KU-798-093

LADY MARGARET HALL LIBRARY
OXFORD

**STRUCTURE AND BONDING
IN SOLID STATE CHEMISTRY**

## ELLIS HORWOOD SERIES IN CHEMICAL SCIENCE



**Kinetics and Mechanisms of Polymerization Reactions:** Applications and Physico-chemical Systematics
P. E. M. ALLEN, University of Adelaide, Australia and
C. R. PATRICK, University of Birmingham

**Electron Spin Resonance**
N. M. ATHERTON, University of Sheffield

**Metal Ions in Solution**
J. BURGESS, University of Leicester

**Organometallic Chemistry** – A Guide to Structure and Reactivity
D. J. CARDIN and R. J. NORTON, Trinity College, University of Dublin

**Structures and Approximations for Electrons in Molecules**
D. B. COOK, University of Sheffield

**Liquid Crystals and Plastic Crystals**
Volume I, Preparation, Constitution, and Applications
Volume II, Physico-Chemical Properties & Methods of Investigation
Edited by G. W. GRAY, University of Hull and
P. A. WINSOR, Shell Research Ltd.

**Polymers and their Properties:** A Treatise on Physical Principles and Structure
J. W. S. HEARLE, University of Manchester Institute of Science and Technology

**Biochemistry of Alcohol and Alcoholism**
L. J. KRICKA and P. M. S. CLARK, University of Birmingham

**Structure and Bonding in Solid State Chemistry**
M. F. C. LADD, University of Surrey

**Metal and Metalloid Amides:** Synthesis, Structure, and Physical and Chemical Properties
M. F. LAPPERT, University of Sussex
A. R. SANGER, Alberta Research Council, Canada
R. C. SRIVASTAVA, University of Lucknow, India
P. P. POWER, Stanford University, California

**Biosynthesis of Natural Products**
P. MANITTO, University of Milan

**Adsorption**
J. OSCIK, Head of Institute of Chemistry, Marie Curie Sladowska, Poland

**Chemistry of Interfaces**
G. D. PARFITT, Tioxide International Limited and
M. J. JAYCOCK, Loughborough University of Technology

**Metals in Biological Systems:** Function and Mechanism
R. H. PRINCE, University Chemical Laboratories, Cambridge

**Applied Electrochemistry:** Electrolytic Production Processes
A. SCHMIDT

**Chlorofluorocarbons in the Environment**
Edited by T. M. SUGDEN, C.B.E., F.R.S., Master of Trinity Hall, Cambridge and
T. F. WEST, former Editor-in-Chief, Society of Chemical Industry

**Handbook of Enzyme Biotechnology**
Edited by ALAN WISEMAN, Department of Biochemistry, University of Surrey

# STRUCTURE AND BONDING IN SOLID STATE CHEMISTRY

M. F. C. LADD
Department of Chemical Physics,
University of Surrey

**ELLIS HORWOOD LIMITED**
Publishers Chichester

Halsted Press: a division of
**JOHN WILEY & SONS**
New York - Chichester - Brisbane - Toronto

First published in 1979 by

**ELLIS HORWOOD LIMITED**

Market Cross House, Cooper Street, Chichester, West Sussex, PO19 1EB, England

*The publisher's colophon is reproduced from James Gillison's drawing of the ancient Market Cross, Chichester*

**Distributors:**

*Australia, New Zealand, South-east Asia:*
Jacaranda-Wiley Ltd., Jacaranda Press,
JOHN WILEY & SONS INC.,
G.P.O. Box 859, Brisbane, Queensland 40001, Australia.

*Canada:*
JOHN WILEY & SONS CANADA LIMITED
22 Worcester Road, Rexdale, Ontario, Canada.

*Europe, Africa:*
JOHN WILEY & SONS LIMITED
Baffins Lane, Chichester, West Sussex, England.

*North and South America and the rest of the world:*
HALSTED PRESS, a division of
JOHN WILEY & SONS
605 Third Avenue, New York, N.Y. 10016, U.S.A.



**British Library Cataloguing in Publication Data**
Ladd, Marcus Frederick Charles
Structure and bonding in solid state chemistry. –
(Ellis Horwood series in Physical Chemistry).
1. Chemical bonds 2. Solid state chemistry
I. Title
541'.224 QD461 78–41289
ISBN 0–85312–095–1 (Ellis Horwood Ltd., Publishers)
ISBN 0–85312–103–6 Pbk. (Ellis Horwood Ltd., Publishers)
ISBN 0–470–26597–3 (Halsted Press)

Typeset in Press Roman by Ellis Horwood Ltd., Publishers
Printed in Great Britain by Cox & Wyman, Fakenham.

Learning in old age is writing on sand but learning in youth is engraving on stone.

*Arabian proverb*

# Table of Contents

**Chapter 2 Ionic Compounds**

**Chapter 3 Covalent Compounds**

# Author's Preface

This book has been developed from an undergraduate course for students of chemistry and chemical physics, and it is hoped that it will prove useful to those reading other chemical and physical sciences. The mathematical treatments are in general not difficult, and should lie within the scope of any chemistry degree student. Some mathematical arguments are developed in Appendices, because their inclusion in the text might distract readers from the development of the subject.

Each chapter has been provided with a set of problems of varying degrees of difficulty, which the reader is encouraged to solve because they will assist in gaining familiarity with the themes of the book and in testing one's ability to apply these themes to new situations. A suggested scheme for solving problems, and a listing of values of fundamental physical constants used in the text, appear before the opening chapter. Solutions to the problems are provided at the end of the book. Those problems marked with an asterisk might be omitted at a first reading.

The SI system of units has been used and, in several situations, conversions between the SI and cgs systems have been indicated. It is not yet, and will not be for some time, possible to neglect the cgs system. Competency in more than one system should not be despised: indeed, such ability may enhance a chemistry student's appreciation of both his subject and its literature. Some conversion factors are tabulated in the introductory matter which appears immediately before Chapter 1.

In studying solids, one needs an understanding of the three-dimensional character of crystals and structures. This aspect of the solid state is not so much intrinsically difficult as it is unfamiliar, despite the nature of the world in which we live. In order to assist in an appreciation of three-dimensional structure, many of the illustrations are provided as stereoscopic pairs, and directions for viewing them are provided.

It is my pleasure to record acknowledgement to my colleagues Dr. R. S. B. Chrystall, Mr. R. J. R. Hayward and Dr. G. A. Webb for critical readings of all parts of the manuscript and for making valuable suggestions, to Dr. S. Motherwell of Cambridge University for the use of the program PLUTO, with which the stereoscopic illustrations have been prepared; to publishers and authors for permission to reproduce those figures which carry appropriate dedications; to numerous students who have worked through problems to my advantage and, I hope, to theirs; to Miss M. R. King for assistance with typing; to Dr. T. M. Sugden, C.B.E., F.R.S., the Series Editor, who made helpful suggestions; and to Ellis Horwood Limited, the publishers, for enabling this work to be brought to a state of completion.

Department of Chemical Physics
University of Surrey

M. F. C. Ladd
31 July 1978

# Solution of Numerical Problems

## INTRODUCTION

Numerical problems are essential to a study of the physical sciences, because they relate experimental observations to theoretical models. The insertion of magnitudes into a given equation is a common scientific activity: it should be mastered and, however trivial, never despised.

The solving of problems leads to an appreciation of several important features:

(a) the orders of magnitude in physical and chemical quantities;
(b) the need for an understanding of units;
(c) the value of checking dimensional homogeneity;
(d) the sources of physical and chemical data;
(e) the precision of the data and its transmission to the result.

Most problems involve algebraic manipulation. It is essential to obtain a clear picture of the chemistry and physics involved in the problem before embarking on a series of mathematical processes. It is often useful to obtain an explicit algebraic expression before inserting numerical values. There are several advantages in so doing:

(f) the expression can be checked dimensionally;
(g) the possible cancellation of terms may improve the precision of the result;
(h) the chemical or physical significance of the result may be more important;
(i) similar problems with other magnitudes can be solved with little additional effort;
(j) if the result is erroneous, it is easy to check whether the error is in the deduction or in the arithmetic;
(k) in examinations, the derivation of a correct explicit expression will score marks, even though the arithmetic may be in error.

If the data is inserted into an expression in the form of numbers between 1 and 10, multiplied by the appropriate powers of 10, it is easy to estimate an approximate answer: calculators and other aids can go wrong; they can even be manipulated incorrectly. Suppose that we have for a relative permittivity, $\epsilon_r$,

$$(\epsilon_r - 1) = Np^2/9\epsilon_0 kT \qquad (1)$$

| | | | |
|---|---|---|---|
| $N$ (number of molecules per unit volume) | = | $2.461 \times 10^{25}$ | $\mathrm{m^{-3}}$ |
| $p$ (dipole moment) | = | $5.11 \times 10^{-30}$ | C m |
| $\epsilon_0$ (permittivity of a vacuum) | = | $8.8542 \times 10^{-12}$ | $\mathrm{F\,m^{-1}}$ |
| $k$ (Boltzmann constant) | = | $1.3807 \times 10^{-23}$ | $\mathrm{J\,K^{-1}}$ |
| $T$ (absolute temperature) | = | 298.15 | K |

We can see that the expression is dimensionally correct; the right-hand side of (1) has the units

$$\frac{\mathrm{m^{-3} \times C^2\,m^2}}{\mathrm{F\,m^{-1} \times J\,K^{-1} \times K}} = 1$$

Inserting the magnitudes, we have

$$(\epsilon_r - 1) = \frac{2.461 \times 10^{25} \times (5.11)^2 \times 10^{-60}}{9 \times 8.8542 \times 10^{-12} \times 1.3807 \times 10^{-23} \times 298.15} \qquad (2)$$

We can see that $(\epsilon_r - 1) \approx \dfrac{60}{100 \times 300}$, or $2 \times 10^{-3}$. Thus, when the expression is evaluated, we can write with confidence, $(\epsilon_r - 1) = 1.96 \times 10^{-3}$.

## APPROACH TO PROBLEMS

There are different ways of tackling problems, so that these notes are offered only as a guide. Sometimes a recommended stage may be changed or bypassed. Elegant derivations are often concise: the converse is not necessarily true, and failure to justify a stage in a derivation may indicate a lack of judgement or a lack of confidence. On the other hand, over-elaboration of trivial detail or of arithmetic manipulation may be equally unacceptable in a polished answer to a problem. Some degree of subjective judgement is involved in the solution of problems, and in the marking of such solutions in examinations. Few examiners would give high marks for a completely correct numerical answer in the absence of satisfactory evidence of the method used.

## PROCEDURE

(a) Read the problem carefully. If you think that it contains an ambiguity (which can happen sometimes), assume the simplest interpretation of the ambiguity, and comment on it.

(b) Summarise the given information by appropriate means, such as:
  (i) labelled drawings;
  (ii) energy-level diagrams;
  (iii) sketch-graphs, correctly labelled;
  (iv) defining symbols used in diagrams and formulae;
  (v) listing numerical values with units.

(c) State the answers required, defining quantities involved together with their units and symbols.

(d) Indicate relevant laws and equations which are to be used in developing the problem, at least initially.

(e) State the method to be used, for example 'take $\log_e$ of both sides of equation (1)'.

(f) Where appropriate, attempt to formulate an explicit equation before inserting numerical data. Look for cancellations of terms, and indicate any physical or functional approximations.

(g) Do *not* make needless numerical approximations, but state any approximations which are made, and include an estimate of the probable error as far as you are able.

(h) For convenience, substitute a new symbol for a complex group of symbols in deriving an expression.

(i) Check the dimensions of both quantities and expressions for consistency. Remember that exponents (and log terms) are dimensionless.

(j) Insert numerical values into expressions carefully. Determine an approximate result by 'gross cancellation', as in the example above.

(k) Think about the answer in terms of your knowledge of the physical sciences, and comment on it in the light of the question. If you feel doubtful about the validity of the result, check your arithmetic and deductions. If you still have some reservations about your answer, indicate their nature.

(l) Keep a neat format in your answer. In an examination, allocate a numerical question only its fair share of time.

## EXAMPLE PROBLEM

The diffusion coefficient ($D$) of carbon in $\alpha$-iron, as a function of temperature, follows the equation

$$D = D_0 \exp(-U/RT) \quad . \qquad (3)$$

Values of $D$ are as follows:

| $T/K$ | 300 | 500 | 700 | 900 | 1100 |
|---|---|---|---|---|---|
| $D/\,m^2\,s^{-1}$ | $4.73 \times 10^{-21}$ | $3.35 \times 10^{-15}$ | $1.08 \times 10^{-12}$ | $2.66 \times 10^{-11}$ | $2.05 \times 10^{-10}$ |

Verify the above equation, and find values for both the activation energy ($U$) for the diffusion process, and $D_0$. Comment briefly on the results.

$$(R = 8.316 \text{ J K}^{-1} \text{ mol}^{-1}).$$

The following solution attempts to illustrate some of the points discussed above:

**SOLUTION**

The presence of $R$ in the exponent and its units indicate that $U$ is expected in J mol$^{-1}$. $D_0$ is the limiting value of $D$ as $T \to \infty$. Taking $\log_e$ of both sides of (3) gives

$$\ln D = \ln D_0 - U/RT \tag{4}$$

If in the graph of $\ln D$ against $1/T$ is a straight line, (3) would be verified. The slope of the line, $\Delta(\ln D)/\Delta(1/T)$, is $-U/R$, and the intercept, $\ln D$ at $1/T = 0$, is $\ln D_0$.

| $D/\text{m}^2\,\text{s}^{-1}$ | $\ln D$ | $T/K$ | $(10^3/T)/\text{K}^{-1}$ |
|---|---|---|---|
| $4.73 \times 10^{-21}$ | −46.800 | 300 | 3.333 |
| $3.35 \times 10^{-15}$ | −33.330 | 500 | 2.000 |
| $1.08 \times 10^{-12}$ | −27.554 | 700 | 1.429 |
| $2.66 \times 10^{-11}$ | −24.350 | 900 | 1.111 |
| $2.05 \times 10^{-10}$ | −22.308 | 1100 | 0.909 |

*Note:* $\ln D$ does not have the units m$^2$ s$^{-1}$; $\ln(D/D_0)$ is dimensionless, and the separation of $\ln D$ from $\ln D_0$ is a convenience.

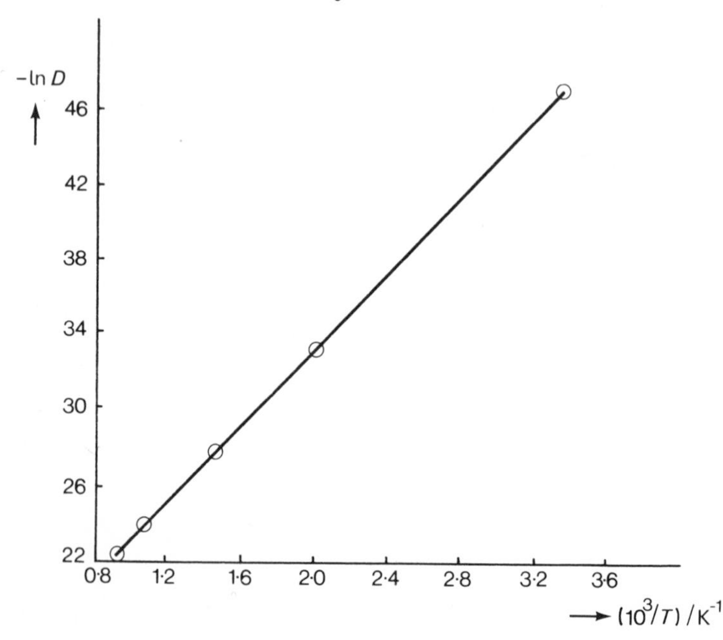

The straight-line graph verifies (3) for the diffusion of carbon in α-iron. From a least-squares fit to (4), the slope is −10104 K, giving $U = 84.0$ kJ mol$^{-1}$; $D_0 = 2.00 \times 10^{-6}$ m$^2$ s$^{-1}$. The sign and magnitude of $U$ seem reasonable, since work must be done on the system to cause diffusion, and the energies of such processes are usually of the order of 1 eV per atom. $D$ increases with $T$, and if (3) continues to hold, would tend to a limiting value of $2.00 \times 10^{-6}$ m$^2$ s$^{-1}$. However, at such high temperatures the material would melt or even vaporize. The usefulness of $D_0$ here is mainly in connection with the evaluation of $D$ from (3).

# Physical Constants

The following data form a self-consistent, least-squares adjusted set, and details of their compilation may be found in *Atomic Masses and Fundamental Constants 5,* edited by J. H. Sanders and A. H. Wapstra (Plenum Press, N.Y., 1976) and in references therein, particularly E. R. Cohen and B. N. Taylor, *Journal of Physical and Chemical Reference Data,* Vol. 2, pp. 663-734 (1973). Each datum is quoted to its known precision although rarely shall we need to use the full precision of any constant: the figure in parentheses is the uncertainty in the final digit of the constant.

| | | | |
|---|---|---|---|
| Speed of light in a vacuum | $c$ | $2.997924590(8) \times 10^{8}$ | m s$^{-1}$ |
| Permittivity of a vacuum | $\epsilon_0$ | $8.85418782(7) \times 10^{-12}$ | F m$^{-1}$ |
| Atomic mass unit | $u$ | $1.660566(9) \times 10^{-27}$ | kg |
| Mass of hydrogen atom | $m_{\mathrm{H}}$ | $1.673560(8) \times 10^{-27}$ | kg |
| Rest mass of proton | $m_{\mathrm{p}}$ | $1.672649(9) \times 10^{-27}$ | kg |
| Rest mass of neutron | $m_{\mathrm{n}}$ | $1.674954(9) \times 10^{-27}$ | kg |
| Rest mass of electron | $m_{\mathrm{e}}, m$ | $9.10953(5) \times 10^{-31}$ | kg |
| Elementary charge | $e$ | $1.602189(5) \times 10^{-19}$ | C |
| Classical electron radius | $r_{\mathrm{e}}$ | $2.817938(7) \times 10^{-15}$ | m |
| Boltzmann constant | $k$ | $1.38066(4) \times 10^{-23}$ | J K$^{-1}$ |
| Planck constant | $h$ | $6.62618(4) \times 10^{-34}$ | J s |
| Bohr radius | $a_0$ | $5.294664(4) \times 10^{-11}$ | m |
| Rydberg constant | $R_{\infty}$ | $1.09737314(1) \times 10^{7}$ | m$^{-1}$ |
| Rydberg constant for hydrogen | $R_{\mathrm{H}}$ | $1.09677582(1) \times 10^{7}$ | m$^{-1}$ |
| Bohr magneton | $\mu_{\mathrm{B}}$ | $9.27408(4) \times 10^{-24}$ | A m$^{2}$ |
| Avogadro constant | $L, N_A$ | $6.022094(6) \times 10^{23}$ | mol$^{-1}$ |
| Gas constant | $R$ | $8.3160(2)$ | J K$^{-1}$ mol$^{-1}$ |
| Ice-point temperature | $T_{\mathrm{ice}}$ | $273.1500(1)$ | K |
| Farady | $F$ | $9.64865(3) \times 10^{4}$ | C mol$^{-1}$ |
| Reduced mass of (proton + election) | $\mu$ | $9.10457(5) \times 10^{-31}$ | kg |

# Symbols and other Literary Impedimenta

In scientific literature, physical quantitites are represented, traditionally, by certain symbols. As a result, a given symbol will have differing meanings in different contexts. While it would be possible to choose, or invent, a unique symbol for every quantity, there is little to recommend such a procedure, and generally no confusion need arise. The following sections introduce common symbols, both English and Greek, superscripts and subscripts, and other common terminologies.

## SYMBOLS

| | |
|---|---|
| $a$ | activity; constant of van der Waals' equation of state; periodicity; unit-cell dimension parallel to the $x$ axis. |
| $a(X)$ | activity of species $X$. |
| $a_{\pm}$ | mean ionic activity. |
| $a_0$ | Bohr radius. |
| aq | hydrated, generally at infinite dilution. |
| $A$ | 'box' length; constant of Debye-Hückel limiting law; cross-sectional area; Madelung constant. |
| Å | angstrom unit. |
| $A_r$ | relative atomic mass. |
| AO | atomic orbital. |
| $b$ | constant of the van der Waals' equation of state; unit-cell dimension parallel to the $y$ axis. |
| $B$ | constant of repulsion energy. |
| Bp | boiling point. |
| $c$ | concentration; speed of light in a vacuum; unit-cell dimension parallel to the $z$ axis. |
| c | centi (prefix). |
| $c_{\pm}$ | mean concentration |
| $c_i$ | LCAO constants. |
| $c_{++}$, $c_{+-}$, $c_{--}$ | constants in the dipole-dipole potential energy term. |

|  |  |
|---|---|
| $C$ | capacitance; resultant dipole-dipole constant. |
| C | Celsius (temperature scale, as in °C); coulomb. |
| $C_P$ | heat capacity (molar) at constant pressure. |
| $C_V$ | heat capacity (molar) at constant volume. |
| $CN$ | coordination number. |
| cos | cosine operator. |
| $\cos\alpha$, $\cos\beta$, $\cos\gamma$ | direction cosines with respect to cartesian axes. |
| $d$ | distance; interplanar spacing. |
| d | deci (prefix); differential operator; orbital descriptor. |
| $d^*$ | distance in reciprocal space. |
| $D$ | determinant; diffusion coefficient; dissociation energy (enthalpy); interatomic spacing. |
| D | differential operator (e.g. $\partial/\partial x$) |
| $e$ | elementary charge. |
| $E$ | electric field intensity; electron affinity; energy. |
| $E_X$ | excitation energy. |
| $E_{\mathrm{F}}$ | Fermi energy. |
| $\mathbf{E}$ | electric field intensity (vector). |
| $\mathcal{E}$ | Rayleigh ratio. |
| eV | electronvolt. |
| Ebs | electrostatic bond strength. |
| exp, e | exponential operator. |
| $f$ | activity coefficient; force constant. |
| $f_\pm$ | mean ionic activity coefficient. |
| $f(E)$ | Fermi-Dirac distribution function. |
| $F$ | Faraday constant; force |
| F | farad. |
| $\mathbf{F}$ | force (vector). |
| $g$ | gravitational acceleration. |
| $g_i$ | $i$th energy state. |
| g | gas; gram. |
| $g(E)$ | density of states function. |
| $G$ | Gibbs' free energy function. |
| $h$ | Miller index; Planck constant. |
| $H$ | coulomb integral; enthalpy (heat content). |
| $H_T$ | enthalpy at temperature $T$. |
| $H^\ominus$ | standard enthalpy. |
| $\Delta H_f^\ominus$ | standard enthalpy of formation |
| $\mathcal{H}$ | Hamiltonian operator. |
| i | operator $\sqrt{-1}$. |
| $\mathbf{i}$ | unit vector parallel to the $x$ axis. |
| $I$ | current; ionic strength; ionization energy (potential). |

| | |
|---|---|
| $j$ | current density. |
| $\mathbf{j}$ | unit vector parallel to the $y$ axis. |
| J | joule. |
| $k$ | Boltzmann constant; Miller index; modulus of wave vector. |
| k | kilo (prefix). |
| $\mathbf{k}$ | unit vector parallel to the $z$ axis; wave vector in **k** space. |
| $k_F$ | radius of sphere in **k** space on the surface of which the energy is $E_F$. |
| K | energy shell descriptor; equilibrium constant; Kelvin (temperature scale). |
| $l$ | azimuthal quantum number; distance. |
| l | liquid. |
| L | Avogadro constant; 'box' length; energy shell descriptor. |
| $\mathcal{L}$ | Langevin function. |
| ln | $\log_e$. |
| LCAO | linear combination of atomic orbitals. |
| $m$ | mass of particle or electron; integer. |
| $m^*$ | effective mass of electron. |
| m | milli (prefix); metre |
| $m_e$ | mass of electron. |
| $m_H$ | mass of hydrogen atom. |
| $m_p$ | mass of proton. |
| $m_n$ | mass of neutron. |
| $m_u$ | atomic mass unit. |
| $m_l$ | magnetic quantum number. |
| $m_s$ | spin quantum number |
| $M$ | metallic element. |
| $M_r$ | relative molecular mass. |
| MO | molecular orbital. |
| Mp | melting point. |
| Min | minimum value. |
| mol | mole. |
| $n$ | defect concentration integer; number of items or particles; number of moles; principal quantum number; refractive index. |
| $N$ | normalization constant; number of items or particles (larger than $n$); number per unit volume. |
| $n_x, n_y, n_z$ | quantum numbers (equivalent to $n$, $l$, $m_l$). |
| $N_A$ | Avogadro constant. |
| $N/V$ | electron concentration. |
| $O$ | origin of $x$, $y$, $z$ axes. |
| $p$ | dipole moment; momentum; polarizing power. |
| p | orbital descriptor |
| $\mathbf{p}$ | dipole moment (vector). |

| | |
|---|---|
| $P$ | polarization; pressure. |
| $P_\mathrm{m}$ | molar polarization. |
| $P(z)$ | Legendre polynomial. |
| $q$ | fraction of electronic charge. |
| $r$ | distance; radial coordinate. |
| $\mathbf{r}$ | distance (vector). |
| $\dot{r}$ | d$v$/d$t$ (velocity). |
| $r_i$ | ionic radius. |
| $r_+$ | cationic radius. |
| $r_-$ | anionic radius. |
| $r_e$ | equilibrium distance. |
| $R$ | distance; gas constant; radial function; radius ratio; resistance. |
| $R_n$ | radius ratio for $n$ coordination. |
| $R_\mathrm{H}$ | Rydberg constant for hydrogen. |
| $R_\infty$ | Rydberg constant for infinite mass. |
| Ry | rydberg. |
| s | orbital descriptor; second; solid. |
| $s$ | specific heat capacity. |
| $\mathbf{s}$ | distance (vector). |
| $S$ | entropy; overlap integral; screening constant. |
| $S_M$ | sublimation energy (enthalpy) of a metal. |
| $S_{++}, S_{+-}, S_{--}$ | lattice sums. |
| sin | sine operator. |
| stp | standard temperature and pressure (0 °C and 1 atm pressure). |
| $t$ | temperature in °C; time. |
| $T$ | absolute temperature (in K); kinetic energy. |
| $T_\mathrm{ice}$ | ice-point absolute temperature. |
| $T_\mathrm{F}$ | Fermi temperature. |
| tan | tangent operator. |
| $u$ | ion mobility. |
| $u_k(x)$ | one-dimensional Bloch function. |
| $U$ | activation energy; internal energy |
| $U_C$ | cohesive, or crystal, energy. |
| $U_\mathrm{D}$ | Debye energy. |
| $U_E$ | electrostatic energy. |
| $U_\mathrm{K}$ | Keesom energy; kinetic energy. |
| $U_\mathrm{L}$ | London energy. |
| $U_S$ | electrostatic self-energy. |
| $U_\mathrm{V}$ | vibrational energy. |
| $v$ | speed. |
| $\mathbf{v}$ | velocity (vector). |
| $\mathbf{v}_d$ | drift velocity (vector). |
| $V$ | volume; molar volume; potential energy; voltage |

| | |
|---|---|
| V | volt |
| $V_c$ | critical volume of gas. |
| $V_m$ | molar volume. |
| VB | valence bond. |
| $w$ | probability; weight. |
| $W$ | total probability; work |
| $x$ | axial direction; electronegativity; fractional coordinate; general quantity (independent variable); mole fraction. |
| $X$ | non-metallic element. |
| $y$ | axial direction; fractional coordinate; general quantity (dependent variable). |
| $Y$ | separable angular function of $\theta, \phi$. |
| $z$ | axial direction; fractional coordinate; vertical height. |
| $z_+$, $z_1$, $z_i$ | cationic charge. |
| $z_-$, $z_2$, $z_j$ | anionic charge. |
| $Z$ | atomic number; partition function. |
| $Z_{eff}$, $Z'$ | effective atomic number. |
| $\alpha$ | angle between $y,z$ axes; coulomb integral; Lagrange multiplier; linear expansivity; polarizability; polymorph descriptor; spin factor. |
| $\beta$ | angle between $x,z$ axes; Lagrange multiplier; polymorph descriptor; resonance integral; spin factor; volume expansivity. |
| $\gamma$ | angle between $x,y$ axes. |
| $\delta$ | path difference; quantum defect; small quantity. |
| $\delta_{ij}$ | Kronecker delta. |
| $\partial$ | partial differential operator. |
| $\epsilon$ | permittivity. |
| $\epsilon_0$ | permittivity of a vacuum. |
| $\epsilon_r$ | relative permittivity. |
| $\epsilon_i$ | energy of the $i$th state. |
| $\zeta$ | AO exponent (the effective atomic number). |
| $\theta$ | Bragg angle; co-latitude. |
| $\kappa$ | compressibility. |
| $\lambda$ | fractional ionic character; Lagrange multiplier; wavelength. |
| $\mu$ | chemical potential; electron mobility; reduced mass. |
| $\mu_B$ | Bohr magneton. |
| $\nu$ | frequency; total number of ions in electrolyte. |
| $\tilde{\nu}$ | wavenumber. |
| $\tilde{\nu}_\infty$ | wavenumber at series limit. |
| $\pi$ | 3.141592654 . . .; bond type; MO descriptor. |
| $\rho$ | density; electrical resistivity; electron density; repulsion parameter. |

| | |
|---|---|
| $\rho_c$ | critical density of a gas. |
| $\sigma$ | bond type; charge density; electrical conductivity; MO descriptor; molecular diameter; standard deviation (estimated). |
| $\tau$ | relaxation time; volume. |
| $\phi$ | azimuthal angle. |
| $\phi_M$ | electronic work function of a metal $M$. |
| $\chi$ | MO; wave function. |
| $\psi$ | AO; wave function. |
| $\psi^*$ | conjugate of $\psi$. |
| $\omega$ | solid angle. |
| $\Gamma(n)$ | gamma function of the argument $n$. |
| $\Delta$ | change in a property between two states; determinant value; difference of two quantities. |
| $\Theta$ | Debye temperature; angular part of Schrödinger wave equation. |
| $\Lambda$ | mean free path. |
| $\Pi$ | product operator. |
| $\Sigma$ | sum operator. |
| $\Phi$ | VB molecular wave function; angular part of Schrödinger wave equation. |
| $\nabla$, grad | $\mathbf{i}\,\partial/\partial x + \mathbf{j}\,\partial/\partial y + \partial/\partial z$ |
| $\nabla^2$ | $\partial^2/\partial x^2 + \partial^2/\partial y^2 + \partial^2/\partial z^2$. |

## SUBSCRIPTS AND SUPERSCRIPTS

| Subscript | (as in) | property | means |
|---|---|---|---|
| $a$ | | $P_{\mathrm{m},a}$ | atom |
| $d$ | | $U_d$ | diffusion |
| $e$ | | $P_{\mathrm{m},e}$ | electron |
| $f$ | | $\Delta H_f$ | formation |
| $g$ | | $\sigma_g$ | gerade (even) |
| $h$ | | $\Delta S_h$ | hydration |
| $hkl$ | | $\theta_{hkl}$ | for the plane ($hkl$) |
| $i$ | | $r_i$ | $i$th species |
| $o$ | | $P_{\mathrm{m},o}$ | orientation |
| $u$ | | $\sigma_u$ | ungerade (odd) |
| $x$ | | $p_x$ | with respect to the $x$ direction |
| d | | $\Delta G_{\mathrm{d}}$ | dissolution |
| e | | $\Delta H_{\mathrm{e}}$ | evaporation |
| f | | $\Delta H_{\mathrm{f}}$ | fusion |
| m | | $P_{\mathrm{m},a}$ | molar |
| s | | $\Delta H_{\mathrm{s}}$ | sublimation |
| t | | $\Delta H_{\mathrm{t}}$ | transition |
| $C$ | | $U_C$ | crystal |

| | | |
|---|---|---|
| I | $a_{\mathrm{I}}$ | state I |
| deloc | $E_{\mathrm{deloc}}$ | delocalization |
| $\pi$ | $E_{\pi}$ | $\pi$-bond |

| Superscript | (as in) | property | means |
|---|---|---|---|
| $\ominus$ | | $\Delta H^{\ominus}$ | standard |
| $\circ$ | | $5^{\circ}$ | degrees (angular measure) |
| $+(-)$ | | $q^{+(-)}$ | positive (negative) |

*Note*

Modern terminology dictates that molar properties, such as the molar heat capacity at constant pressure, should be symbolised with a subscript m, leading to $C_{P,\mathrm{m}}$. For simplicity, we shall often omit the subscript m; confusion is unlikely to arise in the context of this book.

## OTHER TERMINOLOGY

| | |
|---|---|
| [ ] | complex ion; concentration, usually in mol $\mathrm{dm}^{-3}$; crystallographic direction; brackets; sums in least-squares normal equations. |
| <> | average; crystallographic form of directions; crotchets. |
| ( ) | crystallographic plane; parentheses. |
| { } | crystallographic form of planes; braces |
| ! | factorial |
| $\bar{X}$ | average value of $X$ |
| $G(X)$, $G_X$ | property $G$ of species $X$ |
| $\Delta H_f^{\ominus}(MX,\mathrm{s})$ | property $\Delta H_f^{\ominus}$ of species $MX$ in the solid state |
| $\mathrm{Lt}_{h \to 0}\, E$ | limiting value of $E$ as $h$ tends to zero. |
| $\infty$ | infinity |
| $\propto$ | proportional to |

# Conversion of Units

There will be, for a very long time, a need to convert from cgs to SI units and *vice versa.* While the conversion factors can always be determined from first principles, and it is good practice to do so occasionally, a table of conversion factors is useful for quick reference. We set out here a selection of such data.

| Physical quantity | cgs unit | × | conversion factor | → | SI unit |
|---|---|---|---|---|---|
| Length, $l$ | cm | | $10^{-2}$ | | m |
| | Å | | $10^{-10}$ | | m |
| Volume*, $V$ | $cm^3$, ml | | $10^{-6}$ | | $m^3$ |
| Molar volume, $V_m$ | $cm^3\ mol^{-1}$ | | $10^{-6}$ | | $m^3\ mol^{-1}$ |
| Velocity, $v$ | $cm\ sec^{-1}$ | | $10^{-2}$ | | $m\ s^{-1}$ |
| Wavelength, $\lambda$ | cm | | $10^{-2}$ | | m |
| Wavenumber $\nu$ | $cm^{-1}$ | | $10^2$ | | $m^{-1}$ |
| Mass, $m$ | g | | $10^{-3}$ | | kg |
| Density $\rho$ | $g\ cm^{-3}$ | | $10^3$ | | $kg\ m^{-3}$ |
| Force, $F$ | dyne | | $10^{-5}$ | | N |
| Pressure†, $P$ | $dyne\ cm^{-2}$ | | $10^{-1}$ | | $N\ m^{-2}$ |
| Energy, $U$ | cal | | 4.184 | | J |
| Work, $W$ | erg | | $10^{-7}$ | | J |
| Quantity of electricity, $q$ | esu | | $3.3356 \times 10^{-10}$ | | C |
| Current, $I$ | $esu\ sec^{-1}$ | | $3.3356 \times 10^{-10}$ | | A |
| Charge density, $\sigma$ | $esu\ cm^{-2}$ | | $3.3356 \times 10^{-6}$ | | $C\ m^{-2}$ |
| Electric field intensity, $E$ | $dyne\ esu^{-1}$ | | $2.9979 \times 10^4$ | | $V\ m^{-1}$ |
| Electric potential, $V$ | $erg\ esu^{-1}$ | | 299.79 | | V |
| Capacitance, $C$ | $esu^2\ erg^{-1}$ | | $1.1127 \times 10^{-12}$ | | F |
| Permittivity $\epsilon_0$ | 1 | | $8.8542 \times 10^{-12}$ | | $F\ m^{-1}$ |
| Dipole moment¶, $p$ | esu cm | | $3.3356 \times 10^{-12}$ | | C m |
| Resistance, $R$ | ohm | | 1 | | Ω |
| Resistivity, $\rho$ | ohm cm | | $10^{-2}$ | | Ω m |
| Conductivity, $\sigma$ | $ohm^{-1}\ cm^{-1}$ | | $10^2$ | | $\Omega^{-1}\ m^{-1}$ |

| | | | |
|---|---|---|---|
| Molar conductivity, $\Lambda$ | $ohm^{-1}\ cm^2\ mol^{-1}$ | $10^{-4}$ | $\Omega^{-1}\ m^2\ mol^{-1}$ |
| Ionic mobility, $u$ | $cm^2\ V^{-1}\ sec^{-1}$ | $10^{-4}$ | $m^2\ V^{-1}\ s^{-1}$ |
| Specific heat capacity, $s$ | $cal\ g^{-1}\ deg^{-1}$ | 4184 | $J\ kg^{-1}\ K^{-1}$ |
| Molar refraction, $R_m$ | $cm^3\ mol^{-1}$ | $10^{-6}$ | $m^3\ mol^{-1}$ |
| Molar polarization, $P_m$ | $cm^3\ mol^{-1}$ | $10^{-6}$ | $m^3\ mol^{-1}$ |

*1 $l = 1\ dm^3 = 10^{-3}\ m^3$

† 1 atm = 101325 $N\ m^{-2}$; 1 mmHg = 133.322 $N\ m^{-2}$

¶ 1 D (Debye unit) = $10^{-18}$ esu cm.

Chapter 1

# Preamble

## 1.1 ATOMIC NATURE OF MATTER

The fundamental particles of matter are, for our purposes, the electrons, neutrons and protons of which atoms are composed. Atoms may be regarded as spherical in shape, having diameters of approximately $10^{-10}$ m. The nucleus of an atom has a diameter of $10^{-15}$ to $10^{-14}$ m and contains positively charged protons and uncharged neutrons; these particles constitute most of the mass of an atom, being approximately $1.67 \times 10^{-27}$ kg each. The nucleus is surrounded by negatively charged electrons, each of mass approximately $9.11 \times 10^{-31}$ kg, in number equal to that of the protons, so that the atom is, as a whole, electrically neutral.

Electrons, considered as particles, are located most probably within certain regions of space called atomic orbitals, which surround the atom and are associated with certain energy states for the electrons. When we are thinking of the electrons spread out to form a cloud of electron density, we may use the term atomic orbital to mean the wave function, $\psi$, describing the electron. Then, $|\psi|^2 d\tau$† represents the probability of finding the electron in the volume element $d\tau$, and $|\psi|^2$ has the character of electron density.

The chemistry of a substance is determined mainly by those of its electrons that lie farthest from its nuclei, and are thus least strongly bound to them. We may be concerned to study chemistry through processes that involve single species, as with spectroscopic or certain theoretical techniques. Alternatively we may be investigating atoms in aggregation, in which case we need to consider the interactions between atoms.

Atoms may combine to form a substance that can exist, generally, in three different states. Sodium chloride, common salt, is encountered usually as a colourless, crystalline, ionic solid. If pure, it is a non-conductor of electricity, but at 1074 K it melts to form a colourless liquid that conducts electricity by ion transport. At the higher temperature of approximately 1690 K the liquid

†See page 152.

boils, and the vapour consists almost entirely of molecules of NaCl in which the atoms are bonded in a mainly covalent manner. The changes solid → liquid → gas may occur for other substances at much lower temperatures, and with less change in the bonds between the atoms. Thus, in the sequence

$$\text{ice} \underset{}{\overset{273\text{ K}}{\rightleftharpoons}} \text{water} \underset{}{\overset{373\text{ K}}{\rightleftharpoons}} \text{water vapour}^{\dagger} \;,$$

the covalent molecule $H_2O$ is preserved in all three situations.

## 1.2 STATES OF MATTER

The state, solid, liquid or gas, of a substance is determined by the result of a competition between the interatomic forces acting upon its components and the thermal energy that the substance contains, at any given temperature.

Gases may be characterised by the large volume changes consequent upon variations in their temperature or pressure, and by their ability to flow into spaces available to them. Gases, therefore, are fixed neither in shape nor in volume. They are miscible with one another in all proportions, and differ markedly from solids and liquids in that many of their properties are independent of their chemical nature and may be described by general laws.

A liquid is fixed in volume, at a given temperature, but, like a gas, has no definite form: it takes up the shape of its containing vessel, although not filling it completely because its boundary surface restricts the movement of the liquid. Evidently, and unlike a gas, the atoms in a liquid are close enough to one another for interatomic forces to influence their relative spatial distribution. However, the internal energy of liquids is commensurate with their intermolecular binding energies, and order does not exist over distances of more than a few atomic dimensions, often called short-range order.

A solid is a material of fixed volume and shape at a given temperature. In atomic terms we could say that the mean, or equilibrium, positions of the atoms are invariant with time, at a given temperature. However, the atoms are not static: they are vibrating about their mean positions, and their vibrational energy makes a major contribution to the heat capacity of the solid. The atomic vibrations themselves are anharmonic. Figure 1.1 illustrates the variation of the potential energy, $U$, of a pair of atoms as a function of the interatomic distance, $r$. At absolute zero, and ignoring the zero energy,‡ the potential energy, $D_e$, at the equilibrium distance $r_e$, is represented by the point $A$. As the temperature is increased successively to, say $T_1$ and $T_2$, the energy increases to $U_1$ and $U_2$, respectively. Largely because of the asymmetric nature of the potential energy curve, or the anharmonicity of the atomic vibrations, the mean atomic separation

†A vapour is a gas below its critical temperature (Suggested Reading: *Physical Chemistry*).

‡Often called zero-point energy, through a mistranslation of the German word *nullpunktsenergie*. See also page 154.

increases with increasing temperature along the path $AB$. In an assembly of atoms in a solid there are added complications: the problem should be considered in terms of free energy, which introduces an entropy factor, and it may be shown‡ that even if the atomic vibrations were perfectly harmonic, the increase in free energy of a crystal with increasing temperature would produce an increase in volume. This effect is often ignored: certainly it is small compared with that arising from the anharmonicity of the vibrations (see also Problem 5.2). The simple potential energy curve gives a qualitatively correct picture that is useful in understanding certain physical properties of solids. In all states of matter, we shall find that an energy equilibrium is determined by a balance between forces of attraction and repulsion.

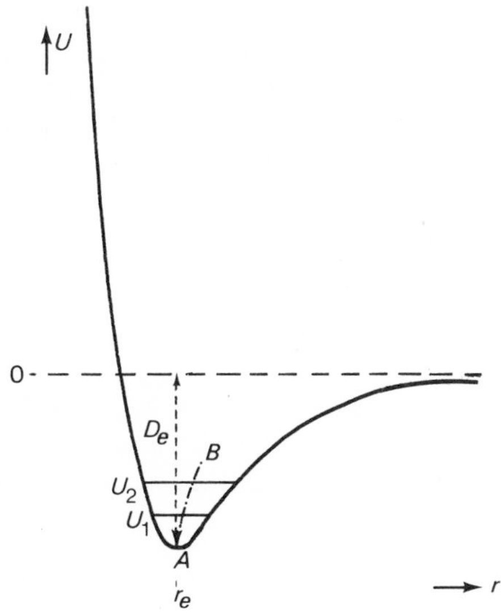


Figure 1.1 – Variations of potential energy $U$, with interatomic distance $r$, for a pair of atoms. Note the steep rise as $r$ decreases below $r_e$.

A qualification is needed about the invariance with time of mean atomic positions, at a constant temperature. In some solids, certain groups of atoms may behave as though their symmetry were greater than that implied by their structure. Thus, potassium cyanide, KCN, at room temperature has the sodium chloride, NaCl, structure type (Figure 1.2¶): although the cyanide ion is linear, it behaves as though it had spherical symmetry. This situation can arise if the group is in a state of free rotation about its mean position, so that the envelope of its motion is a sphere. Alternatively, the group may exhibit orientational disorder, so that when averaged over many hundreds of unit cells having different

‡Suggested Reading: *Mathematics* (Guggenheim, Chapter 8).

¶ Stereoviewing is described in Appendix 1.

orientations of the cyanide group, the ion is again effectively spherical. Recent investigations support the orientationally disordered model for the potassium cyanide cubic structure, with the axes of the cyanide ions lying mainly normal to the (100) and (111) planes† in the cubic unit cells. Thus, we may say that the mean positions of atoms in a solid, averaged over time or space, are constant at a given temperature. At a temperature of about 233 K, potassium cyanide develops another crystalline modification in which the cyanide ions are oriented regularly in the unit cell (Figure 1.3).

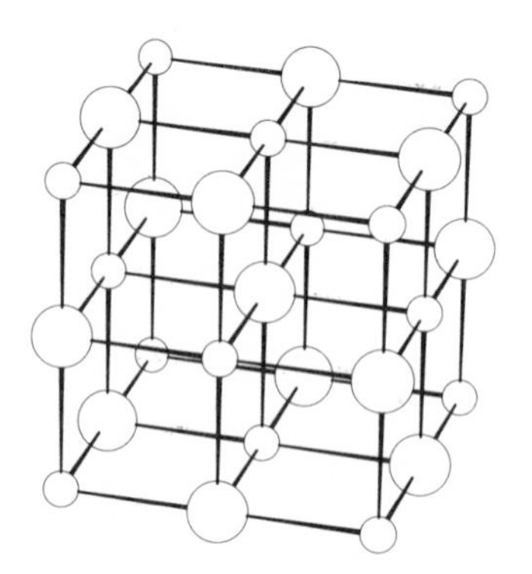
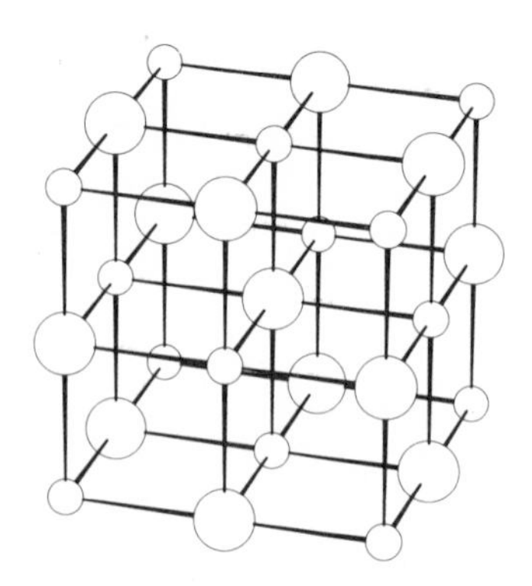

Figure 1.2 – Unit cell of the NaCl structure type: KCN at room temperature; circles in decreasing order of size are (CN) and K.

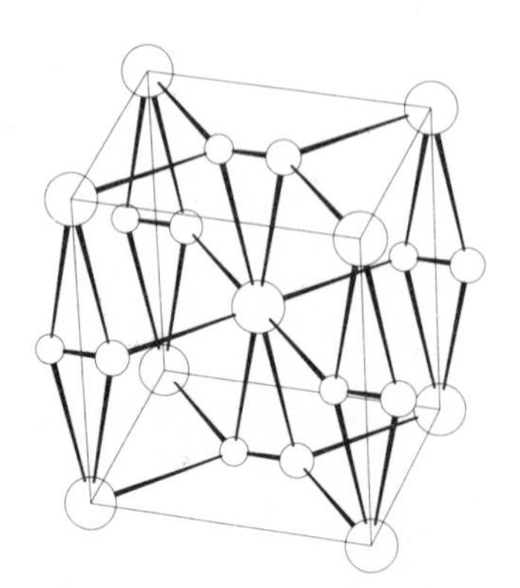
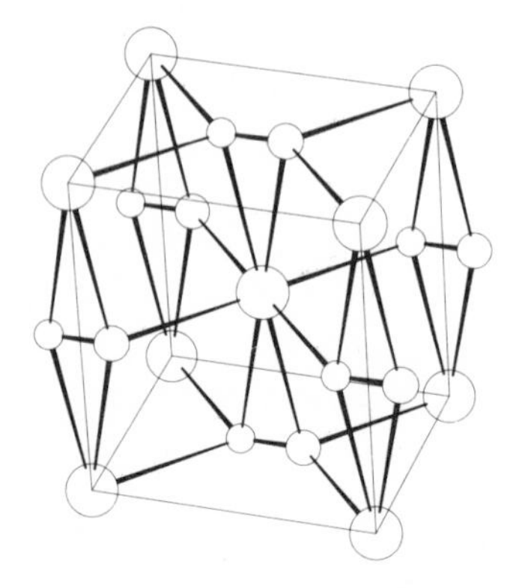

Figure 1.3 – Unit cell of the orthorhombic structure of KCN at 233 K; circles in decreasing order of size are K, N and C. Can you see how this structure is related to that in Figure 1.2?

†Miller indices of planes are discussed in Appendix 2.

## 1.3 CRYSTALLINE AND AMORPHOUS SOLIDS

The essence of crystallinity is the regular repetition of structural units, atoms or groups of atoms, over distances equivalent to many thousands of atomic dimensions, often called long-range order. If the order is in three dimensions, we speak of crystals. Their regular appearance is well known, and is the external manifestation of the periodicities of the crystal structure; Figure 1.4 is an example of a silicate crystal.

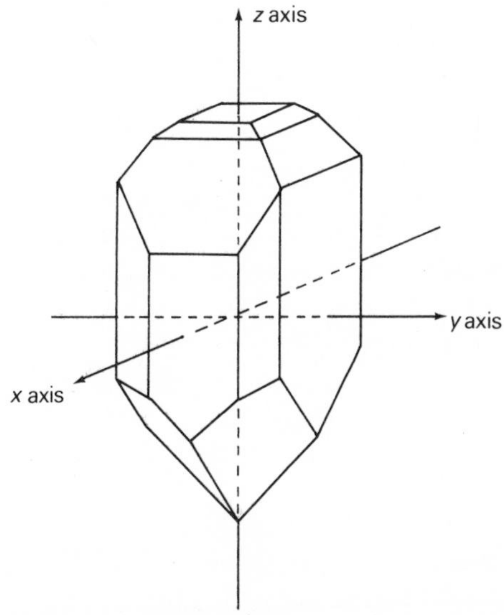


Figure 1.4 – Crystal of hemimorphite $Zn_4Si_2O_7(OH)_2{\cdot}H_2O$ with the crystallographic reference axes drawn in the conventional orientation.

Lower degrees of crystallinity are recognised: certain stretched polymer sheets exhibit two-dimensional order, in the plane of the sheets, and many natural and synthetic fibres exhibit long-range order in only one dimension, the fibre axis. Mammalian hair contains the fibrous protein, α-keratin, with a periodicity to X-rays of about 5 Å† along the fibre axis. If the hair is extended

$R_1$ H N — CO $R_2$
CH CH $R_3$
NH CO - - - - - - HN CO CH NH
5·2Å

† 1Å = $10^{-10}$ m.

by about 100% in steam, $\beta$-keratin, with a periodicity of about 3.5 Å in the fibre direction, is developed. The transformation between the two forms is

$R_2$

H H

CO N CH CO N

N CH CO N CH

H H

$R_1$ $R_3$

← 3·5Å →

reversible, but if the stretched hair is held in steam for some time the reversibility is lost. This transformation is used in the 'permanent' waving of hair.

Most crystalline materials exhibit optical anisotropy under examination with a polarizing microscope. Anisotropy refers to the variation with direction of the value of physical properties; in the case of optical anisotropy, the property is the refractive index of the material. When light travels through a substance the electric vector of the light wave interacts with the electrons of that substance. The degree of interaction will in general vary with the structural complexity in different directions. Thus, anisotropy in physical properties is to be expected for crystalline materials. Crystals belonging to the cubic crystal system, because of their high degree of symmetry, do not show this variation of refractive index; they are optically isotropic.

Optical anisotropy can be induced by stress. Thus, a drawn fibre may often show this phenomenon unless annealed after drawing. One must be careful not to deduce the presence of crystallinity from this type of optical observation, but rather to seek further evidence through an X-ray fibre diagram (Figure 1.5).

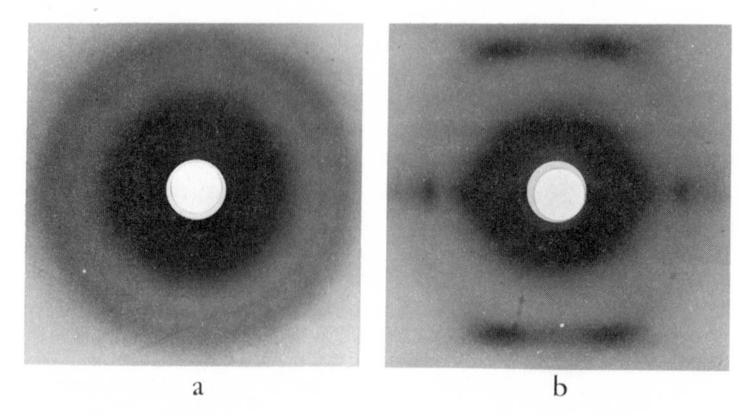

a b

Figure 1.5 – Fibre X-ray diffraction patterns from ethylene-propylene-diene terpolymer: (a) Zero extension, 10% crystallinity (b) 220% extension, 70% crystallinity. (By courtesy of Dr. E. J. Wheeler).

Amorphous solids, like liquids, exhibit only short-range order. Figure 1.6 is a schematic diagram for X-ray diffraction from powdered aluminium (crystalline) enclosed by Sellotape (amorphous). The sharply-defined lines, of which only the first two are shown, belong to the diffraction pattern aluminium; the diffuse band arises from the Sellotape.

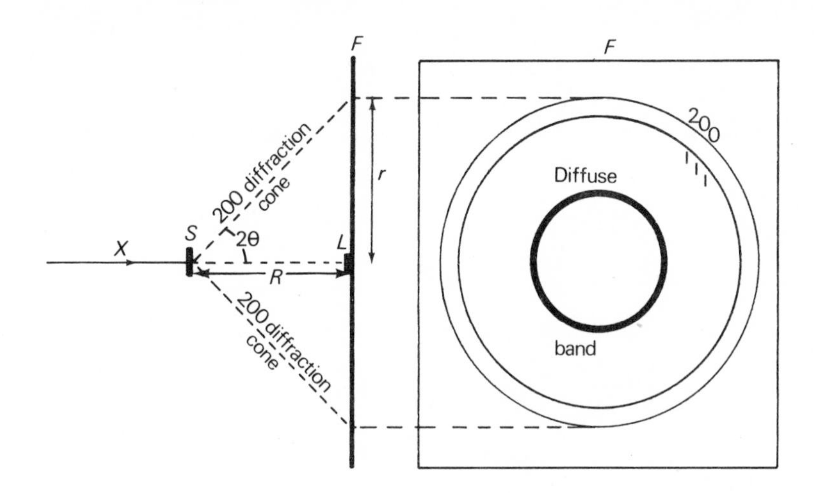


Figure 1.6 – Schematic diagram for X-ray diffraction: $X$ X-rays; $S$ Specimen; $F$ Film; $L$ Leas trap for direct beam.

We shall make a geometrical interpretation of this diagram, which is drawn to scale for Cu $K\alpha$ X-radiation ($\lambda = 1.5418$ Å). From the figure,

$$r/R = \tan(2\theta_{hkl}) \ , \tag{1.1}$$

where $\theta_{hkl}$ is the Bragg reflexion angle for the ($hkl$) planes. In Appendix 3 we discuss the Bragg equation and show that

$$\sin\theta_{hkl} = (\lambda/2a)\,(h^2 + k^2 + l^2)^{1/2} \ , \tag{1.2}$$

where $a$ is the cubic unit-cell dimension. Hence,

$$a = \lambda(h^2 + k^2 + l^2)^{1/2} / \{2\sin(\tfrac{1}{2}\tan^{-1}[r/R])\} \, . \tag{1.3}$$

In an actual experiment, $R$ was 30.00 mm and $r$ for the 200 diffraction ring was 29.5 mm. Hence, from (1.3), $a = 4.08$ Å.

The diffuse band cannot be examined through the Bragg equation. It is the strongest part of the radial distribution of scattered X-radiation from the Sellotape, and it arises from a prominent interatomic spacing ($D$) given, without proof here, by

$$D = 0.61\,\lambda/\sin\theta \ . \tag{1.4}$$

Using (1.1)

$$D = 0.61\,\lambda/\{\sin(\tfrac{1}{2}\tan^{-1}[r'/R])\} \ . \tag{1.5}$$

By experiment, the mean radius ($r'$) of the diffuse band was 11.5 mm, and from (1.5), $D$ = 5.17 Å. Now, Sellotape is based on a polysaccharide: if we construct a model of a 1,4-glucosidic monomer fragment, we find that the overall length of this major structural unit is, in fact, about 5.2 Å.

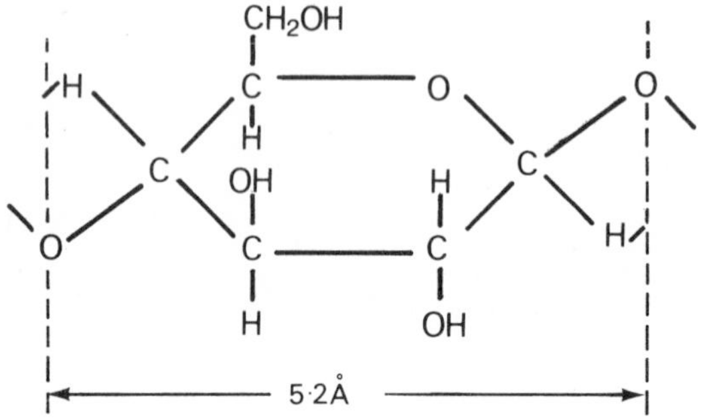

The physical properties of amorphous solids are isotropic. The absence of long-range order means that the structural character in different directions, when averaged over a macroscopic amount of a substance, is effectively the same. Cubic crystals are isotropic for some physical properties, such as refractive index, electrical and thermal conductivities and dielectric polarization, but anisotropic for elasticity and photoelasticity. All crystals are isotropic with respect to density measurements, but amorphous substances are isotropic in all of their physical properties.

Crystalline solids may be differentiated from amorphous solids also from a consideration of their behaviour on melting. Pure crystalline solids generally have a clearly defined melting point, and the range of melting is very short, generally less than 1 K. Amorphous solids do not have a sharp melting point. Instead, they soften over a wide range of temperature; glass is a notable example. Amorphous solids are not common: glasses, plastic sulfur, some borates, silicates and phosphates, alkanes of large molecular weight and polyvinyl chloride are examples of this class of solid.

## 1.4 LIQUID CRYSTALS

Certain organic crystals, on heating carefully, pass into a state which is intermediate between those of solid and liquid. This state is called the mesomorphic state, or liquid crystal. Liquid crystals form the basis of devices which are sensitive to small changes in temperature and pressure. Thus, they are of considerable interest in the field of medical diagnosis.

Liquid crystals were discovered by Reinitzer in 1888: if cholesteryl benzoate, $C_{34}H_{50}O_2$, is heated, it melts sharply at 419 K forming an opaque liquid crystal; at 452 K, there is a sudden clearing to produce an isotropic liquid.

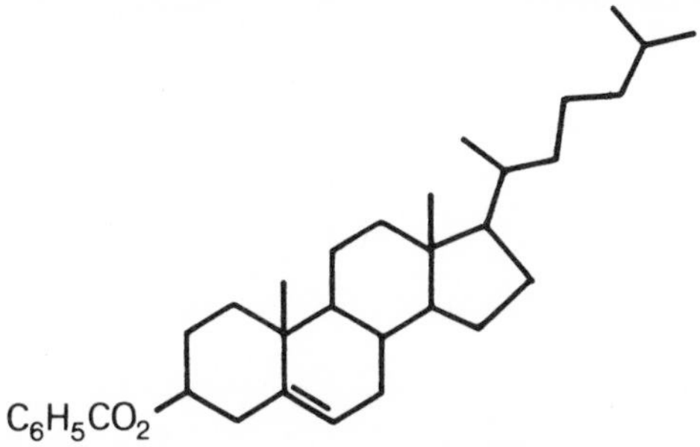

$C_6H_5CO_2$

Liquid crystals usually consist of large, elongated molecules, possessing one or more polar groups such as $-NH_2$ or $\rangle CO$. In the crystalline state, these substances pack with their molecules aligned parallel to one another, and linkages arise through both attractions between the polar groups and van der Waals' induced dipolar forces. On heating, the weaker van der Waals' forces are overcome first by the thermal energy supplied to the crystal, and relative movement can occur. Further heating breaks also the dipolar bonds, and the substance passes into true liquid state.

Several phases of liquid crystal are recognised (Figure 1.7). Not all liquid crystals exhibit each phase, however. In the nematic phase, shown, for example, by ammonium oleate, $CH_3(CH_2)_7CH{=}CH(CH_2)_7CO_2NH_4$, the long molecules are arranged with their lengths parallel to one another, but without any periodicity, rather like an army of descending parachutists. In the smectic phase, shown for example by *p*-azoxyanisole, $H_3CO{-}\langle\bigcirc\rangle{-}N(O){=}N{-}\langle\bigcirc\rangle{-}OCH_3$, molecules are arranged on equally spaced planes, but without any lateral periodicity, rather like a crowd of shoppers in a department store. The transitions between the mesophases are reversible, occurring at definite transition temperatures which vary with pressure according to the Clausius-Clapeyron equation:†

$$dP/dT = \Delta H_t/T(V_2-V_1) \,, \tag{1.6}$$

where $V_1$ and $V_2$ are the molar volumes of the two phases and $\Delta H_t$ is the enthalpy change of the transition.

†Suggested Reading: *Thermodynamics* (Smith).

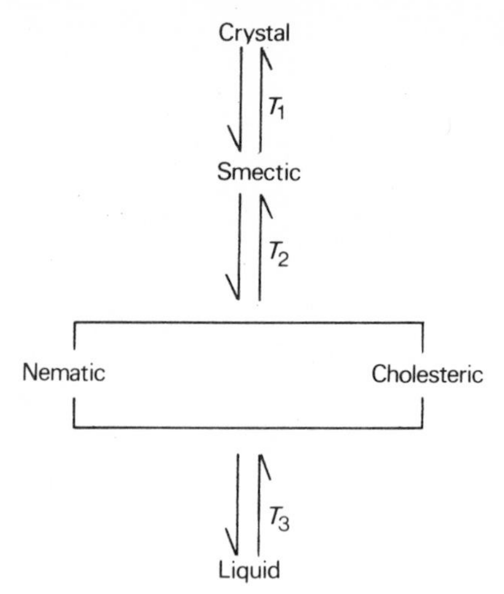


Figure 1.7 – Phases of liquid crystals; $T_1$ to $T_3$ represent transition temperatures.

The cholesteric liquid crystals also exhibit optical activity.

## 1.5 ISOMORPHISM AND POLYMORPHISM

According to Mitscherlich (1819), substances showing similar crystal shape would be expected to correspond in their chemical compositions. This early law of isomorphism may be illustrated through data on the axial ratios $(a:b:c)$† of sulphates and selenates of a number of singly-charged cations (Table 1.1). The results indicate that the components of the different crystalline substances pack together in the solid state with Rb taking the place of K, Se taking the place of S, and so on.

**Table 1.1 – Crystallographic axial ratios**

| | $a$ | : | $b$ | : | $c$ |
|---|---|---|---|---|---|
| $K_2SO_4$ | 0.573 | | 1 | | 0.742 |
| $Rb_2SO_4$ | 0.572 | | 1 | | 0.749 |
| $Cs_2SO_4$ | 0.571 | | 1 | | 0.753 |
| $Tl_2SO_4$ | 0.564 | | 1 | | 0.732 |
| $(NH_4)_2SO_4$ | 0.556 | | 1 | | 0.733 |
| $K_2SeO_4$ | 0.573 | | 1 | | 0.732 |
| $Rb_2SeO_4$ | 0.571 | | 1 | | 0.739 |
| $Cs_2SeO_4$ | 0.570 | | 1 | | 0.742 |
| $Tl_2SeO_4$ | 0.555 | | 1 | | 0.724 |

†See Appendix 2.

A wider definition of isomorphism is now recognised. Chemical resemblance is not wholly essential; it is a similarity between structural units that is most important. Thus, KCl and RbCl are isomorphous in the NaCl structure type (Figure 1.2) at stp, but CsCl (Figure 1.8) has a different structure. Again, $KNO_3$ and $NaNO_3$ are not isomorphous at stp, but $NaNO_3$ is isomorphous with $CaCO_3$ (calcite) under these conditions of temperature and pressure.

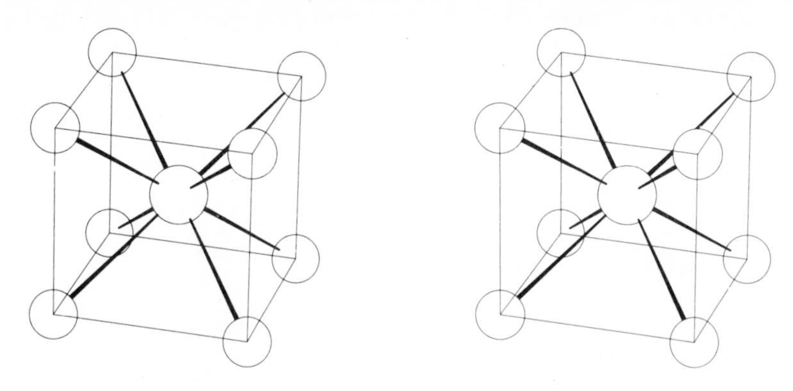

Figure 1.8 – Unit cell of the CsCl structure type; circles in decreasing order of size are Cl and Cs.

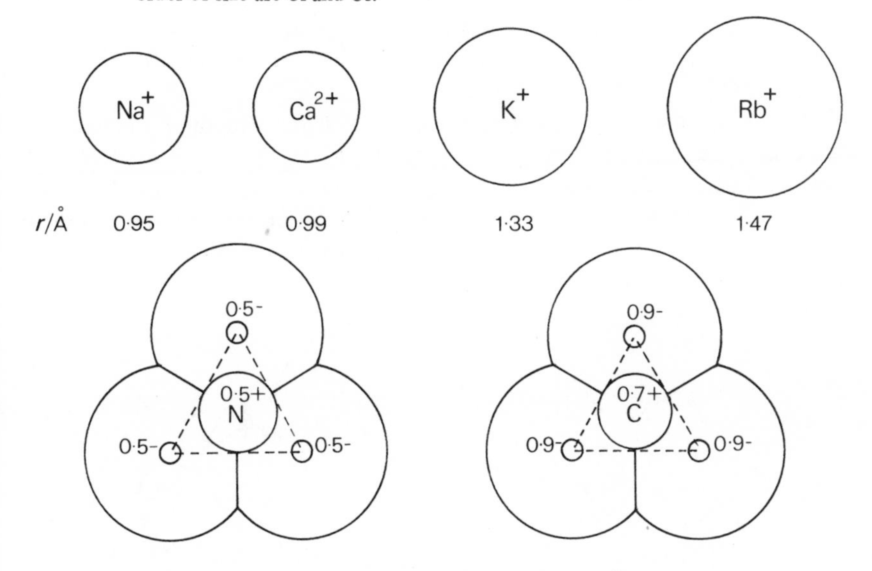


Figure 1.9 – Sizes and shapes of some structural units. In $[NO_3]^-$ and $[CO_3]^{2-}$, the fractional charges quoted were obtained by molecular orbital calculations; the dashed lines emphasise that the shape of these ions is a prism on an equilateral triangle as base.

In these examples, the structural units are simple or complex ions, and their similarities are shown in Figure 1.9. The size and shape of the cations $Na^+$ and $Ca^{2+}$, and of the anions $[NO_3]^-$ and $[CO_3]^{2-}$, enable $NaNO_3$ and $CaCO_3$ to be isomorphous in the calcite structure type (Figure 1.10). The $K^+$ ion is too large to be accommodated in this way. Generally, if size differences are less than about 15%, isomorphism may occur, subject to the condition of overall electrical neutrality of the structure. This rule is not inviolable; NaCl, KCl and RbCl are isomorphous at stp.

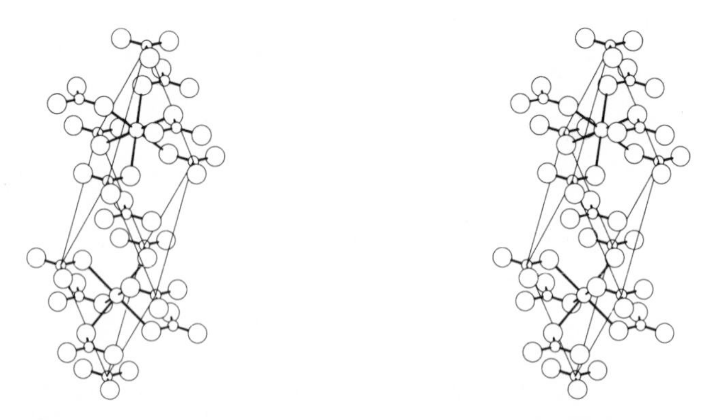

Figure 1.10 – Unit cell of the calcite ($CaCO_3$) structure type: circles in decreasing order of size are O, Ca and C. This structure is isomorphous with that for $NaNO_3$ at stp.

Two solids are said to be anti-isomorphous if the structural arrangements are geometrically similar, but with the positions of the atoms interchanged. Fluorite, $CaF_2$, is illustrated in Figure 1.11. $Li_2O$, in which Li and O replace F and Ca respectively, is sometimes called an anti-fluorite structure. Both of the terms anti-isomorphous and anti-fluorite are unnecessary, possibly misleading, and will not be used again in this book.

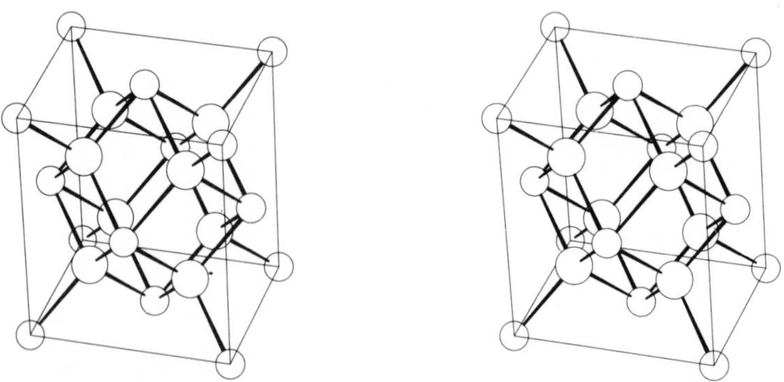

Figure 1.11 – Unit cell of the $CaF_2$ structure type; circles in decreasing order of size are F and Ca. Compare with Figure 1.8.

Solids which exhibit more than one structure type are called polymorphic. The different structures may involve quite different interatomic forces and, therefore, show very different physical properties. White tin ($\beta$-Sn, Figure 1.12) is metallic and conducts electricity, whereas grey tin ($\alpha$-Sn, Figure 1.13) is powdery and non-metallic. The number of equidistant or near-equidistant neighbours of a given atom is called its coordination number. Thus, the co-ordination numbers of $\alpha$-Sn and $\beta$-Sn are 4 and 6 (irregular), respectively. Another good example of polymorphism is afforded by the diamond and graphite structures (Figures 1.14 and 1.15): polymorphism in elements is sometimes called allotropy. Among compounds, ammonium nitrate has at least five polymorphs, and calcium carbonate crystallizes as aragonite and vaterite, as well as calcite. In the aragonite structure type (Figure 1.16), $CaCO_3$ and $KNO_3$ are isomorphous at stp.

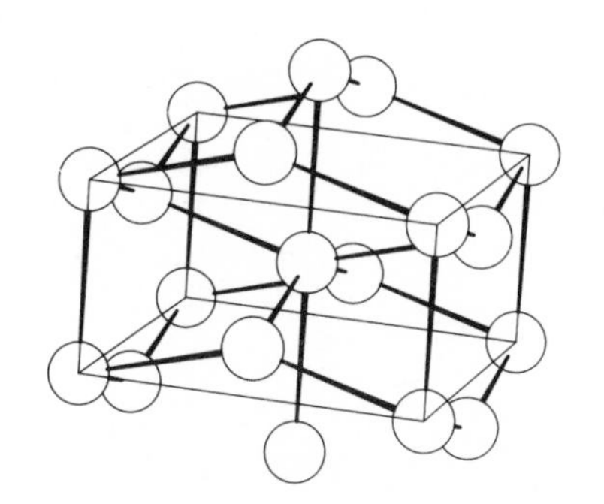
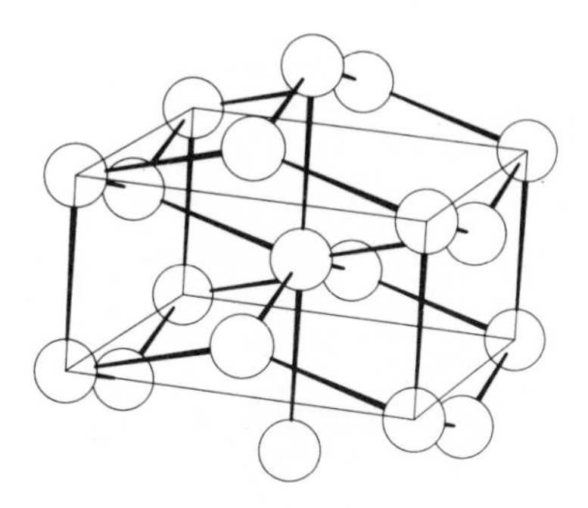

Figure 1.12 – Unit cell of the structure of white tin ($\beta$-Sn).

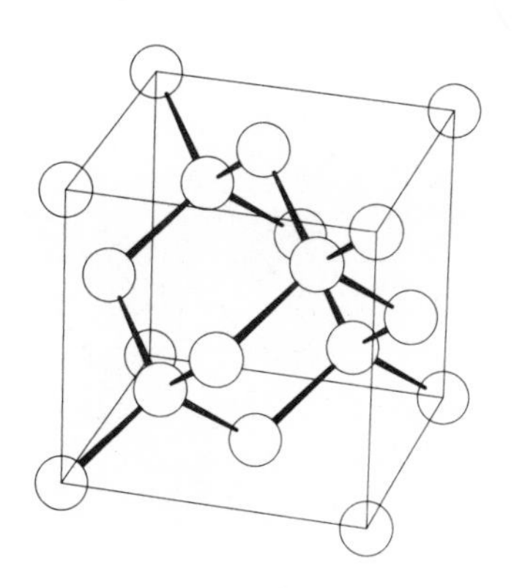
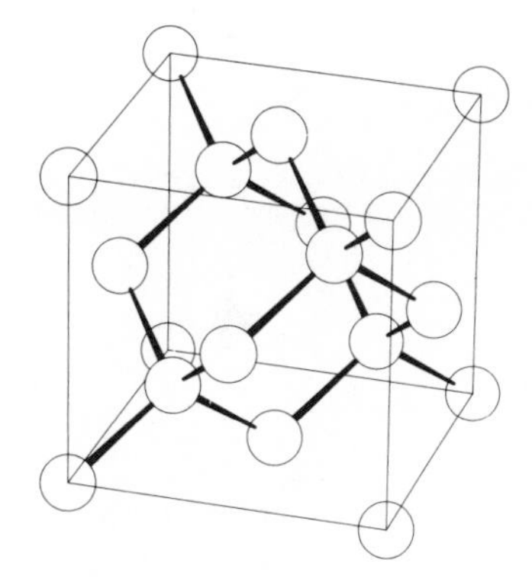

Figure 1.13 – Unit cell of the structure of grey tin ($\alpha$-Sn). Compare with Figure 1.14.

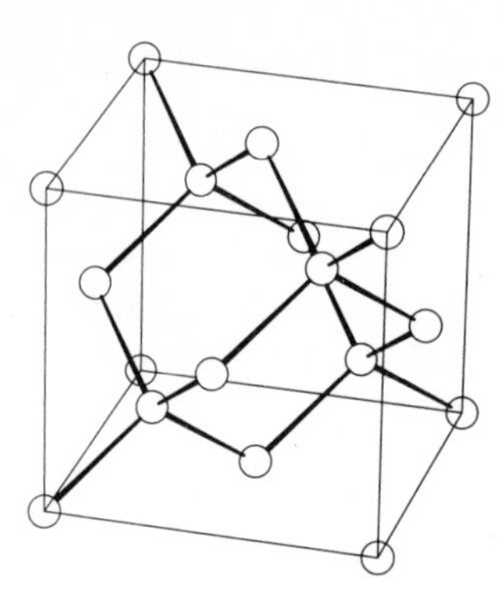
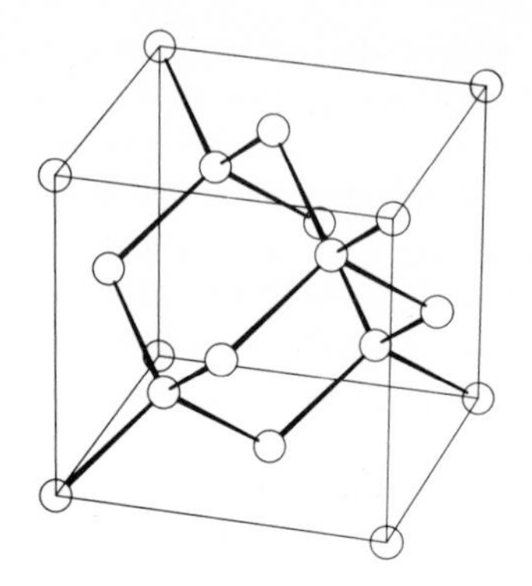

Figure 1.14 – Unit cell of the structure of diamond (C). Observe the three-dimensional configuration of similar bonds.

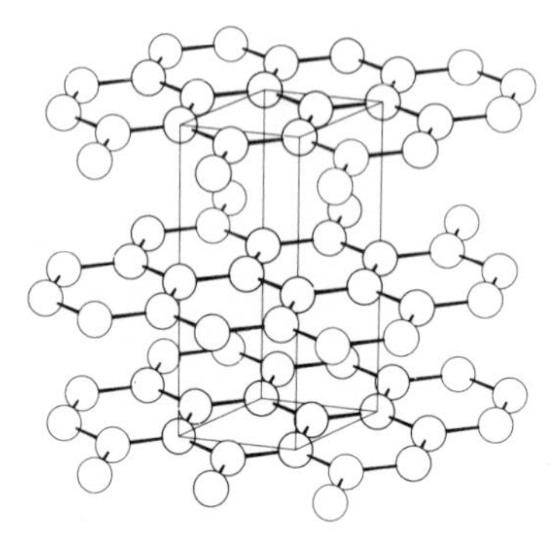
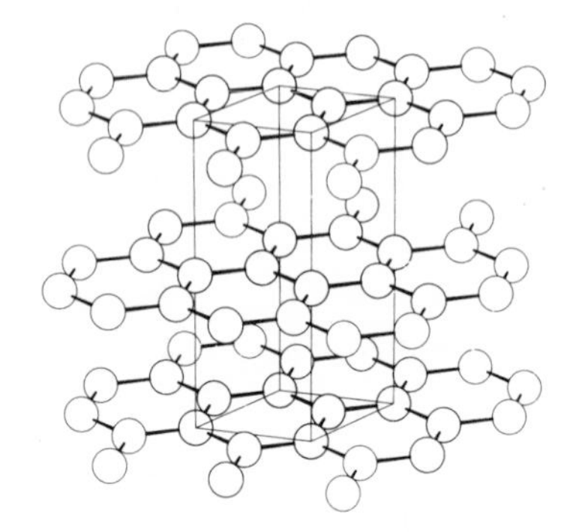

Figure 1.15. – Unit cell of the graphite (C) structure type; note the longer distances between the layers than between adjacent atoms in any layer.

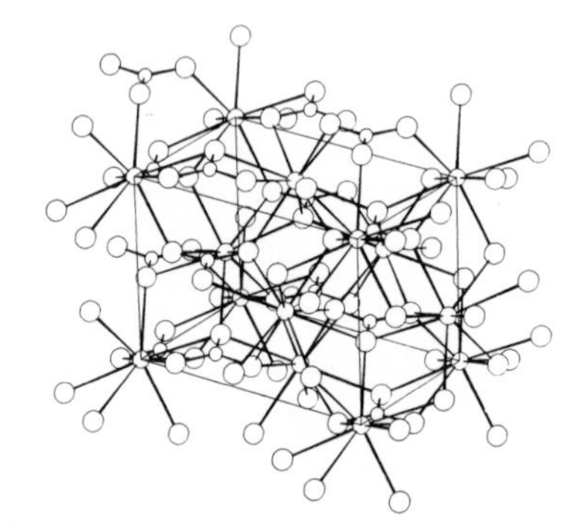

Figure 1.16 – Unit cell of the aragonite ($CaCO_3$) structure type; circles in decreasing order of size are O, Ca and C. This structure is isomorphous with that for $KNO_3$ at stp.

Generally, each polymorph of a given compound is stable over a given range of temperature and pressure. The change from one polymorph to the other may not take place rapidly. Thus, at atmospheric pressure, α-Sn is stable below 286 K and β-Sn is stable above this temperature, but the $\beta \rightarrow \alpha$ transformation becomes rapid only in the neighbourhood of 233 K or less.

## 1.6 TRANSITIONS IN SOLIDS

Much of the data on the solid state that are available to us, particularly crystal structures, refer to a temperature of about 290 K and a pressure of 1 atmosphere (101325 N $m^{-2}$). The temperature range of availability of solids is from approximately 4000 K (tantalum carbide) to absolute zero: above 4000 K all known solids will have metled, vaporized or decomposed.

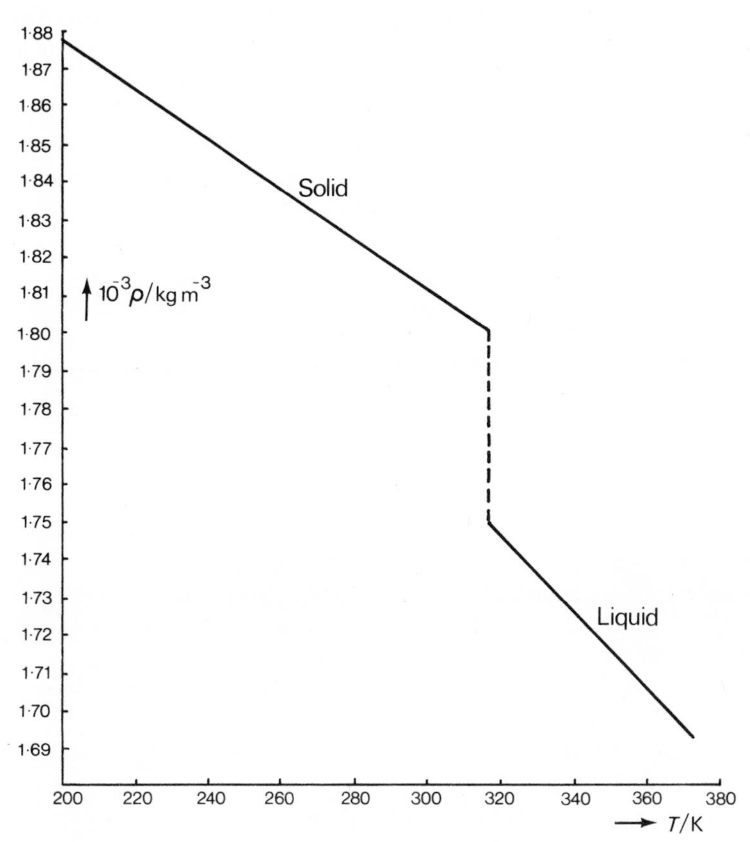


Figure 1.17 – Variation of density with temperature for white phosphorus (from Sapper and Biltz, 1931; Biltz and Meisel, 1931; Dobinski 1933, 1935). Note the discontinuity at the melting point.

It is important to consider the conditions of temperature and pressure when examining structures for isomorphism. At a temperature greater than 500 K, potassium nitrate transforms from the aragonite to the calcite structure type; under these conditions, $NaNO_3$ and $KNO_3$ are isomorphous. Polymorphic transitions are well established, and transition temperatures and pressures are known from experimental studies. Transitions may be classed as sharp or gradual, and each type possesses certain important characteristics.

A sharp transition occurs at a precise temperature and pressure, and the Clausius-Clapeyron equation (1.6) is obeyed at the transition point. Many physical properties, such as density, heat capacity and entropy, show discontinuities at transition points. Figure 1.17 shows the variation of density with temperature for white phosphorus. The discontinuity at the change from solid to liquid at 317 K shows that melting is a sharp transition.

When a solid undergoes a transition, the packing of atoms and, hence, the molar volume, changes considerably and abruptly. Since the mass is constant, there is a consequent alteration in density which reflects the change in atomic or molecular structure. Generally the solid phase of a substance is more dense than its liquid form; ice is a notable exception.

In gradual transitions, the temperature of the transformation is not clearly defined, and the change may take place over an appreciable range of temperature. Discontinuities in physical properties are not observed, but maxima occur in the temperature or pressure variations of, for example, heat capacity or compressibility. The Clausius-Clapeyron equation is not obeyed, and some gradual transitions may be accompanied by hysteresis in a physical property (Figure 1.18).

The change in heat capacity with temperature for ammonium chloride shows an interesting feature known as a $\Lambda$-type transition (Figure 1.19). Ammonium chloride has the CsCl structure type (Figure 1.8). At very low temperatures, the hydrogen atoms on the $[NH_4]^+$ ion at the centre of a unit cell are directed tetrahedrally to four of its eight corners, in an ordered manner throughout the crystal. As the temperature is raised, a new structure is taken up with a unit cell of twice the size of that of the low-temperature form in order to accommodate $[NH_4]^+$ ions in both of the possible orientations, again in an ordered manner. At 243 K, the $\Lambda$-type transition in the heat capacity is observed: it is related to the formation of a new structure, having the normal, smaller size of unit cell, but in which twofold orientational disorder occurs, because the two possible tetrahedral coordinations of $[NH_4]^+$ are taken up randomly among the unit cells in a crystal of macroscopic dimensions.

The molar entropy change of the $\Lambda$-type transition, $\Delta S_t$, is known from experiment to be 4.6 J $K^{-1}$ $mol^{-1}$. This value may be compared with that obtained through Boltzmann's statistical equation

$$\Delta S = k\ln(W_2/W_1) , \qquad (1.7)$$

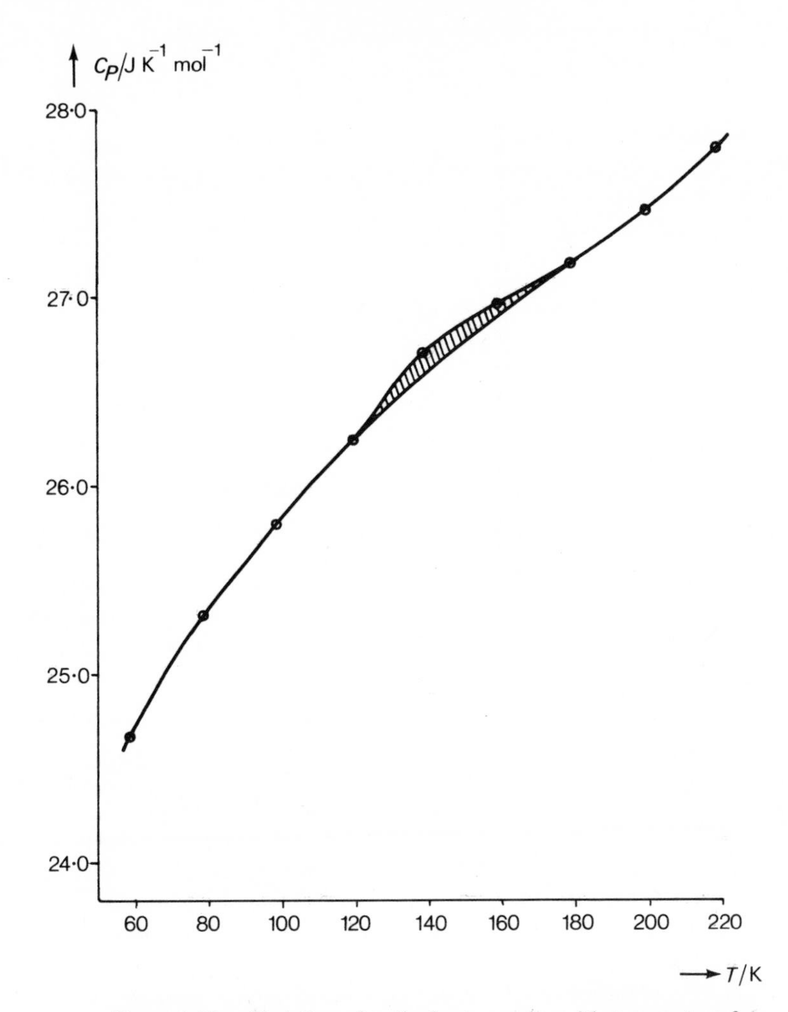


Figure 1.18 – Variation of molar heat capacity with temperature for cesium, showing hysteresis between 110 and 190 K (from Dauphinee, Martin and Preston-Thomas, 1955); the return path (cooling) is the lower curve.

where $k$ is the Boltzmann constant and $(W_2/W_1)$ is the ratio of the numbers of arrangements of the two systems. Hence, in molar terms,

$$\Delta S_t = Lk\ln(W_2/W_1) \ , \tag{1.8}$$

where $L$ is the Avogrado constant. Since $(W_2/W_1)$ is 2 for the disordering of the

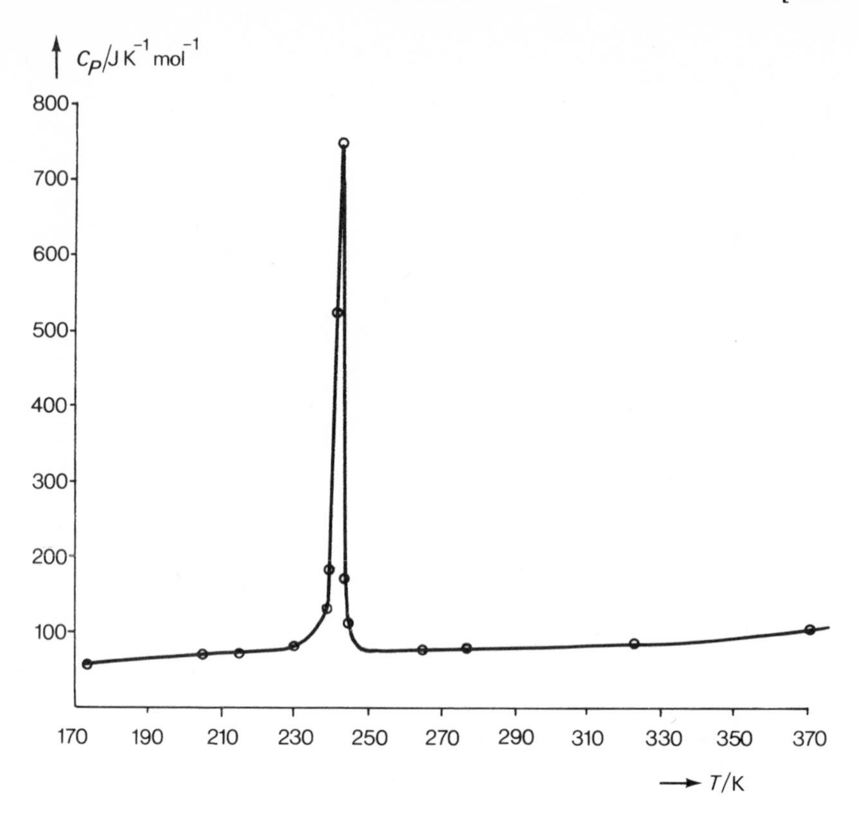


Figure 1.19 – Λ-type transition at 243 K in the molar heat capacity of ammonium chloride (from Simon, Simson and Ruhemann, 1927; Klinkhardt, 1927); a perfect example of its name.

$NH_4Cl$ structure, $\Delta S_t$ is 5.8 J K$^{-1}$ mol$^{-1}$, which compares favourably* with the experimental result. Equation (1.7) is related to the Boltzmann equation for the distribution of energies: in a system of molecules at equilibrium, the number $n_\epsilon$ of molecules of energy greater than $\epsilon$ is given by

$$n_\epsilon = n_o \exp(-\epsilon/kT) \; , \tag{1.9}$$

where $n_o$ is constant, equal to the number of molecules in the lowest energy state. Multiplying the numerator and denominator of the exponent in (1.9) by $L$ leads to a similar result, in molar terms:

$$N_E = N_o \exp(-E/RT) \; . \tag{1.10}$$

*In measuring $\Delta H_t$ it is necessary that the transition is approached and carried out slowly so that a complete conversion from one form to the other is achieved ($\Delta S_t = \Delta H_t/T_t$).

A simplified derivation of this distribution equation is given in Appendix 4, and the form of (1.10) is shown in Figure 1.20.

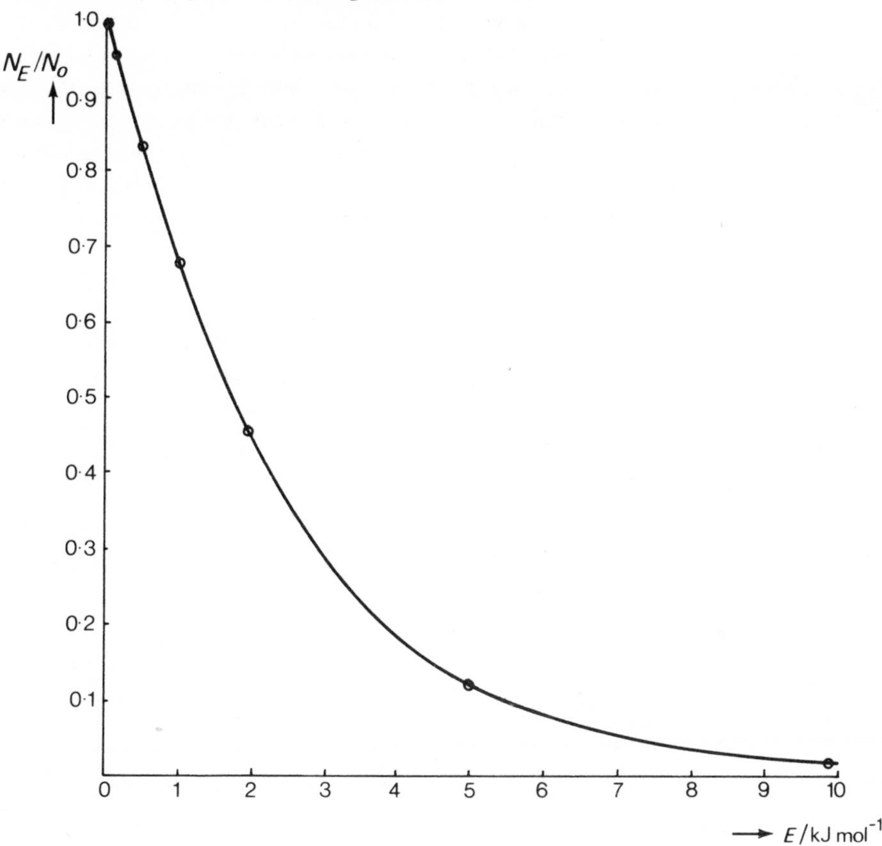


Figure 1.20 – Variation of $N_E/N_0$ with $E$ at 298 K according to the Boltzmann distribution equation; notice the initial rapid decrease with increasing $E$.

As the temperature of a solid is increased, the thermal motion of its component atoms is enhanced. The consequent changes in the structure and properties of the solid depend upon the balance between this thermal energy and the interatomic forces. The thermal vibrations may bring about polymorphic transformation (KCN at 233 K), melting (Na at 371 K), sublimation ($I_2$‡ at 458 K) or decomposition ($Pb[NO_3]_2$ at 743 K). In each of these examples, the entropy change for the transition is positive, indicating that it is accompanied by a decrease in the degree of order of the substance. The 3rd law of thermodynamics states that the entropy of an infinite crystal of a pure substance† is

‡Provided that the partial pressure of iodine vapour is less than 90 mmHg.

†Suggested Reading: *Thermodynamics* (Denbigh)

zero at 0 K, indicating a situation of perfect order. This law does not preclude complete order being attained at temperatures above 0 K: indeed, it is possible that it exists in a superconductor, for example Mo at 0.9 K, although the ordering process may be more elaborate than that associated with the third law.

When the temperature of a solid is decreased, atomic vibrations decrease in amplitude and any rotational motion ceases, until finally the atoms remain with only their zero energy of vibration. The changes in atomic motion are accompanied by a decrease in the heat capacity which, from the 3rd law, must be zero at 0 K. The loss of vibrational modes in the solid can be detected through the decrease in intensity, and finally the absence of, infrared spectral lines. The electrical resistivities of solids change markedly with temperature. Those of ionic solids tend to very high values with decrease in temperature, whereas the resistivities of metals tend to zero near 0 K. We shall discuss these effects after studying ionic and metallic solids in later chapters.

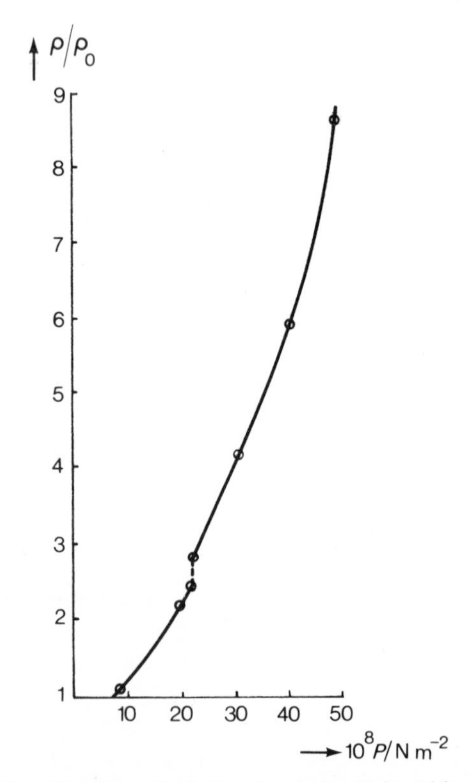


Figure 1.21 – Variation of relative electrical restivity with pressure for cesium at ambient temperature; $\rho_0$ is the resistivity at zero pressure (from Bridgman, 1952). A discontinuity accompanies the polymorphic transition.

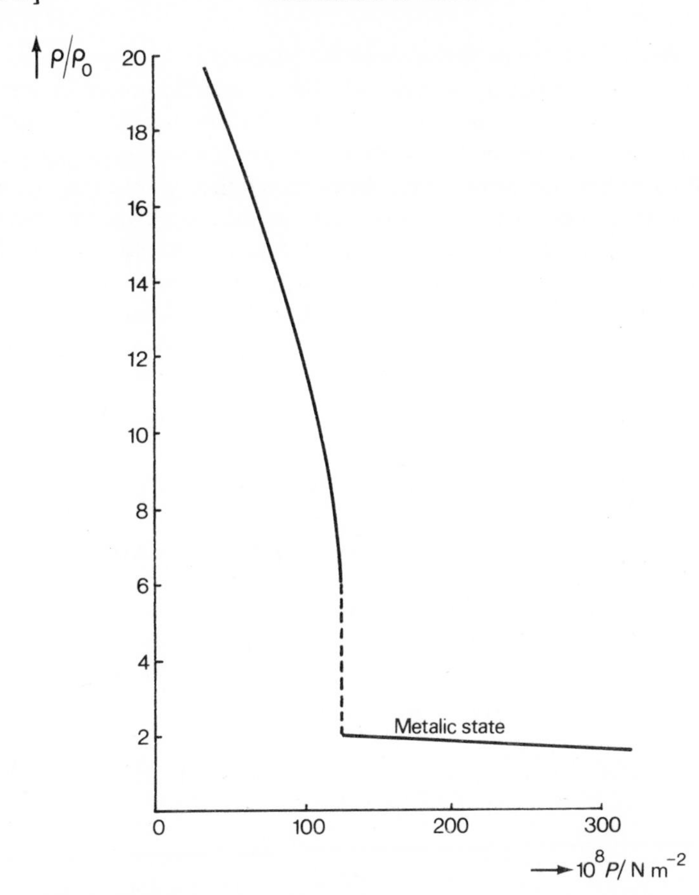


Figure 1.22 – Variation of relative electrical resistivity with pressure for selenium at ambient temperature (from Balchan and Drickamer, 1961) showing a discontinuity at high pressure.

The effects of pressure on solids, though important, are often less dramatic than those produced by changes in temperature. Generally, an increase in pressure produces changes in solids akin to those obtained by a decrease in temperature: crystals undergo polymorphic transitions producing closer-packed structures (higher coordination number); atomic vibrations decrease in amplitude, with a weakening in the intensity of infrared spectral lines. The changes in electrical resistivity vary considerably among different classes of solid. Figures 1.21 and 1.22 illustrate the change in electrical resistivity with pressure for cesium and selenium, respectively. The discontinuity in the resistivity curve for cesium at $21.5 \times 10^8$ N m$^{-2}$ (20 000 atm) corresponds to the transition from a body-centred cubic (8-coordinate) to a face-centred cubic (12-coordinate) structure. In the case of selenium, a metallic structure develops at a pressure of about 120 000 atmospheres.

### 1.6.1 Thermodynamic properties at transition points

Thermodynamic data on substances at their transition points are informative; a selection of data is given in Table 1.2. We see that all $\Delta H$ values are positive: liquids are in higher energy states than are the corresponding solids at their melting points, and gases are of higher energy than are the corresponding liquids at their boiling points. All $\Delta S$ values are also positive: the degree of randomness increases in the sequence solid → liquid → gas. Excluding helium, the enthalpies of fusion range from about 0.1 to 46 kJ $mol^{-1}$, and the corresponding entropy changes range from about 8 to 44 J $K^{-1}mol^{-1}$. For vaporization, $\Delta H$ ranges from about 9 to 300 kJ $mol^{-1}$, but the corresponding $\Delta S$ changes cluster around 100 J $K^{-1}mol^{-1}$ to within about 20%, excluding helium and hydrogen.

**Table 1.2 – Thermodynamic data of transition points**

| | | | $\Delta H$/kJ $mol^{-1}$ | | $\Delta S$/J $K^{-1}mol^{-1}$ | |
|---|---|---|---|---|---|---|
| | Mp/K | Bp/K | Fusion | Vaporization | Fusion | Vaporization |
| $MgCl_2$ | 985 | 1691 | 43.1 | 137 | 43.9 | 81 |
| KCl | 1043 | 1680 | 25.5 | 162 | 24.4 | 96 |
| NaCl | 1074 | 1738 | 28.5 | 171 | 26.5 | 98 |
| $BeCl_2$ | 678 | 793 | 12.6 | 105 | 18.6 | 132 |
| Ge | 1210 | 3103 | 34.7 | 285 | 28.7 | 92 |
| Si | 1683 | 2953 | 46.4 | 297 | 27.6 | 101 |
| He | 1.0¶ | 4.2 | 0.021 | 0.084 | 6.3 | 21 |
| $H_2$ | 14.0 | 20.0 | 0.13 | 0.92 | 9.3 | 46 |
| $CH_4$ | 91 | 112 | 0.96 | 9.20 | 10.5 | 82 |
| $H_2O$ | 273 | 373 | 5.86 | 47.3 | 21.5 | 127 |
| Hg | 234 | 630 | 2.43 | 64.9 | 10.4 | 103 |
| Na | 371 | 1165 | 2.64 | 103 | 7.1 | 88 |

¶ At 26 atmospheres.

These data express in thermodynamic terms some of the ideas which we have discussed on the states of matter and transitions between them. The regularity in solids leads to their greater energetic stability compared with liquids. When the temperature of a substance is high enough for the thermal energy of the liquid state to dominate the cohesive energy of the solid, melting occurs. We recall from thermodynamics that the tendency for a reaction to take place is expressed by the free energy change, $\Delta G$, of the process under consideration, a negative value of $\Delta G$ indicating the spontaneous, or preferred, direction of reaction. The equation

$$\Delta G = \Delta H - T\Delta S \,, \tag{1.11}$$

shows that the determining quantity $\Delta G$ represents a sort of compromise between its often opposing enthalpic and entropic components. Of particular interest is the liquid → gas transition. Because of the much greater randomness of gases compared to liquids, these entropy changes are dominated by the entropies in the gaseous state, which are of similar value, since gases have, from Avogadro's hypothesis, approximately equal molar volumes; hence, $\Delta S$ (vaporization) tends to a constant value. This trend is sometimes known as Trouton's law, but we can see from Table 1.2 that it is only an approximate relationships: other factors, such as hydrogen bonding in water and its vapour, modify the Trouton value of $\Delta S$ (vaporization) because they constrain the degree of randomness of the molecules in the liquid and gas phases.

## 1.7 CLASSIFICATION OF SOLIDS

No method of classifying solids is entirely free from ambiguity. Yet it is desirable that the vast body of available structural information be discussed over a framework which attempts to group solids according to a few chosen parameters. We can classify solids into four groups, dependent upon the bond type mainly responsible for cohesion in the solid state. Some authors define five, or even seven, classes of solid. The differences among some of these groups are, however, more a matter of degree than of kind; since uniqueness is still not attained, the simpler grouping seems preferable. The four main classes of solids are based on ionic, covalent, van der Waals' and metallic bonding, and a scheme is exemplified through Table 1.3.

### 1.7.1 Ionic bonding

Ionic bonding involves an electron donor-receptor mechanism among the participating atoms, leading to the formation of ions in which electrons are essentially localized in the atomic orbitals of the charged species. The electronic structures of ions are often similar to those of inert gases,† which are known to be particularly stable configurations. The attractive forces between oppositely charged ions are largely coulombic in nature, and they are balanced by repulsions both between similarly charged species and, more strongly, especially under compressive stress, between electrons in closed inner shells, or energy levels, of all species. A good example of an ionic solid is potassium fluoride, KF; it has the NaCl structure type (Figure 1.2).

†The reader is referred to the Periodic Table inside the back cover of this book.

**Table 1.3 – Classification of Structures:** I, class bond type predominant; II, overlap with other areas of the classification; *Ebs*. the electrostatic bond strength, is the oxidation number of a species divided by the coordination number; $z_X$ is the charge on an anion species (of negative oxidation number, like oxygen −2).

| | IONIC | | | COVALENT | |
|---|---|---|---|---|---|
| Structure Class | I $Ebs < \lvert z_X \rvert/2$ | II $Ebs = \lvert z_X \rvert/2$ | I $Ebs > \lvert z_X \rvert/2$ | I | II |
| Close-packed, or nearly close-packed | Halides and oxides of metals: $MX$, $MX_2$, $MX_3$ types. Perovskites Spinels | Borates Silicates Germanates | Salts of inorganic acids | Diamond, Si, Ge. Compounds of *B*-group elements amongst themselves, such as ZnS. | Compounds of *B*-Group metals with P, As.NiAs, Pyrite. |
| Chain | – | Pyroxenes Amphiboles | – | – | – |
| Layer | – | Micas | Gypsum | – | – |
| Framework | – | Felspars and zeolites | – | – | – |
| | VAN DER WAALS' | | | METALLIC | |
| | I | II | | I | II |
| Close-packed or nearly close-packed | Inert gases | Molecular gases: $O_2$, $N_2$, HCl. Sulphur Organic compounds | | True metals and their alloys | Zn, Cd, Hg, Sn. Alloys of more metallic *B*-group metals with one another or with true metals |
| Chain | – | Rubber. Cellulose Fibrous proteins | | – | Se, Te, $Sb_2S_3$ |
| Layer | – | Layer structures, such as graphite | | – | As, Sb, Bi, $MoS_2$ |
| Framework | – | Globular proteins | | Interstitial compounds | – |

### 1.7.2 Covalent bonding

Covalent bonding involves an electron-sharing mechanism by means of which an inert-gas configuration is again attained by the individual atoms present. We may say that the atomic orbitals overlap one another leading to electron sharing (electron density fluctuating with time) and exhibiting the strongly directional character inherent in the orbitals themselves. A balancing repulsion arises mainly from the inner closed electron shells. The electron-pair shared by two atoms often remains essentially localized to the orbitals of these atoms. Diamond (Figure 1.14) is the best example of a covalent solid. The bonds between the atoms in the molecules of van der Waals' compounds and between the atoms in complex ions are mainly covalent too.

### 1.7.3 Van der Waals' bonding

At low temperatures, the inert gases can be solidified. We can envisage a closed-shell repulsion between these atoms when they are brought close together, but what is the mechanism of attraction? The electron density of an atom is changing with time. At any instant, the effective centres of gravity of the positive nucleus and the negative electrons may not coincide. This separation of charges defines a dipole, and it can induce a dipole in a neighbouring atom and so lead to a form of electrostatic attraction, albeit very weak compared with ionic or covalent bonding, which binds the inert gas atoms in the solid state. This type of binding exists between all species, but it is of paramount importance among van der Waals' compounds. Molecules may possess permanent dipoles, which increases the degree of intermolecular attraction. There are a number of dipolar-type interactions; they may be described collectively as van der Waals' bonding forces. A good example of a van der Waals' or molecular, solid is argon, Ar (Figure 1.23); the majority of organic molecules from molecular solids.

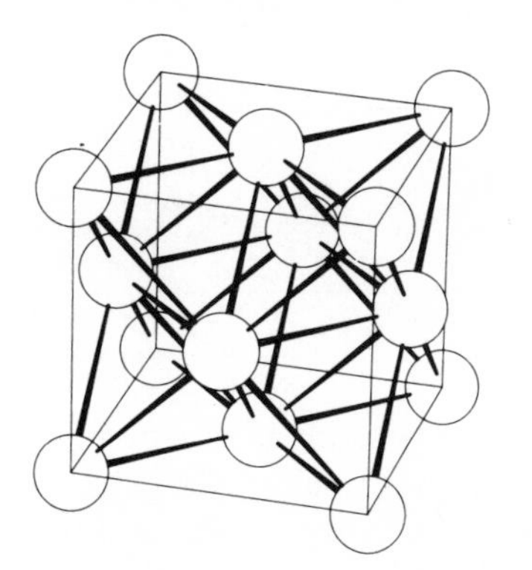

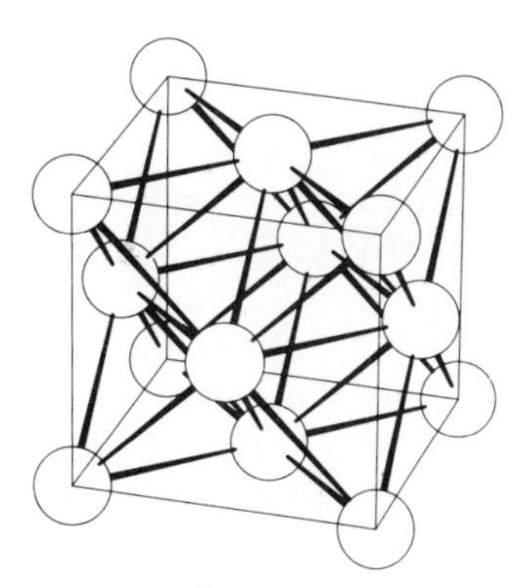

Figure 1.23 – Unit cell of the structure of argon, Ar; an example of a close-packed cubic structure. See Figure 5.13 for a space-filling diagram of the same structure type.

1.7.3.1 *Hydrogen bonding*

Hydrogen bonding is a notable example of a dipolar-type interaction. Hydrogen is able to form a bridge between two atoms, most strongly with F, O and N, which enhances the normal intermolecular attractions. We can find evidence of the strength of this bond among the hydrides of the Group V*B*, VI*B* and VII*B* elements (Table 1.4). Normally, melting temperatures increase with increasing molecular mass; the important exception of water, is well known Hydrogen bonds enhance the attractive forces in both ionic and molecular solids: in gypsum $CaSO_4{\cdot}2H_2O$, the cohesion in one direction is governed by hydrogen bonds (Figure 1.24).

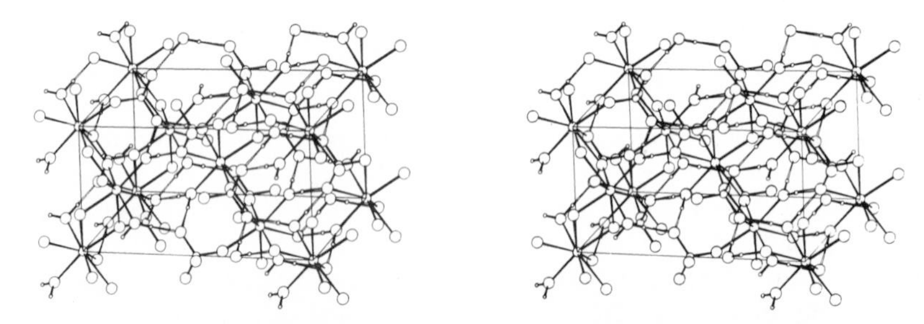

Figure 1.24 – Unit cell of structure of gypsum, $CaSO_4{\cdot}2H_2O$; circles in decreasing order of size are O, Ca, S and H. Note the importance of the hydrogen bonds (double lines) to the cohesion of the structure.

**Table 1.4 – Melting temperatures of some hydrides/K**

| | | | | | |
|---|---|---|---|---|---|
| $H_3N$ | 195 | $H_2O$ | 273 | HF | 190 |
| $H_3P$ | 140 | $H_2S$ | 190 | HCl | 159 |
| $H_3As$ | 157 | $H_2Se$ | 207 | HBr | 186 |
| $H_3Sb$ | 185 | $H_2Te$ | 224 | HI | 222 |

**1.7.4 Metallic bonding**

We know that in certain molecules, such as benzene or naphthalene, some of the bonding electrons are delocalized; they belong to the molecule as a whole. In metallic solids, this delocalization is extended to the crystal as a whole; a regular three-dimensional array of positive ions is bound by a sea of electrons. Gold is a good example of a metallic solid; it has the same close-packed structure as argon (Figure 1.23).

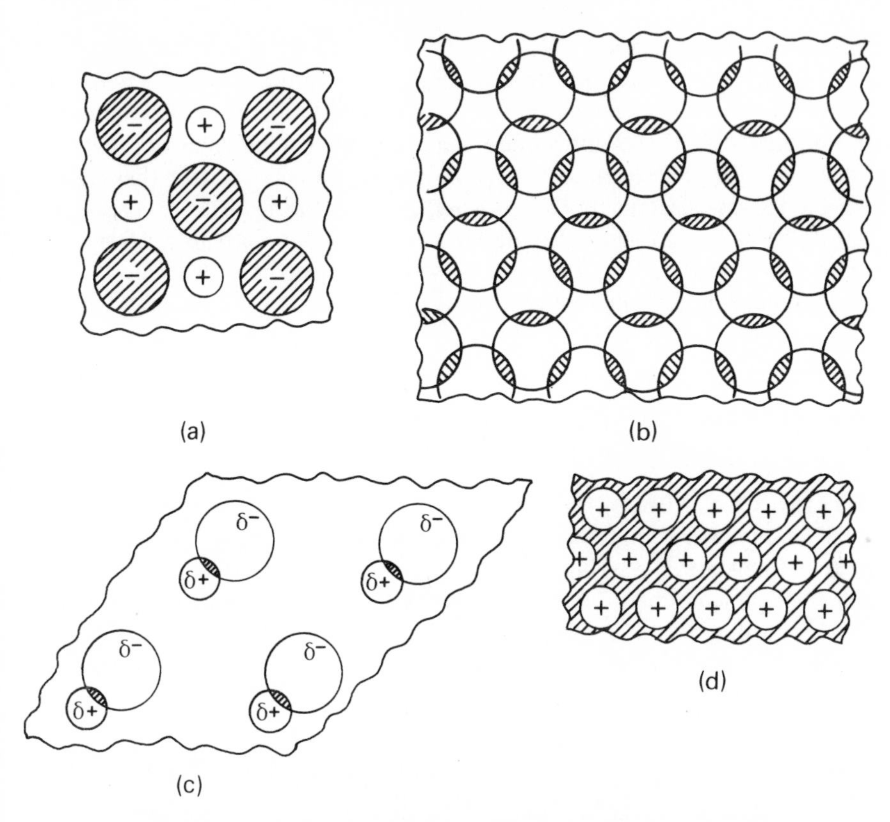


Figure 1.25 – Pictorial representation of the four main bonding forces in solids: (a) ionic, (b) covalent, (c) van der Waals' (dipolar), (d) metallic.

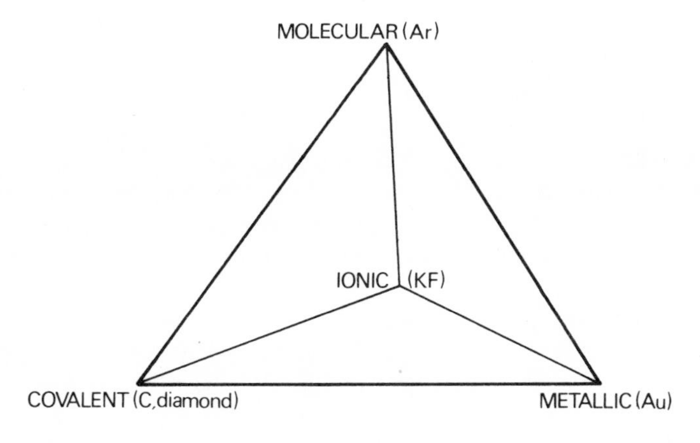


Figure 1.26 – Schematic representation of bond type, with typical solids indicated.

### 1.7.5 Concluding remarks

Figure 1.25 is schematic representation of the forms of bonding which we have discussed. We must remember that no solid exists (with the possible exception of the 'solid' inert gases) with a single bond type; some solids are better representatives of their class than are others. In general, we may consider that a given bond type lies, diagramatically, on a tetrahedron (Figure 1.26): the bond has something of the character of more than one of the four extreme types, yet is a single entity and not a mixture of these components. We shall develop our study of bonding and solid state chemistry around the four main bond types.

## APPENDIX 1 STEREOVIEWING

The representation of crystal and molecular structures by stereoscopic pairs of drawings is a valuable aid to their full understanding. Sophisticated computer programs exist whereby stereoviews may be prepared from structural data. Two diagrams of a given object are needed, and they must correspond to the views seen by the eyes in normal vision. Their correct viewing requires that each eye sees only the appropriate drawing, and there are several ways in which stereopsis may be accomplished.

1. A stereoscope can be purchased for a modest sum. Two suppliers are:

(a) C. F. Casella and Company Limited, Regent House, Britannia Walk, London N1 7ND. This maker supplies two grades of stereoscope.

(b) Taylor-Merchant Corporation, 25 West 45th Street, New York, N.Y. 10036, U.S.A.

2. The unaided eyes can be trained to defocus, so that each eye sees the appropriate diagram. The eyes must be relaxed and be directed straight ahead. A thin white card placed edgeways between the stereoviews may aid direct viewing. The stereoscopic image appears in the centre of the two diagrams but will be very slightly out of focus.

3. A stereoscope can be constructed with a little effort. Two biconvex or planoconvex lenses, each of focal length about 10 cm and diameter 2 to 3 cm, are mounted in a frame such that the centres of the lenses are about 6 cm apart. The frame must be shaped so that it can be held close to the eyes. Two pieces of thin card, prepared as shown in Figure A1.1, and glued together with the lenses in position, form a satisfactory stereoscope.

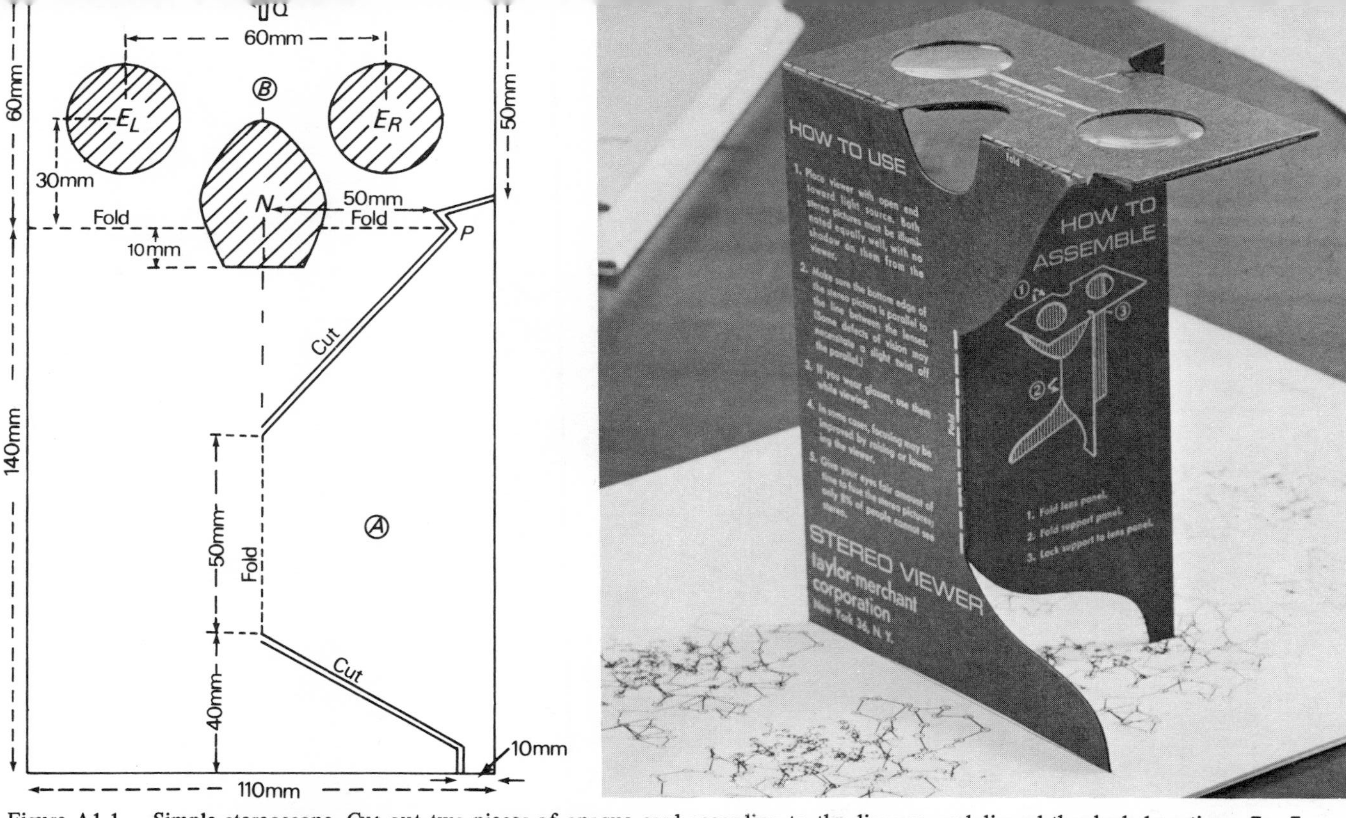


Figure A1.1 – Simple stereoscope. Cut out two pieces of opaque card according to the diagram, and discard the shaded portions, $E_L$, $E_R$ and $N$. Make cuts along the three double lines. Glue the front of one card, as seen in diagram, to the back of the other, with the lenses, $E_L$ and $E_R$, in position. Fold the portions $A$ and $B$ backwards, and engage $P$ in the cut at $Q$. View from the side marked $B$. (Photograph shows a similar stereoscope marketed by the Taylor-Merchant corporation, New York).

## APPENDIX 2 MILLER INDICES

In studying a crystal, it is convenient to refer it to three non-coplanar axes. Figure A2.1 represents the unit cell of a crystal. The origin of the crystallographic references axes is at $O$, and the unit cell has sides of length $a$, $b$ and $c$ along the right-handed axes $x$, $y$ and $z$ respectively. Let $PQR$ be any plane which makes intercepts $(1/h)a$, $(1/k)b$ and $(1/l)c$ along the $x$, $y$ and $z$ axes respectively, where $h$, $k$, and $l$ are integers. The Miller indices of the plane $PQR$ are defined as $(hkl)$: they are, thus, the reciprocals of the fractional intercepts made by the plane on the corresponding unit-cell edges. In other words, $h = 1/[(1/h)a/a]$, and so on.

If a plane is parallel to a given axis, its intercept may be said to be at infinity. Thus, the plane $BDFG$ is (010), and $BCED$ is (011). If a plane intercepts an axis on the negative side of the origin, the corresponding Miller index is negative; thus, $ABH$ is $(11\bar{2})$, read as one-one-bar-two.

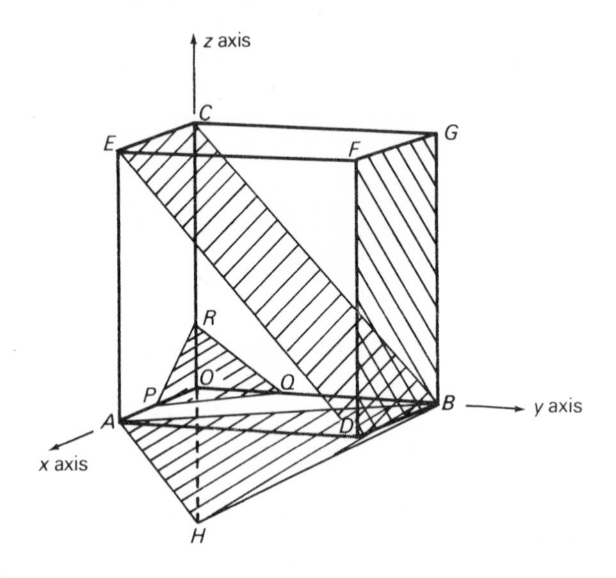


Figure A2.1 – Development of Miller indices of crystal planes.

In a morphological description of a crystal, the origin of the crystallographic axes is taken as its centre and $x$, $y$ and $z$ are chosen so that they are parallel to prominent edges of the crystal (Figure 1.4). Miller indices of crystal faces do not have, and do not need, a common factor. In other words $(hkl)$ and $(nh,nk,nl)$ are indistinguishable morphologically. If the crystallographic axes are chosen in the conventional manner described above, $h$, $k$ and $l$ must be integral (Figure A2.2).

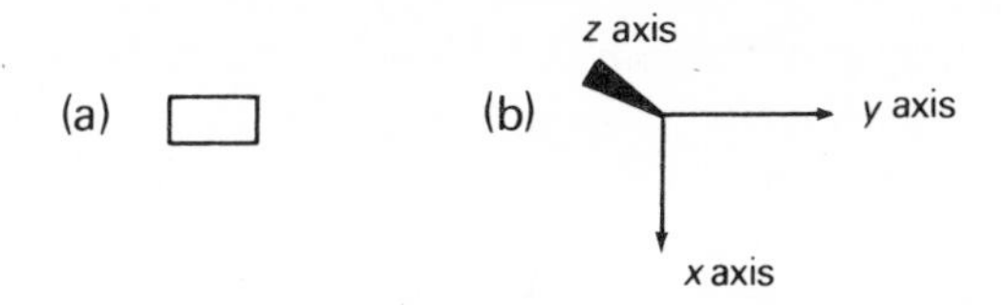


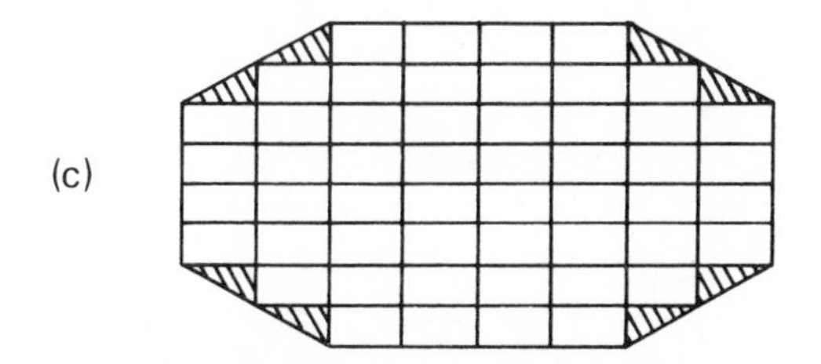


Figure A2.2 – Rational character of Miller indices
(a) Unit cell, viewed along the $z$ axis
(b) Crystallographic axes
(c) Possible crystal shape, viewed along the $z$ axis, obtained by stacking unit cells. The bounding planes are, in sequence, (100), (110), (010), ($\bar{1}$10), ($\bar{1}$00), ($\bar{1}\bar{1}$0), (0$\bar{1}$0) and (1$\bar{1}$0). The size of the unit cell and the crystal are, in this representation, disparate: the shaded steps are of the order of $10^{-6}$ mm whereas the crystal may have an average dimension of 1 mm.

A crystal volume 1 mm$^3$ will contain about $10^{18}$ unit cells of average side 10 Å stacked in a manner similar to that indicated by Figure A1.2. In studying such a crystal by X-rays, numerous equally-spaced planes of the type (100), (110) and in general ($hkl$) will be involved. The same Millerian notation is used to describe these families of parallel, equidistant planes in a crystal. The faces of a crystal are, then, the terminations of some of the possible families of planes.

The descriptor ($hkl$) applies specifically to the first plane in a family nearest to the origin in a chosen unit cell. Thus, from Figure A2.1, the plane (234) in the family of these indices makes intercepts of $a/2$, $b/3$ and $c/4$ along the $x$, $y$ and $z$ axes, respectively. The second plane in this family has the corresponding intercepts $a$, $2b/3$ and $c/2$, measured from the same origin. If these intercepts are used to define the family, we obtain ($1\frac{3}{2}2$). Clearing the fraction results in (234), as before.

## APPENDIX 3 BRAGG REFLEXION OF X-RAYS FROM CRYSTALS

Bragg's treatment of X-ray diffraction from crystals as though it were reflexion from planes of electron density was based on experimental observation. If a crystal in a diffracting position was rotated through an angle $\phi$ to the next diffracting position, the direction of the diffracted beam was found to have rotated through $2\phi$. A similar situation is experienced with the reflexion of light from a plane mirror (Figure A3.1), and in the optical lever.

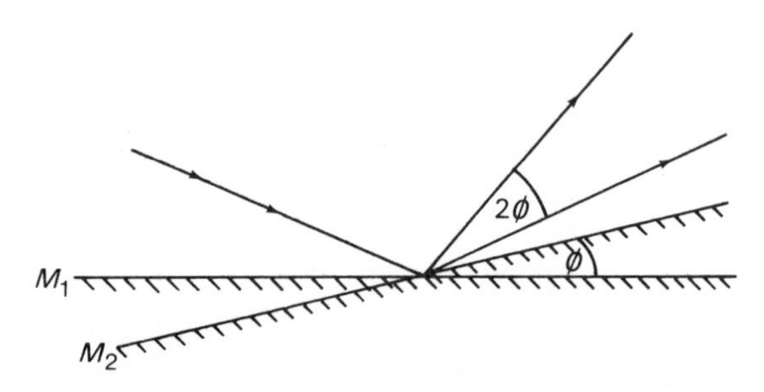


A3.1 – Reflexion of light from a plane mirror, initially at $M_1$, and after rotation through $\phi$ to $M_2$; the reflected beam has been turned through an angle $2\phi$.

We shall begin the derivation of the Bragg equation with X-rays incident at an angle $\theta$ on a family of parallel, equidistant planes, and being reflected (diffracted) from them at an equal angle (Figure A3.2). We need to determine the condition that X-rays reflected from successive planes interfere constructively, or reinforce one another, in the given direction of reflexion.

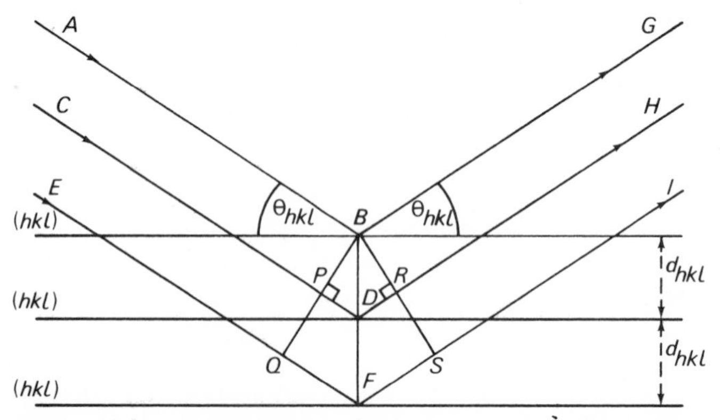


Figure A3.2 – Bragg reflexion of X-rays from crystal planes; three planes of an $(hkl)$ family are shown.

In Figure A3.2, $AB$, $CD$ and $EF$ represent typical parallel rays in an X-ray wave incident on the $(hkl)$ planes of interplanar spacing $d_{hkl}$ at an angle $\theta_{hkl}$. The rays are in phase both along the normal $BQ$ to the incident wavefront and, after reflexion, along the normal $BS$ to the reflected wavefront. For the first two planes, the excess path $\delta$ of $CDH$ over $ABG$ is given by

$$\delta = PD + DR \ . \tag{A3.1}$$

From the construction,

$$PD = DR = d_{hkl} \sin \theta_{hkl} \tag{A3.2}$$

Thus,

$$\delta = 2d_{hkl} \sin \theta_{hkl} \ . \tag{A3.3}$$

For constructive interference between reflected rays, the path difference must be equal to the wavelength $\lambda$ of the radiation. Hence,

$$2d_{hkl} \sin \theta_{hkl} = \lambda \ , \tag{A3.4}$$

which is the Bragg equation. A simple extension of the construction shows that if (A3.4) is satisfied for the first and second planes in a family, it will be satisfied also for the second and third, third and fourth, and so on for the whole family.

Sometimes the Bragg equation is written as

$$2d \sin \theta_n = n\lambda \ . \tag{A3.5}$$

However, in X-ray crystallography all families of planes are define uniquely by their Miller indices, including multiples such as $(nh,nk,nl)$. Thus $d$ in (A3.5), referring to the fundamental spacing ($h = k = l = 1$) of a family, is equal to $nd_{hkl}$ in (A3.4) with $\theta_n = \theta_{hkl}$. In particular examples, $d_{200} = d_{100}/2$, $d_{330} = d_{110}/3$, $d_{12,84} = d_{321}/4$, and so on. The reader should sketch the (110), (200), (110), (220) and (330) families of planes for a crystal, in projection on the $xy$ plane, given $a = 6, b = 9, c = 10$ Å.

It is convenient to relate $d_{hkl}$ to the values of $h$, $k$ and $l$. Consider Figure A3.3; for simplicity the axes will be assumed to be orthogonal; the plane $ABC$ is the first plane from the origin in the $(hkl)$ family. In the right-angled triangle $ONA$.

$$ON = (a/h) \cos \alpha \ . \tag{A3.6}$$

Since, by definition of Miller indices, $OA = a/h$,

$$d_{hkl} = (a/h) \cos \alpha \ . \tag{A3.7}$$

or

$$(h^2/a^2) d_{hkl}^2 = \cos^2 \alpha \ . \tag{A3.8}$$

Similarly, from triangles $ONB$ and $ONC$,

$$(k^2/b^2)d^2_{hkl} = \cos^2 \beta \,, \tag{A3.9}$$

and

$$(l^2/c^2)d^2_{hkl} = \cos^2 \gamma \;. \tag{A3.10}$$

Summing (A3.8) to (A3.10), remembering that the sum of the squares of the direction cosines of a line is unity,

$$d^2_{hkl} = (a^2/h^2) + (b^2/k^2) + (c^2/l^2) \;. \tag{A3.11}$$

In a cubic unit cell, $a = b = c$. Hence,

$$1/d^2_{hkl} = (h^2 + k^2 + l^2)/a^2 \;, \tag{A3.12}$$

From (A3.4) and (A3.12).

$$\sin \theta_{hkl} = (\lambda/2a)\,(h^2 + k^2 + l^2)^{1/2} \;. \tag{A3.13}$$

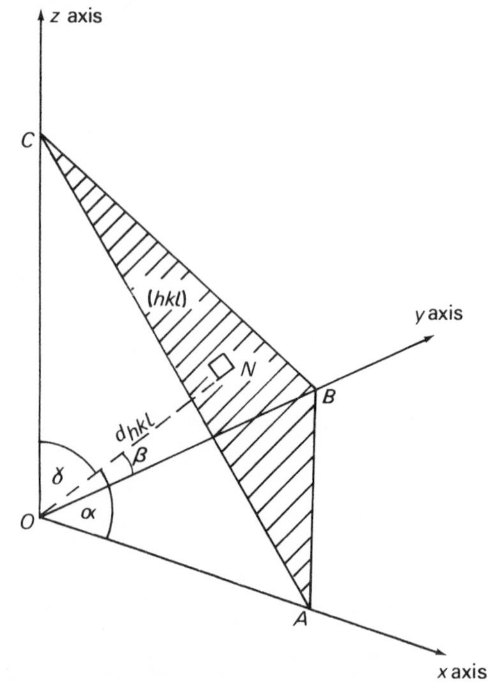


Figure A3.3 – Relationship between $d_{hkl}$ and $hkl$.

## APPENDIX 4 THE HYPSOMETRIC FORMULA – A SPECIAL CASE OF THE BOLTZMANN DISTRIBUTION

Consider a rectangular column of an ideal gas of cross-sectional area $A$ and height $z$, with reference to an origin, $O$, at ground level, at a uniform temperature $T$ (Figure A4.1). The mass of a gas molecule is $m$ and, at the height $z$, let the pressure of the gas be $P$ and let there be $N$ molecules per unit volume. We need to determine how $N$ varies with $z$. Since the gas is assumed to be ideal, no intermolecular attractions need be considered. Hence, from the gas laws,

$$PV = RT . \qquad (A4.1)$$

The gas constant $R$ is $Lk$, where $L$ is the Avogardro constant and $k$ is the Boltzmann constant, and $V$ is the volume containing 1 mole of gas. Since $L/V = N$,

$$P = NkT . \qquad (A4.2)$$

At a height $(z + \mathrm{d}z)$ the pressure is $(P + \mathrm{d}P)$. The gravitational force on the segment of width $\mathrm{d}z$ is $A\rho g\mathrm{d}z$, where $\rho$ is the density of the gas and $g$ is the gravitational acceleration. The pressure difference across the segment is $(-\mathrm{d}P/\mathrm{d}z)\mathrm{d}z$, the negative sign indicating that the gas pressure decreases in the positive direction of $z$. The hypsometric force on the segment is $-A(\mathrm{d}P/\mathrm{d}z)\mathrm{d}z$, and at equilibrium the two forces are balanced.

$$A(\mathrm{d}P/\mathrm{d}z)\mathrm{d}z = A\rho g\mathrm{d}z \qquad (A4.3)$$

or

$$\mathrm{d}P = \rho g\mathrm{d}z . \qquad (A4.4)$$

Equation (A4.4) might be obtained also from a definition of pressure.

From (A4.2)

$$\mathrm{d}P = kT\mathrm{d}N , \qquad (A4.5)$$

and using the fact that $\rho = mN$ we obtain,

$$\frac{\mathrm{d}N}{N} = -mg\mathrm{d}z/kT . \qquad (A4.6)$$

On integration, we obtain

$$\ln N = -mgz/kT + \text{constant} . \tag{A4.7}$$

At $z = 0$ let $N = N_0$. Thus the constant becomes $N_0$ and

$$N = N_0 \exp(-mgz/kT) . \tag{A4.8}$$

Now, $mgz$ is the gravitational potential energy per molecule of gas. Let $U$ represent this potential energy per mole of gas. Then $U = Lmgz$, and

$$N = N_0 \exp(-U/RT) . \tag{A4.9}$$

As an example, consider air, of mean molar mass 0.030 kg $mol^{-1}$, at a height of 30 km above ground level at 298 K. From (A4.9), $N/N_0$ is approximately 1/35 times its value at ground level; at 300 km it is only about $3 \times 10^{-16}$ times its value at ground level.

A more general derivation of (A4.9) may be obtained through statistical mechanics but this topic is beyond the scope of this book. We shall need this equation in studying both ion mobility in solids, and dipoles.

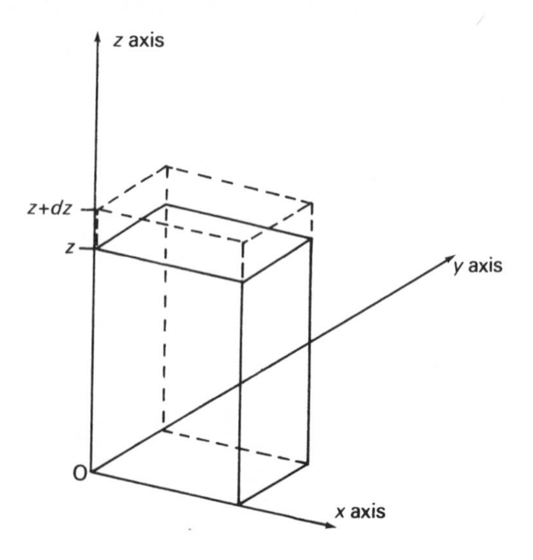


Figure A4.1 – Construction for the hypsometric formula.

## PROBLEMS TO CHAPTER 1

1. At 420 K $NH_4NO_3$ has the CsCl structure type. What is an important structural implication of this observation?
2. Refer first to Problem 1. In an X-ray diffraction experiment, similar to that shown in Figure 1.6, but with $NH_4NO_3$ at 420 K, the first two sharp rings had diameters of 22.1 and 32.7 mm; these reflexions have indices 100 and 110 respectively. If the specimen-to-film distance was 30.00 mm and the X-ray wavelength 1.542 Å, calculate an average value for the unit-cell dimension of $NH_4NO_3$ from the data on the two reflexions.
3. Which of the following pairs of substances are structurally isomorphous at stp?

| | | | | | |
|---|---|---|---|---|---|
| NaCl | KCl | $BaSO_4$, | $PbSeO_4$ | $CaF_2$, | $\beta$-$PbF_2$ |
| RbCl | CsCl | $SrSO_4$, | $CaSO_4$ | NaBr, | MgO |
| $KNO_3$, | $CaCO_3$ (calcite) | $CaF_2$, | $MgF_2$ | CaO | BeO |

(*Crystal Structures*, Vols. I, II and III, by R. W. G. Wyckoff, may prove useful).
4. The entropy of vaporization of $BeCl_2$ is 36 J $K^{-1}mol^{-1}$ greater than that for KCl. What does this result suggest about the liquid state of $BeCl_2$?
5. The heats of combustion of orthorhombic ($\alpha$) and monoclinic ($\beta$) sulphur at 298 K to form $SO_2$ are −297.0 and −297.3 kJ $mol^{-1}$ respectively. The $\alpha \rightarrow \beta$ transition temperature is 386 K and the entropy of the $\alpha$ form at this temperature is 31.88 J $K^{-1}mol^{-1}$. Draw an enthalpy-level diagram to illustrate the changes involved. Assuming that the heat contents of $\alpha$- and $\beta$-sulphur do not change between 298 and 386 K, calculate the entropy of $\beta$-sulphur at the transition temperature.
6. Classify the following substances according to the principal type of bonding responsible for cohesion in the *solid* state.

| | | | |
|---|---|---|---|
| RbF | $CO_2$ | $C_6H_6$ | $Cu_3Au$ |
| $P_4$ | AlN | Ne | $Na_2SO_4$ |
| $P_2Cl_{10}$ | Pb | $KClO_3$ | SiC |

7. What are the Miller indices of planes which make the following intercepts on the crystallographic $x$, $y$ and $z$ axes?

| | | |
|---|---|---|
| $a/2$ | $b$ | $-c/3$ |
| $a$ | // to $b$ | $2c/3$ |
| $-a/3$ | // to $b$ | // to $c$ |
| $a$ | $-2b/3$ | $3c/4$ |
| // to $a$ | $b/2$ | $-3c/4$ |
| // to $a$ | $-b$ | $2c$ |

*8. Consider the NaCl structure type in Figure 1.2. The path difference for X-rays reflected from successive (010) planes is $2d_{010} \sin \theta_{010}$. State and explain the effect of the interleaving, identical planes of atoms on the reflexion of X-rays when the crystal is in the correct geometrical position for the 010 reflexion?

Chapter 2

# Ionic Compounds

## 2.1 INTRODUCTION

Among our four principal classes of compounds, ionic materials provide a convenient starting point for a detailed study, because it is possible to discuss them in a fairly quantitative manner on a classical basis. We shall see in ionic structures a clear embodiment of the general electrostatic rule, 'unlike charges attract, like charges repel'. This simple principle will enable us often to form a useful mental picture of bonding.

## 2.2 IONIC BOND

The ionic, or electrovalent, bond is formed between atoms of widely differing electron configurations. One of the species, typically a metal, becomes ionized, and the other, typically a non-metal, acquires an excess of electrons: the ions may then attract one another by means of coulombic forces, which are inversely proportional to the square of the distance between the ions and directly proportional to the product of their charges.

The atom of an element is, in general, a stable entity. It does not ionize spontaneously, because energy is required in order to remove electrons from an atom. This energy, the ionization energy, must be supplied to the atom if an electron is to be expelled. Two possible stages of ionization are

$$M(\mathrm{g}) \longrightarrow M^{+}(\mathrm{g}) + e^{-} \tag{2.1}$$

and

$$M^{+}(\mathrm{g}) \longrightarrow M^{2+}(\mathrm{g}) + e^{-} \, . \tag{2.2}$$

We speak of the first and second ionization energies corresponding to the processes (2.1) and (2.2), respectively. Sometimes the less desirable term ionization potential is used instead of ionization energy: the unit of ionization energy is the joule, J, but it is often quoted in electronvolt, eV. One eV is the energy acquired by a single electron moving through a potential difference of one volt:

we may note that 1 eV = (1.602189±0.000005)$\times 10^{-19}$ J, and 1 eV atom$^{-1}$ = 96486.5±0.3 J mol$^{-1}$.

Evidently, we shall be concerned with energy changes in our study of ionic solids, and we will discuss these compounds first from this point of view.

## 2.3 THERMODYNAMIC DESCRIPTION OF IONIC COMPOUNDS

The stages in the formation of an ionic compound may be considered initially in terms of enthalpy changes. Subsequently, we shall introduce a relationship between the crystal enthalpy and the crystal energy.

Sodium chloride is a typical ionic compound, and we are familiar with the equation

$$\mathrm{Na} + 1/2\mathrm{Cl_2} \longrightarrow \mathrm{NaCl} \ . \tag{2.3}$$

It is both more descriptive and more precise to write

$$\mathrm{Na(s)} + 1/2\mathrm{Cl_2(g)} \longrightarrow \mathrm{NaCl(s)} \ , \tag{2.4}$$

and the corresponding enthalpy of formation $\Delta H_f$(NaCl, s) is −412.5 kJ mol$^{-1}$: but even this equation contains, implicitly, several stages which we must examine in turn.

Solid sodium is sublimed:

$$\mathrm{Na(s)} \longrightarrow \mathrm{Na(g)} \ . \tag{2.5}$$

The enthalpy of sublimation $S_M$(Na) is 110.2 kJ mol$^{-1}$ at 298 K. The gas is next ionized:

$$\mathrm{Na(g)} \longrightarrow \mathrm{Na^+(g)} + e^- \ , \tag{2.6}$$

and the corresponding enthalpy change $I$(Na) at 0 K is 495.8 kJ mol$^{-1}$. The total enthalpy change for these two processes is, at 298 K, 612.2 kJ mol$^{-1}$: $I$(Na) has been increased by $5RT/2$ to take into account the enthalpy of one mole of electrons at 298 K. Clearly, there is no tendency for (2.5) and (2.6) to occur spontaneously. Considering next chlorine, we have

$$1/2\mathrm{Cl_2(g)} \longrightarrow \mathrm{Cl(g)} \tag{2.7}$$

and

$$\mathrm{Cl(g)} + e^- \longrightarrow \mathrm{Cl^-(g)} \ , \tag{2.8}$$

and the enthalpy changes are one-half the dissociation enthalpy $1/2D_0(\mathrm{Cl_2})$ 121.7 kJ mol$^{-1}$ at 298 K, and the electron affinity $E$(Cl) −348.6 kJ mol$^{-1}$ at

0 K respectively. The total for (2.5) and (2.6) is −233.1 kJ mol$^{-1}$ at 298 K, allowing again for the enthalpy of one mole of electrons at 298 K. For the four processes (2.5) to (2.8), the total enthalpy change is 379.1 kJ for the production of one mole of each of the $Na^+$ and $Cl^-$ gaseous ions; this enthalpy change does not favour the formation of the gaseous species. However, the driving force for (2.4) may be said to lie in the stability conferred on the NaCl(s) species by the enthalpy decrease, the crystal enthalpy, when the gaseous ions condense to form the sodium chloride structure.

The crystal, or cohesive, energy, also called, less desirably, lattice energy, is the quantity of energy liberated when one mole of the crystal is formed from its component ions in the gaseous state. The crystal enthalpy is the corresponding enthalpy change; at 0 K these energy and enthalpy changes are equal. Conventionally, they are negative quantities and, thermodynamically, they are concerned with the relative stabilities of two states, the crystal and the gaseous ions, both at the same temperature and pressure. Thus, for example,

$$\Delta U_C(MX) = U_C(MX) - [U(M^+,\text{g}) + U(X^-,\text{g})] \; ; \qquad (2.9)$$

we shall use the subscript $C$ to refer to the crystal in the thermodynamic discussion of energetics.

It is necessary to choose a reference state for the crystal energy, and the ideal ion-gas at 0 K is defined as the zero of crystal energy; there are no forces of attraction between the ions in the reference state. Hence for any solid† at 0 K

$$\Delta U'_C = U'_C \; , \qquad (2.10)$$

so that, although we are concerned with a change between two states, the crystal energy is often given the symbol $U$.

The processes which we have considered may be represented diagrammatically by a thermochemical cycle which was put forward, independently, by Born, Haber and Fajans (Figure 2.1). Although it is convenient to be able to equate the crystal enthalpy $\Delta H'_C$ with the crystal energy $U'_C$ by considering processes such as (2.5) to (2.8) at 0 K, it is sometimes more practical to consider them at some other temperature $T$, where $T$ is normally 298 K.‡ If we apply (2.11) to the process ion-gas ⟶ crystal, at constant temperature and pressure,

$$\Delta U_C = \Delta H_C + P\Delta V \; ; \qquad (2.11)$$

we have

$$\Delta U_C = \Delta H_C + PV_{\text{ion-gas}} \; , \qquad (2.12)$$

†The prime on $\Delta U_C$ or $\Delta H_C$ refers here to absolute zero.
‡Strictly, 298.15 K.

since $V_{\text{ion-gas}}$ is very much greater than $V_C$ under the given conditions. Because the ion-gas is, by definition, ideal, we may replace the $PV$ term in (2.12) by $nRT$, where $n$ is the number of moles of gaseous species. Hence,

$$\Delta U_C \text{ (or } U_C) = \Delta H_C + nRT \; . \tag{2.13}$$

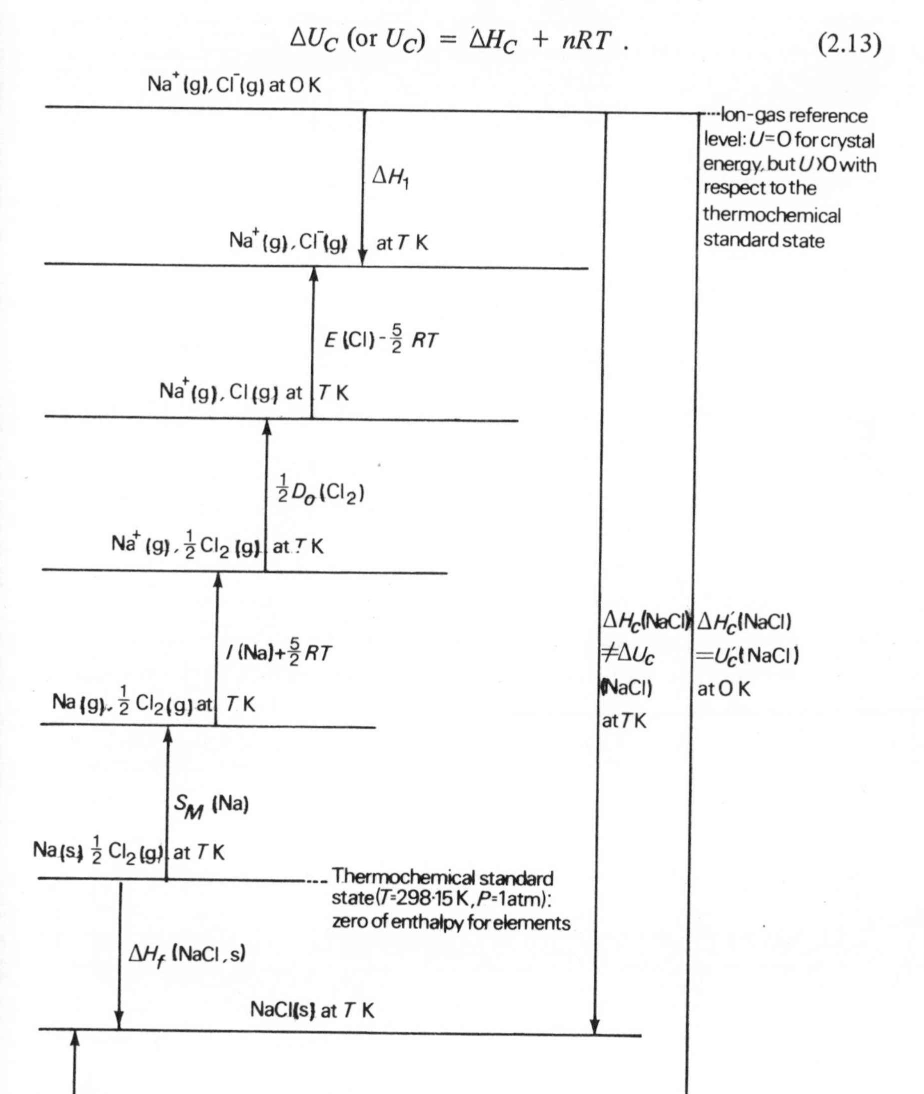


Figure 2.1 – Born-Haber-Fajans thermochemical cycle illustrated for NaCl; be careful to distinguish between the two reference levels shown.

In the cycle (Figure 2.1), the terms $\Delta H_1$ and $\Delta H_2$ are given by

$$\Delta H_1 = \int_0^T C_P(g)\, dT \,, \tag{2.14}$$

and

$$\Delta H_2 = \int_0^T C_P(s)\, dT \,, \tag{2.15}$$

where $C_P$ represents the molar heat capacity at constant pressure. From the cycle, at $T$ K, and using (2.13), we may write

$$U_C = \Delta H_f - S_M - I - \tfrac{1}{2}D_0 - E + nRT + \Delta H_1 \,; \tag{2.16}$$

the terms $\pm 5RT/2$ associated with (2.6) and (2.8) at $T$ K cancel. Inserting the quantities for NaC1, we have

$$U_C(\text{NaCl}) = -412.5 - 110.2 - 495.8 - 121.7 + 348.6 + 5.0 + 12.4 \,; \tag{2.17}$$

hence

$$U_C(\text{NaCl}) = -774 \text{ kJ mol}^{-1} \,. \tag{2.18}$$

This value represents the crystal energy of sodium chloride at 298 K. Its precision is determined by those of the quantities in (2.16): in NaCl, the precision of $U_C$ is about $\pm 2$ kJ mol$^{-1}$, but this quality is not often attainable.

The exact nature of the ionic binding energy is not explicit in the thermodynamic description which we have used. The bonding interactions are effectively bulked together in the practical quantity $\Delta H_f$, and, of course, in $U_C$. We shall consider next the terms in the Born-Haber-Fajans cycle in detail, so as to form an idea of both how they are measured and their probable precision.

### 2.3.1 Enthalpy of sublimation

We are concerned with the process

$$M(\text{s}) \longrightarrow M(\text{g}) \,, \tag{2.19}$$

at a temperature $T$ of 298 K. Enthalpies of sublimation may be obtained by applying the Clausius-Clapeyron equation to measurements of vapour pressure. Vapour pressures of heated metals are small, but they may be measured by an effusion technique. The vapour is streamed through a hole, the diameter of which is small compared with the mean free path of the atoms in the vapour. An atom reaching the hole will pass through it, and we can say that the number passing through the hole in unit time is that number which would strike the same surface area as that of the hole in the same time. In other words, the rate of effusion is proportional to the pressure of the vapour. The pressure of a streaming vapour may be measured, over a range of temperature, by a torsion-fibre apparatus, which must be first calibrated.

The Clausius-Clapeyron equation for the sublimation change of state, at a temperature $T$, is

$$\mathrm{d}P/\mathrm{d}T = \Delta H/\{T[V(\mathrm{g}) - V(\mathrm{s})]\} . \tag{2.20}$$

If we make the reasonable assumptions that $V(\mathrm{g})$ is very much greater than $V(\mathrm{s})$, and that the vapour, at the low pressures involved, behaves ideally, then $V(\mathrm{s}) \approx 0$, and, for 1 mole of vapour, $PV(\mathrm{g}) = RT$. Hence,

$$\mathrm{d}P/\mathrm{d}T = \Delta HP/RT^2 , \tag{2.21}$$

or

$$(1/P)\mathrm{d}P/\mathrm{d}T = \Delta H/(RT^2) . \tag{2.22}$$

We may write (2.22) as

$$\mathrm{d}(\ln P)/\mathrm{d}T = \Delta H/(RT^2) , \tag{2.23}$$

and, since $\mathrm{d}(1/T) = -(1/T^2)\mathrm{d}T$, we obtain

$$\mathrm{d}(\ln P)/\mathrm{d}(1/T) = -\Delta H/R . \tag{2.24}$$

Thus a graph of $\ln P$ against $(1/T)$ should be a straight line of slope $-\Delta H/R$. The best line may be fitted by the method of least squares (Appendix 5), and the value of $\Delta H$ so obtained may be considered to apply at the average temperature of the experiment, conducted over a small range, say 200 K. It is necessary to obtain the value at 298 K; this process is illustrated in Figure 2.2. From the cycle,

$$\Delta H_1 = \Delta H_2 + \Delta H_3 - \Delta H_4 , \tag{2.25}$$

or

$$\Delta H_1 = \Delta H(\text{experiment}) + \int_{T_1}^{T_2} C_P(\mathrm{s})\,\mathrm{d}T - \int_{T_1}^{T_2} C_P(\mathrm{g})\,\mathrm{d}T . \tag{2.26}$$

In some cases, $C_P(\mathrm{s})$ can be expressed as a function of temperature in which case the first integral on the right-hand side of (2.26) can be evaluated analytically; in other cases, it may be determined numerically (Appendix 6). The second integral in (2.26) is $5R(T_2-T_1)/2$, since the vapour was assumed to behave like an ideal monatomic gas; $\Delta H_1$ is the term $S_M$ in (2.16). The enthalpies of sublimation of the alkali metals are listed in Table 2.1.

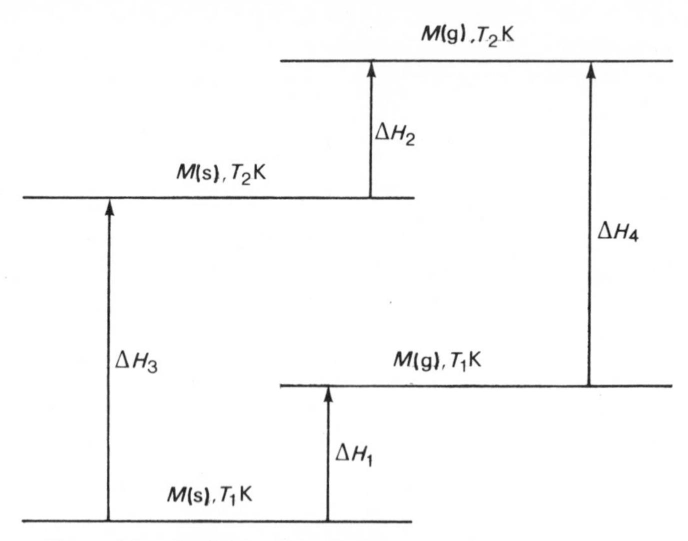


Figure 2.2 – Enthalpy cycle for the extrapolation of $\Delta H$ from $T_2$ to $T_1$: in practice, many metals would be liquid at the upper temperature $T_2$ and the enthalpy of fusion would feature in the cycle.

**Table 2.1**
**Enthalpies of sublimation of the alkali metals**

| Element | $S_M$/kJ mol$^{-1}$ |
|---|---|
| Li | 161.5 ± 5.5 |
| Na | 110.2 ± 1.5 |
| K | 90.0 ± 1.9 |
| Rb | 85.8 ± 2.1 |
| Cs | 78.7 ± 2.1 |

### 2.3.2 Ionization energy

The process under consideration is

$$M(g) \longrightarrow M^{+}(g) + e^{-} . \tag{2.27}$$

Many stages of electron excitation are possible in an atom, whereby an electron may be raised from one energy level to another. The ionization energy, $I$, refers to the complete removal of an electron from an atom in its ground state, that is the lowest energy state (at 0 K). At any other temperature $T$, the enthalpy of the electron, $5kT/2$, where $k$ is the Boltzmann constant, must be added to $I$.

The excitation energies $\Delta E$ of hydrogen-like species, such as $H^+$ or $Li^{2+}$, may be calculated from (2.28):

$$\Delta E = E_2 - E_1 = \mathrm{Ry}[(1/n_1^2)-(1/n_2^2)] \ , \tag{2.28}$$

where Ry represents the rydberg, $(2.17991 \pm 0.00001) \times 10^{-10}$ J: $n_1$ and $n_2$ are the principal quantum numbers of the atomic electron energy levels $E_1$ and $E_2$, respectively, involved in the excitation. For the ionization energy of hydrogen, $n_1 = 1$ and $n_2 = \infty$; hence, $I(\mathrm{H}) = 2.17991 \times 10^{-18}$ J atom$^{-1}$ or 1312.76 kJ mol$^{-1}$.

The first ionization energies of the alkali metals cannot be obtained in this manner. The wavenumber of a line in the absorption spectra of these species is given by the equation

$$\tilde{\nu} = \tilde{\nu}_\infty - R_\infty/(n - \delta)^2 \ . \tag{2.29}$$

The absorption spectrum of sodium (Figure 2.3) consists of a series of lines known as the principal ($P$) series. The first two members of this series are the well known, closely spaced sodium D yellow lines. Subsequent members lie in the ultraviolet region of the spectrum, and converge on a series limit, $\tilde{\nu}_\infty$ in (2.29), at about 41500 cm$^{-1}$ which is followed by an ionization continuum. At the convergence limit, the expelled electron has zero energy at 0 K: at any other temperature, the energy of the electron is represented at some other frequency along the continum.

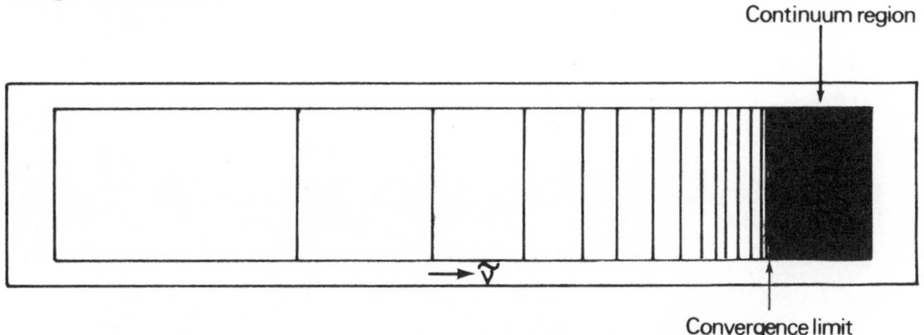


Figure 2.3 – Schematic representation of the absorption spectrum of Na; the convergence limit of the series represents the energy level corresponding to $n = \infty$.

Spectroscopists often speak of the wavenumber ($\tilde{\nu}$) of a spectral line as its 'frequency': $\tilde{\nu} = 1/\lambda = \nu/c$, where $\lambda$ is the wavelength, $\nu$ is the true frequency and $c$ is the speed of light in a vacuum.

**Table 2.2 – Spectral wavenumbers† and the first ionization energy for sodium**

| $n$ | $\tilde{\nu}$/cm$^{-1}$ | $[R_\infty/(n-\delta)^2]$/cm$^{-1}$ | $\tilde{\nu}_\infty$/cm$^{-1}$ | $\delta$ | |
|---|---|---|---|---|---|
| | | *$\delta = 0.81$* | | | |
| 10 | 40137.2 | 1299.3 | | 0.867 | |
| 11 | 40383.2 | 1056.8 | | 0.871 | 0.87 |
| 12 | 40566.0 | 876.4 | | 0.875 | |
| 13 | 40705.7 | 738.5 | | 0.880 | |
| 14 | 40814.5 | 630.8 | | 0.888 | |
| 15 | 40901.1 | 545.0 | | 0.896 | |
| 16 | 40971.2 | 475.6 | | 0.905 | |
| 17 | 41028.7 | 418.7 | | 0.914 | |
| | | | *41452.7* | | |
| | | *$\delta = 0.87$* | | | |
| | | 1316.5 | | 0.853 | |
| | | 1069.4 | | 0.852 | 0.85 |
| | | 885.9 | | 0.850 | |
| | | 745.8 | | 0.848 | |
| | | 636.5 | | 0.847 | |
| | | 549.6 | | 0.845 | |
| | | 479.4 | | 0.842 | |
| | | 421.8 | | 0.838 | |
| | | | *41448.8* | | |
| | | *$\delta = 0.85$* | | | |
| | | 1310.7 | | 0.857 | |
| | | 1065.2 | | 0.858 | |
| | | 882.7 | | 0.859 | |
| | | 743.4 | | 0.859 | 0.86 |
| | | 643.6 | | 0.861 | |
| | | 548.1 | | 0.862 | |
| | | 478.1 | | 0.863 | |
| | | 420.7 | | 0.863 | |
| | | | *41450.1* | | |
| | | *$\delta = 0.856$*‡ | | | |
| | | 1312.4 | | 0.856 | |
| | | 1066.4 | | 0.857 | |
| | | 883.6 | | 0.856 | |
| | | 744.1 | | 0.855 | |
| | | 635.2 | | 0.857 | 0.856 |
| | | 548.5 | | 0.857 | |
| | | 478.5 | | 0.857 | |
| | | 421.0 | *41449.71* | 0.856 | |

† 1 cm$^{-1}$ = 100 m$^{-1}$.
‡ Subjective estimate from iterations 2 and 3.

In (2.29), both $\tilde{\nu}_\infty$ and $\delta$, an arbitrary parameter known as the quantum defect, are unknown initially. The wavenumbers of the $P$ series for sodium are listed in Table 2.2. If we plot $\tilde{\nu}$ against $1/n^2$ and extrapolate to $1/n^2 = 0$, we can determine, through (2.29), a starting value for $\delta$; $R_\infty$ is the Rydberg constant, 109737.3 $cm^{-1}$ and $\delta = 0.81$. Taking this value of $\delta$, $\tilde{\nu}$ is fitted to $R_\infty/(n-\delta)^2$ by the method of least squares. A better estimate of $\tilde{\nu}_\infty$ and $\delta$ are obtained, and the procedure repeated until it converges on constant values, of $\delta$ and $\tilde{\nu}_\infty$; the stages are set out in Table 2.2.

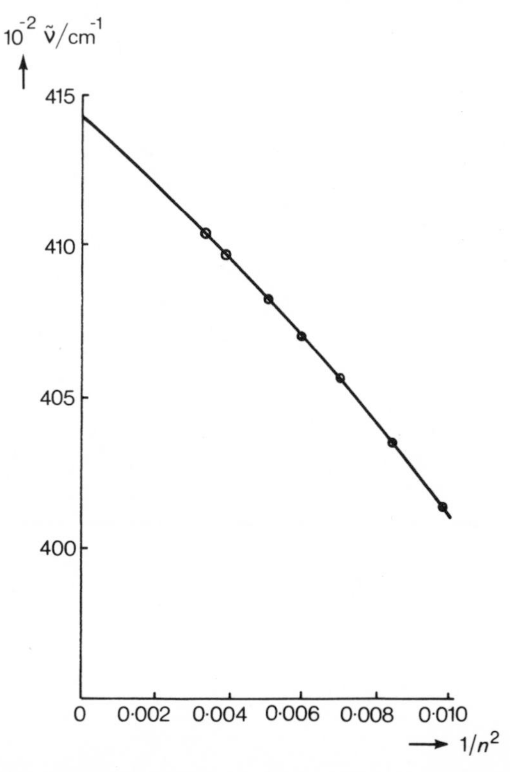


Figure 2.4 – Plot of $\tilde{\nu}$ against $1/n^2$ for the $P$ spectral series of Na; notice that both $1/n^2$ and $1/(n + \delta)^2$ must extrapolate to the same value of $\tilde{\nu}$.

The convergence limit of 41449.7 $cm^{-1}$ may be converted into an ionization energy by the relation $I = hc\ \tilde{\nu}_\infty$, where $h$ is the Planck constant and $c$ is the speed of light in a vacuum. Thus $I$(Na) = 5.1390 eV $atom^{-1}$, or 495.85 kJ $mol^{-1}$. The currently accepted values of the first ionization energies of the alkali metals are listed in Table 2.3; the precision of $I$ is about $\pm 0.001$ kJ $mol^{-1}$.

**Table 2.3 – First ionization energies of the alkali metals**

| | $I$/eV atom$^{-1}$ | $I$/kJ mol$^{-1}$ |
|---|---|---|
| Li | 5.3916 | 520.22 |
| Na | 5.1390 | 495.85 |
| K | 4.3406 | 418.81 |
| Rb | 4.1771 | 403.04 |
| Cs | 3.8938 | 375.70 |

### 2.3.3 Dissociation energy†

The most precise data on dissociation energies are obtained from the electronic absorption spectra of molecules. When a gas is heated, electronic transitions occur in the molecule and radiation is absorbed. The spectrum of a molecule has a complex band structure. For diatomic molecules, however, the banded spectrum ultimately becomes a continuum, indicating dissociation of the molecule into its component atoms. The process under consideration is

$$1/2X_2(g) \longrightarrow X(g) \ . \tag{2.30}$$

Again, a convergence limit must be established, and the frequency at the low-energy end of the continuum represents the minimum energy $D_0(X_2)$ needed to dissociate the molecule; $D_e$ is less than $D_0$ (see page 3) by the zero energy. For chlorine the appropriate wave number is 21189 cm$^{-1}$, corresponding to an energy of 2.627 eV molecule$^{-1}$. However, the dissociated atoms will not both be in the ground state (Figure 2.5); the excitation energy $E_X$ is obtainable from observations on the atomic spectrum of the gas, and for chlorine it is 0.109 eV. Hence the value of $D_0(Cl_2)$ is 2.518 eV, or 243.0 kJ mol$^{-1}$: the results for the halogens are listed in Table 2.4; their precision is better then ±0.01 kJ mol$^{-1}$ except for fluorine.

**Table 2.4 – Dissociation energies for the halogens**

| | $D_0(X_2)$/eV molecule$^{-1}$ | $D_0(X_2)$/kJ mol$^{-1}$ |
|---|---|---|
| $F_2$ | 1.63 | 157. ± 4. |
| $Cl_2$ | 2.518 | 243.0 |
| $Br_2$ | 2.319 | 223.8 |
| $I_2$ | 2.215 | 213.7 |

†Also called dissociation enthalpy.

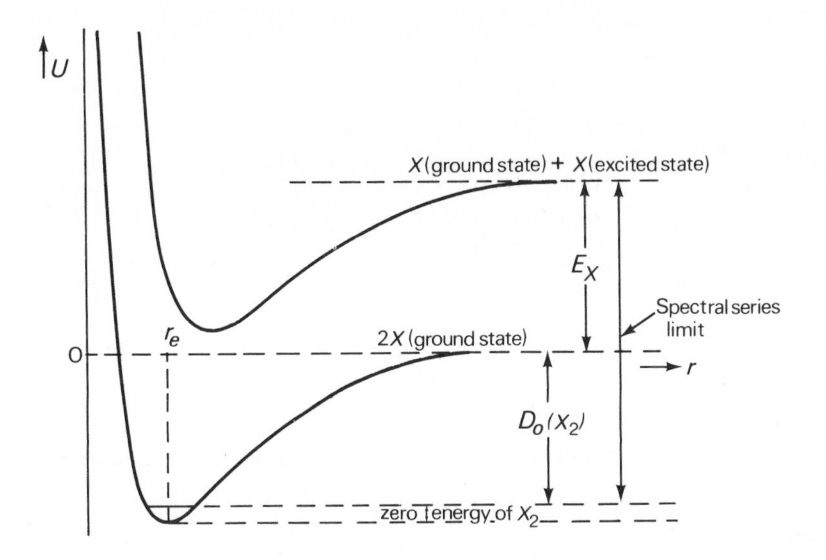


Figure 2.5 – Variation of potential energy $U$ with interatomic distance $r$ for a diatomic molecule $X_2$:$r_e$ is the equilibrium interatomic distance. For clarity the value of $E_X$ has been somewhat enlarged. The difference between $D_e$ (Figure 1.1) and the practical quantity $D_o$ is the zero energy of vibration.

### 2.3.4 Electron affinity

The electron affinity is defined for the process, at 0 K,

$$X(g) + e^- \longrightarrow X^-(g) \ . \tag{2.31}$$

Single electron affinities have been measured experimentally for the halogens and for a few other atomic species; others may be deduced from crystal energies in a manner to be described presently. Alkali-metal halides, such as RbI or CsC1, when heated by shock waves produce a vapour containing $I^-$ of $Cl^-$ species in abundance. Their ultraviolet absorption spectra are continua with sharp low-energy thresholds corresponding to the photodetachment of electrons:

$$Cl^-(g) + h\nu \longrightarrow Cl(g) + e^- \ . \tag{2.32}$$

The threshold wavelengths for the halogens have been ascribed precisely, and the electron affinities can be calculated from them (Table 2.5).

Table 2.5 – Electron affinities for the halogens

| | $E(X)$/eV atom$^{-1}$ | $E(X)$/kJ mol$^{-1}$ |
|---|---|---|
| F | −3.399 ± 0.002 | −328.0 ± 0.2 |
| Cl | −3.613 ± 0.003 | −348.6 ± 0.3 |
| Br | −3.363 ± 0.003 | −324.5 ± 0.3 |
| I | −3.063 ± 0.003 | −295.5 ± 0.3 |

### 2.3.5 Enthalpy of formation

The enthalpy of formation refers to the formation of a compound from its elements:

$$M(\mathrm{s}) + 1/2X_2(\mathrm{g}) \longrightarrow MX(\mathrm{s})\ . \quad (2.33)$$

At 298.15 K and 1 atmosphere pressure, the $\Delta H$ of this reaction is the standard enthalpy of formation $\Delta H_f^{\ominus}(MX,\mathrm{s})$. This quantity may be measured for the direct combination of, say, lithium and iodine, but for cesium and chlorine it would be wise to use an indirect approach. A sequence of reactions may be considered:

| Reaction | | $\Delta H$/kJ mol$^{-1}$ |
|---|---|---|
| $H_2(g) + 1/2O_2(g)$ | $\longrightarrow H_2O(l)$ | −285.9 |
| $Cs(s) + H_2O(l)$ | $\longrightarrow Cs^+(aq) + OH^-(aq) + 1/2H_2(g)$ | −191.8 |
| $1/2H_2(g) + 1/2Cl_2(g)$ | $\longrightarrow HCl(g)$ | − 92.3 |
| $HCl(g) + aq$ | $\longrightarrow H_3O^+(aq) + Cl^-(aq)$ | − 75.1 |
| $Cs^+(aq) + OH^-(aq) + H_3O^+(aq) + Cl^-(aq)$ | $\longrightarrow Cs^+(aq) + Cl^-(aq) + H_2O(l)$ | − 55.8 |
| $CsCl(s) + aq$ | $\longrightarrow Cs^+(aq) + Cl^-(aq)$ | + 18.0 |
| $Cs(s) + 1/2Cl_2(g)$ | $\longrightarrow CsCl(s)$ | $\Delta H_f^{\ominus}$ |

The reader is invited to construct a thermochemical cycle to show that $\Delta H_f^{\ominus}(\mathrm{CsCl, s})$ is −433.0 kJ mol$^{-1}$.

The precision of the available data on $\Delta H_f(\mathrm{s})$ is very variable: among the alkali-metal halides, for example, it ranges from about 1 to 17 kJ mol$^{-1}$. Table 2.6 lists the standard enthalpies of formation for the alkali-metal halides.

**Table 2.6 – Standard enthalpies of formation,** $\Delta H_f^{\ominus}(MX,\text{s})$, for the alkali-metal halides; the precision is given in parentheses. All values are in kJ $\text{mol}^{-1}$.

| | Li | Na | K | Rb | Cs |
|---|---|---|---|---|---|
| F | 612.1(8.4) | 571.1(2.1) | 562.7(4.2) | 549.4(8.4) | 530.9(16.7) |
| Cl | 405.4(8.4) | 412.5(0.8) | 436.0(0.8) | 430.5(8.4) | 433.0(8.4) |
| Br | 348.9(8.4) | 361.7(1.7) | 392.0(2.1) | 389.1(8.4) | 394.6(12.6) |
| I | 271.1(8.4) | 290.0(2.1) | 327.6(1.3) | 328.4(8.4) | 336.8(10.5) |

### 2.3.6 Crystal energy

From Sections 2.3.1 to 2.3.5, it is clear that the terms $I$, $D$ and $E$ are generally known with high precision for the alkali metal and halogen elements, with $S_M$ and $\Delta H_f$ somewhat less so. Hence the precision in $U_C$ varies from about ± 2 kJ $\text{mol}^{-1}$ at best to about ± 25 kJ $\text{mol}^{-1}$. Among other solids a precision of less than 5% may be regarded as good, but not always attainable.

## 2.4 ELECTROSTATIC MODEL FOR IONIC CRYSTALS

The thermodynamic development of the crystal energy has not needed to enquire into the nature of the ionic bond. However, in order to gain a better understanding of this bond, we must approach crystal energy calculations by means of an atomic model. We know from X-ray crystal structure analysis (Appendix 7) both the structural arrangment in the alkali-metal halides and that they contain ions: hence our model will have an electrostatic character.

Consider two ions of charges $z_1e$ and $z_2e$, of opposite sign, treated initially as points, distant $r$ apart. Their electrostatic, or coulombic, potential energy is given by

$$u_E = -z_1z_2e^2/(4\pi\epsilon_0 r) \;, \tag{2.34}$$

where $e$ is the charge on an electron and $\epsilon_0$ is the permittivity of a vacuum: the negative sign arises from the negative ion, and implies that the potential energy is attractive.

In a crystal such as the NaCl type in Figure 1.2, we need to consider the effect on the energy of a pair of ions, given by (2.34), of all the other ions in the crystal. We shall write, for this structure type,

$$U_E = -Az_1z_2e^2/(4\pi\epsilon_0 r) \;, \tag{2.35}$$

where $A$ is a constant for the structure type, known as the Madelung constant, after E. Madelung (1918).

### 2.4.1 Madelung constant

The meaning of the Madelung constant is illustrated first in one dimension, by an infinite row of regularly spaced alternating positive and negative charges, each of magnitude $e$, distant $r$ apart (Figure 2.6). Let a negative charge be taken as a reference point. Its immediate neighbours produce an attractive potential energy of $-2e^2/(4\pi\epsilon_0 r)$. The next nearest neighbours give rise to a repulsion of $+2e^2/(8\pi\epsilon_0 r)$, the next nearest, an attraction of $-2e^2/(12\pi\epsilon_0 r)$, and so on. Thus we obtain a series of terms.

$$-2e^2/(4\pi\epsilon_0 r)\,(1 - 1/2 + 1/3 - 1/4 + - \ldots)\ . \tag{2.36}$$

The series in parentheses is ln (2). Hence, the effect of an infinite row of charges of alternating sign is to modify (2.34) by the factor 2ln(2), or approximately 1.3863; this number is the Madelung constant for the one-dimensional structure under examination.

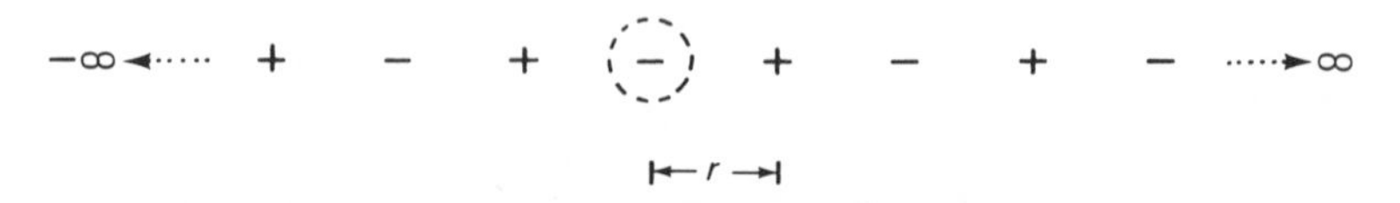


Figure 2.6 – Infinite row of regularly spaced unit charges of alternating sign; a one-dimensional point-charge structure.

In three dimensions the problem is more difficult. Consider again the NaCl structure type (Figure 1.2) and take the central $Cl^-$ ion in the unit cell as a reference point. The six nearest neighbours, at the centres of the unit-cell faces, give rise to an attraction of $-6e^2/(4\pi\epsilon_0 r)$ where $r$ is the interionic distance. The next nearest neighbours, at the centres of the unit-cell edges, produce a repulsion of $+\ 12e/(4\pi\epsilon_0 r\sqrt{2})$; the next nearest, at the corners, an attraction of $-8e^2/(4\pi\epsilon_0 r\sqrt{3})$, and so on. Writing these terms as a series we obtain

$$U_E = -e^2/(4\pi\epsilon_0 r)\,(6.000 - 8.485 + 4.619 - + \ldots)\ ; \tag{2.37}$$

this series has a very slow rate of convergence. H. M. Evjen (1932) showed that the convergence can be improved by working with nearly neutral blocks of structure. Potential energy falls off more rapidly with distance for a neutral group than for a group with an excess of charge (ion-ion $1/r$, ion-dipole $1/r^2$, dipole-dipole $1/r^3$).

Consider the unit cell of the NaCl structure type (Figure 1.2) as a neutral block of structure. A crystal is built up by stacking unit cells face to face. Hence, an ion at the corner of a unit cell is shared by 8 adjacent unit cells, and its contribution to one unit cell is, therefore, 1/8. Similarly, an ion on an edge contributes 1/4, and an ion on a face 1/2. If we add up all these contributions for the unit cell, we obtain a neutral block of $4Na^+$ and $4Cl^-$ species. Now, the first three terms in the series (2.37) are weighted accordingly, and the right-hand side becomes $-1.46e^2/(4\pi\epsilon_0 r)$; in other words, the Madelung constant is calculated as 1.46. If we take a cube twice as large, the value obtained is 1.75, and by much more sophisticated methods, $A = 1.74756$. It is interesting to note that the last two results are an expression of one of Pauling's empirical rules for crystal structures: 'the charge on an ion tends to be neutralized by its immediate neighbours.' An NaCl structure of eight unit cells magnitude behaves towards a central ion like an infinite structure to within about 0.1%.

We may look at the Madelung constant by another model. Imagine an ion of charge $e$ existing in free space. The effect of incorporating it into a crystal structure can be simulated by surrounding it with and earthed, spherical, conducting shell, since in both situations its electric field is neutralized by its surroundings.

The electrostatic potential† due to a charge $e$ at a distance $r$ is $e/(4\pi\epsilon_0 r)$. The work done in decreasing the charge by an amount $\delta e$ is $e\delta e/(4\pi\epsilon_0 r)$. For neutralizing the whole charge, the work done per ion, $W$, is given by

$$W = (1/4\pi\epsilon_0) \int_0^e (e/r)\mathrm{d}e \ , \tag{2.38}$$

which is $0.5e^2/(4\pi\epsilon_0 r)$. The volume $V$ of a spherical shell of radius $r$ is $4\pi r^3/3$. Hence (2.38) may be written as

$$W = 0.5e^2/[(4\pi\epsilon_0)(3V/4\pi)^{\frac{1}{3}}] = 0.8060e^2/(4\pi\epsilon_0 V^{\frac{1}{3}}) \ . \tag{2.39}$$

Let us apply this result to the cesium chloride structure (Figure 1.8), confining our attention to one unit cell. The unit cell volume is $a^3$, where $a$ is the length of the cell edge. But $a\sqrt{3} = 2r_e$ where $r_e$ is the equilibrium interionic distance. Hence, the unit cell volume is $8r_e^3/(3\sqrt{3})$, and that per ion is half this value, since each ion has an identical environment in the structure. Using (2.39), and choosing to identify $r_e$ with the radius of the conducting shell, $V^{\frac{1}{3}}$ becomes $4^{\frac{1}{3}}r_e/\sqrt{3}$ and $W$ is $0.8795e^2(4\pi\epsilon_0 r)$. The work term per $Cs^+Cl^-$ ion pair is $2W$, or $1.759e^2/(4\pi\epsilon_0 r)$. Thus 1.759 is the result for the Madelung constant of the CsCl structure type: the accepted value is 1.7627. A selection of Madelung constants is listed in Table 2.7.

†Electrostatic theorems are considered in Appendix 18.

**Table 2.7 – Madelung constants and approximate crystal energies for some simple structure types**

| Structure type | $z_1$ | $z_2$ | $z_1z_2A$ | $-U$ (for $\rho/r_e = 0.1$ and $r_e = 3$ Å) kJ mol$^{-1}$ |
|---|---|---|---|---|
| CsCl | 1 | 1 | 1.7627 | 734 |
| CsCl | 2 | 2 | 7.0508 | 2939 |
| NaCl | 1 | 1 | 1.7476 | 728 |
| NaCl† | 2 | 2 | 6.9904 | 2913 |
| α-ZnS | 1 | 1 | 1.6407 | 684 |
| α-ZnS | 2 | 2 | 6.5629 | 2735 |
| β-ZnS | 1 | 1 | 1.6381 | 683 |
| β-ZnS | 2 | 2 | 6.5522 | 2731 |
| $CaF_2$ | 2 | 1 | 5.0388 | 2100 |
| $CaF_2$ | 4 | 2 | 20.155 | 8401 |
| $TiO_2$‡ | 2 | 1 | 4.770 | 1988 |
| $TiO_2$‡ | 4 | 2 | 19.080 | 7952 |
| β-$SiO_2$ | 2 | 1 | 4.597 | 1916 |
| β-$SiO_2$ | 4 | 2 | 18.386 | 7664 |

†MgO, for example.

‡There is no unique value of $r_e$ in the rutile structure type and $A$ is calculated in terms of the shortest distance in the structure, 1.945 Å for rutile.

### 2.4.2 Crystal energy calculation

In an ionic crystal, the main interatomic forces are the attraction of oppositely charged ions, and the repulsions between like ions and between the inner core electron shells of atoms. In Figure 1.1, these effects are represented in a qualitative manner, respectively, by the high- and low-$r$ portions of the potential energy curve. It is clear that the repulsion potential energy rises rapidly as $r$ decreases below $r_e$. The Madelung constant takes account of the electrostatic repulsion between like ions, and a satisfactory repulsion energy term takes the form $B\exp(-r/\rho)$, where $B$ is a constant for the structure type, and $\rho$ is a parameter of the particular compound. We express the crystal energy as a function of $r$ by

$$U(r) = Az_1z_2e^2/(4\pi\epsilon_0 r) + B\exp(-r/\rho) \ . \tag{2.40}$$

For simplicity of manipulation, we write

$$U = A'/r + B\exp(ar) \ , \tag{2.41}$$

where $A' = -z_1 z_2 e^2/(4\pi\epsilon_0)$ and $a = -1/\rho$. From Figure 1.1 we see that $U(r)$ is a minimum when $r = r_e$, the equilibrium interionic distance. From (2.41)

$$\mathrm{d}U/\mathrm{d}r = -A'/r^2 + aB\exp(ar) \ . \tag{2.42}$$

At $r = r_e$, $\mathrm{d}U/\mathrm{d}r = 0$ and

$$\mathrm{B} = \frac{A'}{r_e^2\, a\exp(ar_e)} \ . \tag{2.43}$$

Hence, from (2.40) and (2.43), inserting the true values of $A'$ and $a$, and multiplying by the Avogadro constant $L$ to give the result in J mol$^{-1}$, we obtain

$$U(r_e) = -LAz_1z_2e^2(1 - \rho/r_e)/(4\pi\epsilon_0 r_e) \ . \tag{2.44}$$

The parameter $\rho$ is related, not surprisingly, to the compressibility of the particular compound under investigation. The equation of state for an isotropic (cubic) solid under hydrostatic pressure, and at constant temperature, may be written, without sensible error, as (Appendix 8)

$$P = -(\partial U/\partial V)_T \ , \tag{2.45}$$

and it follows that

$$(\partial P/\partial V)_T = -(\partial^2 U/\partial V^2)_T \ . \tag{2.46}$$

The isothermal compressibility, $\kappa$, is defined by

$$\kappa = -(1/V)\,(\partial V/\partial P)_T \ , \tag{2.47}$$

and, from (2.46) and (2.47), we have, at constant temperature,

$$1/\kappa V = \mathrm{d}^2U/\mathrm{d}V^2 \ , \tag{2.48}$$

where $V$, in our analysis, is the volume occupied in the crystal by a pair of oppositely charged ions. For isotropic structures, we can write

$$V = kr^3 \ , \tag{2.49}$$

where $k$ is a constant. Hence,

$$\mathrm{d}V/\mathrm{d}r = 3kr^2 = 3V/r \ , \tag{2.50}$$

and

$$\mathrm{d}^2V/\mathrm{d}r^2 = 6kr = 6V/r^2 \ . \tag{2.51}$$

Now
$$dU/dV = (dU/dr)/(dV/dr) \ , \qquad (2.52)$$

and, from (2.52) we have

$$d^2U/dV^2 = \frac{1}{(dV/dr)}\frac{d}{dr}[(dU/dr)/(dV/dr)]$$
$$= \frac{(dV/dr)(d^2U/dr^2) - (dU/dr)(d^2V/dr^2)}{(dV/dr)^3} \ . \qquad (2.53)$$

From (2.42), we have

$$d^2U/dr^2 = 2A'/r^3 + a^2B\exp(ar) \ . \qquad (2.54)$$

Substituting (2.53) in (2.48), and using (2.42), (2.50), (2.51) and (2.54), we obtain

$$1/\kappa V = \frac{(3V/r)\,[2A'/r^3 + a^2B\exp(ar)] - (6V/r^2)\,[-A'/r^2 + aB\exp(ar)]}{(3V/r)^3} \ . \qquad (2.55)$$

Eliminating $B$ through (2.43), we obtain, with rearrangements, and at the equilibrium distance, $r_e$,

$$9V/\kappa - 2A'/r_e = aA = aAr_e/r_e \ . \qquad (2.56)$$

Introducing the true values of $A'$ and $a$, we obtain

$$\rho/r_e = [Az_1z_2e^2/(4\pi\epsilon_0 r_e)]\,/\,[9V/\kappa + 2Az_1z_2e^2/(4\pi\epsilon_0 r_e)] \ , \qquad (2.57)$$

which is used in (2.44) to obtain the cohesive energy. Let us apply these equations to sodium chloride, using the following data:

$L = 6.0221 \times 10^{23}\ \text{mol}^{-1}$, $A = 1\ 7476$, $e = 1.6022 \times 10^{-19}$ C,
$r_e = 2.82 \times 10^{-10}$ m, $\kappa = 4.1 \times 10^{-11}\ \text{N}^{-1}\ \text{m}^2$, $\epsilon_0 = 8.854 \times 10^{-12}\ \text{F m}^{-1}$,
$V = 2r_e^3$ (NaCl structure type).

From (2.57), $\rho/r_e = 0.113$, and from (2.44) $U(r_e) = -764$ kJ mol$^{-1}$. Comparing this result with that from the thermochemical cycle, (2.18), we see that the latter value is about 1% lower (more negative). The main reason for this difference lies in the fact that the electrostatic model developed so far has treated the ions in the structure as point charges. In fact, they are finite regions of electron density, and a more refined model must take into account other factors, such as the mutual polarization of the ions, and the approximation in (2.45).

### 2.4.3 Polarization

Polarization may be regarded qualitatively as a distortion of the electron density of a species by the electric field of its neighbours. This distortion creates a system of dipoles in each ion which, although they are symmetrical and do not lead to a permanent dipole moment in the solid, do give rise to a dipole-dipole attractive potential energy proportional to $\alpha(Na^+)\alpha(Cl^-)/r_e^6$ (for NaCl) where $\alpha$ is the polarizability of a species (see also Chapter 4). Other terms, proportional to $1/r_e^8$, may also be introduced into a refined model, but they represent only about 10% of the $1/r_e^6$ contribution, which itself is small.

In the case of NaCl, inclusion of these and other refinements into the electrostatic model leads to a crystal energy of $-774$ kJ mol$^{-1}$ (see Table 6.1), in excellent agreement with the thermochemical value. It is notable that among the alkali-metal halides, and, indeed, in many other ionic crystals, the value of $\rho/r_e$ is very close to 0.1. This fact means that about 10% of the total electrostatic energy is negated by the inner core repulsion potential energy.

The polarizability of a species depends upon the ease with which its electron density can be distorted. Hence, it becomes more significant the larger the species and the more loosely bound its electron density. The effectiveness of a species in producing a deformation in another species may be called its polarizing power, $p$, and is measured qualitatively by the value of the electric field of an ion at its boundary, $ze^2/(4\pi\epsilon_0 r)$, where $r$ is the radius of the ion (see page 93 ff). Table 2.8 lists values for the polarizabilities and the polarizing powers of some ions.

**Table 2.8 – Polarizabilities $\alpha$ and polarizing powers $p$ of some ions**

| | $10^{40}\alpha$/F m$^2$ | $10^{18}p$/C$^2$ m$^{-1}$ |
|---|---|---|
| $Li^+$ | 0.03 | 3.4 |
| $Na^+$ | 0.3 | 2.3 |
| $K^+$ | 1.3 | 1.7 |
| $Rb^+$ | 2.0 | 1.5 |
| $Cs^+$ | 3.4 | 1.3 |
| $Be^{2+}$ | 0.01 | 13.2 |
| $Mg^{2+}$ | 0.1 | 5.7 |
| $Ba^{2+}$ | 2.8 | 3.4 |
| $F^-$ | 1.0 | 1.7 |
| $Cl^-$ | 3.4 | 1.2 |
| $Br^-$ | 4.8 | 1.2 |
| $I^-$ | 7.3 | 1.1 |
| $O^{2-}$ | 2.7 | 3.5 |
| $S^{2-}$ | 6.1 | 2.5 |
| $Se^{2-}$ | 7.8 | 2.4 |
| $Te^{2-}$ | 10.0 | 2.2 |

We shall consider polarization again when we study the chapter on van der Waals' compounds.

## 2.5 USES OF CRYSTAL ENERGIES

The crystal energy tells us something about the stability of a solid with respect to its components. It can be used to comment on a range of interesting features connected with the energetics of formation of a crystal or its components: here, we consider three of them.

### 2.5.1 Electron affinities and similar quantities

When electrostatic calculations of the crystal energies of the alkali-metal halides were carried out first, by Born and Landé in about 1918, there were no experimental values for either the dissociation energies or the electron affinities of the halogens. The other terms in (2.16) were known, and were used to calculate $(1/2D_0 + E)$. The constancy of this term for a given halogen was regarded as good evidence for the applicability of the model. As $D_0$ became measurable so $E$ was obtainable directly from the average value of $(D_0 + E)$. For many years $E$(F) was given as approximately −393 kJ mol$^{-1}$. However, later work showed that $D_0(F_2)$ at approximately 266 kJ mol$^{-1}$ was erroneous, and it was revised during the early 1950's to 155 kJ mol$^{-1}$. Hence $E$(F) was immediately amended to −338 kJ mol$^{-1}$. Table 2.4 shows that among the halogens $D_0$ has its greatest uncertainty for fluorine. However, $E$(F) does not now rest on this value, for it has been measured experimentally (Table 2.5). Single electron affinities for a few other species, such as O or OH, have been measured experimentally, but the affinities for O and S, for example, for *two* electrons must be deduced by comparing the thermodynamic and electrostatic models for $U$. In addition, there are some species, such as $[NO_3]^-$ and $[SO_4]^{2-}$, for which the terms $D_0$ and $E$ do not have a clear meaning. In these cases we can refer to $(D_0 + E)$ collectively, for $[NO_3]^-$ say, as $\Delta H_f(NO_3^-,g)$. If we can set up an electrostatic calculation for a crystalline nitrate, then we can obtain $\Delta H_f(NO_3^-,g)$ from a combination of (2.16) and (2.44). The precision of such a quantity will be rather less than that of similar experimentally measured properties.

### 2.5.2 Compound stability

We have discussed NaCl in some detail, and we might wish to consider how it is that we do not find a comparable ionic compound NeCl. If it were to exist, it would be, perhaps, not unreasonable to assume that it might have a simple structure type, such as that of NaCl. From (2.16), we can write

$$\Delta H_f(\mathrm{NeCl,s}) = U_C(\mathrm{NeCl}) + S_M(\mathrm{Ne}) + I(\mathrm{Ne}) + 1/2D_0(\mathrm{Cl_2}) + E(\mathrm{Cl}) - nRT \ . \tag{2.58}$$

Inserting the data, where known, we obtain

$$\Delta H_f(\text{NeCl,s}) = U_C(\text{NeCl}) + 2080 + 0 + 121.5 - 348.6 - 5.0 = U_C(\text{NeCl}) - 1848 \; . \tag{2.59}$$

If we say that for NeCl to exist, $U_C$(NeCl) must be lower than –1848 kJ mol$^{-1}$, then using (2.44) with $r_e$ in Å, and taking $\rho/r_e$ as 0.1, we can show that

$$-2185/r_e < -1848 \text{ kJ mol}^{-1} \; , \tag{2.60}$$

which means that $r_e < 1.18$ Å. Since we know that $r(\text{Cl}^-)$ is $\geqslant 1.7$ Å (Table 2.11), $r_e$ cannot be less than about 2.5 Å. We must conclude that NeCl cannot form a stable ionic compound, because too much energy has to be expanded on the formation of $Ne^+$(g) for the overall cycle of (2.16) to be energetically favourable to the production of NeCl.

In this connexion, an interesting and important topic arises. In 1962, Bartlett prepared, by an accident, a crystalline compound, $O_2[PtF_6]$, from the gas-phase reaction

$$O_2(g) + PtF_6(g) \longrightarrow O_2[PtF_6](s) \; . \tag{2.61}$$

The compound was found to be crystalline, and X-ray analysis showed that it consisted of the species $O_2^+$ and $[PtF_6]^-$; oxygen had been oxidized. The crystal energy was calculated as approximately –502 kJ mol$^{-1}$, a not unreasonable value for a crystal with an average interionic distance of about 4.35 Å. The ionization energies for $O_2(g) \longrightarrow O_2^+ + e^-$ and $Xe(g) \longrightarrow Xe^+(g) + e^-$ were known to be 1182 and 1175 kJ mol$^{-1}$, respectively. From their similarity, it appeared that (2.61) might be performed with Xe in place of $O_2$. It was found that Xe was oxidized by $PtF_6$, forming a crystalline solid of cohesive energy approximately – 460 kJ mol$^{-1}$. These experiments heralded the chemistry of the inert gases, which name now belies their nature, and subsequently a number of interesting compounds such as $NeF_4$, $ArF_4$ and $KrCl_4$ have been prepared and studied.

There is, in the chemical literature, a report by Ebert and Woitinek (1933) on the preparation of CuF. From an estimate of its free-energy of formation, which is positive, its existence has been challenged, because it is thermodynamically unstable with respect to its elements in their standard state. However, this criterion is not all-important. For example, silver cyanide, AgCN, has a positive free-energy of formation of 164 kJ mol$^{-1}$, but it can be prepared by mixing aqueous solutions of silver nitrate and potassium cyanide. Once prepared, AgCN shows little tendency to decompose into its elements: it is thermodynamically unstable with respect to its elements in their standard state, but not with respect to the aqueous ions (Figure 2.7). We must exercise caution in using the word 'stability'; the reference level should always be defined precisely.

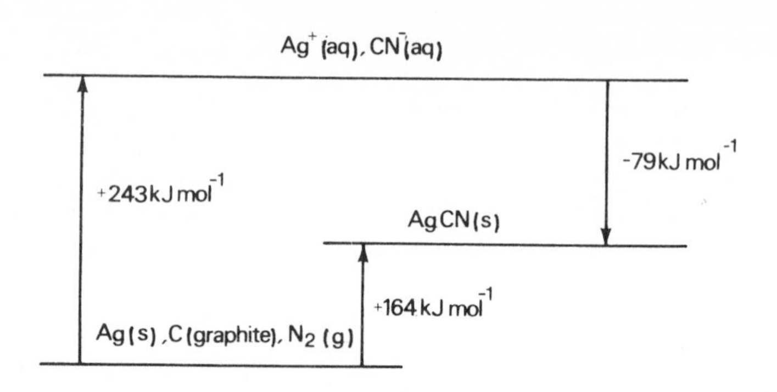


Figure 2.7 – Free-energy level diagram of AgCN.

### 2.5.3 Charge distribution on complex ions

At 298 K, KCN and NaCN, but not LiCN, have the NaCl structure type (Figure 1.2). At 279 K, NaCN and at 233 K, KCN, transform to the orthorhmobic structure (Figure 1.3). The reader may care to consider a possible reason for the differences in these transformation temperatures in the light of Table 2.8.

At a transition temperature, two polymorphs have the same free energy: hence, at constant pressure, the free-energy change at the transition temperature is given by

$$\Delta G = \Delta U - T\Delta S + P\Delta V \ . \tag{2.62}$$

The volume change from one solid polymorph to the other is negligibly small, and, at equilibrium, $\Delta G = 0$: thus,

$$\Delta U = T\Delta S \ . \tag{2.63}$$

In other words, the difference in the crystal energies of the polymorphs (Figure 2.8) may be determined from the entropy change (or enthalpy change†) of the transition; for NaCN and KCN, the $\Delta U$ values are 2.9 and 1.3 kJ mol$^{-1}$, respectively.

The crystal energies of NaCN and KCN, in their NaCl structure type, may be calculated from (2.44), with corrections for polarization, at their transition temperatures. At these temperatures, the values of $\Delta U$ for the cubic ⟶ orthorhombic forms can be obtained from (2.63). Now, in the orthorhombic structures, the $[CN]^-$ ions are fixed, and the electrostatic energy of the crystal must be calculated in two parts. The single negative charge on $[CN]^-$ may be con-

†At the transition temperature, $\Delta H/T = \Delta S$.

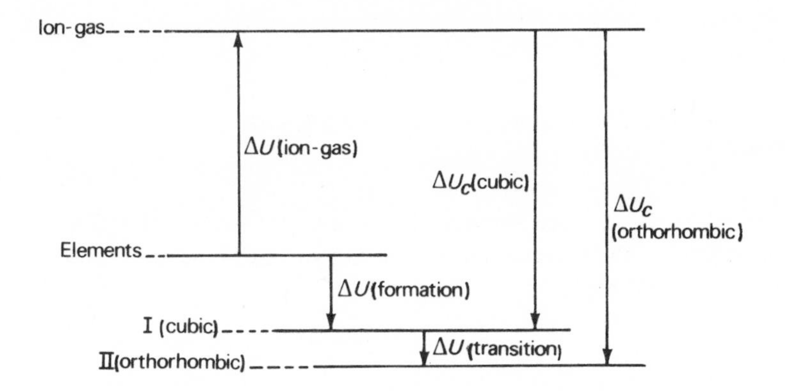


Figure 2.8 – Energy levels for two polymorphs, I and II. Except for $\Delta U$ (transition) the other $\Delta U$ terms could be called simply $U$ values as each refers to a process starting from a reference zero of energy (see Figure 2.9).

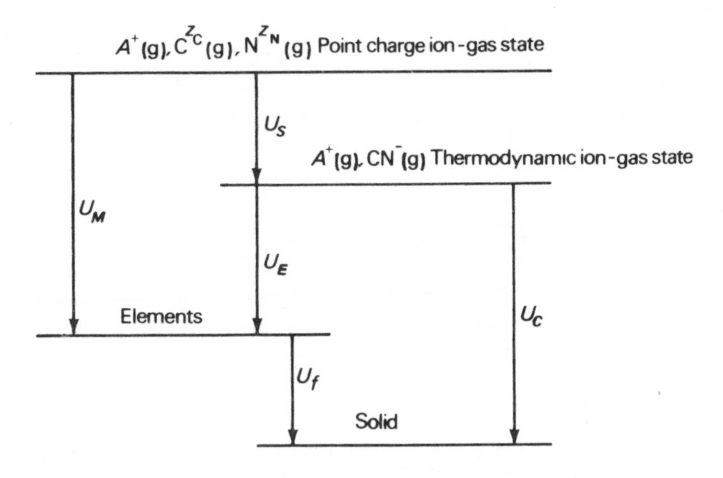


Figure 2.9 – Electrostatic and other energy levels for KCN and NaCN.

sidered, approximately, to be divided as $[C^{z_C}N^{z_N}]$, where $z_C + z_N = -1$. In Figure 2.9, the components of the electrostatic energy for a cyanide, $A$CN, are shown: $U_M$ is the electrostatic energy obtained by the Madelung calculation on the array of charged species $A^+$, $C^{z_C}$ and $N^{z_N}$ in the crystal; $U_E$ is the electrostatic energy of the array of species $A^+$ and $[CN]^-$ in the crystal; the difference is the electrostatic self-energy of the complex ion, $U_S$. This term refers to the process

$$C^{z_C}(g) + N^{z_N}(g) \longrightarrow [CN]^-(g) \; , \qquad (2.64)$$

with the C–N distance taken as that in the crystal. The term $U_S$ may be defined formally by

$$U_S = \sum_{\substack{i,j \\ i \neq j}} z_i z_j e^2/(4\pi\epsilon_0 r_{ij}) \;, \tag{2.65}$$

where $r_{ij}$ is the distance between the $i$th and $j$th species, of charges $z_i$ and $z_j$, respectively, in the complex ion. In the case of NaCN and KCN, (2.65) reduces to, with $r_{ij} = r(\text{CN})$,

$$U_S = z_C z_N e^2/[4\pi\epsilon_0 r(\text{CN})] \;. \tag{2.66}$$

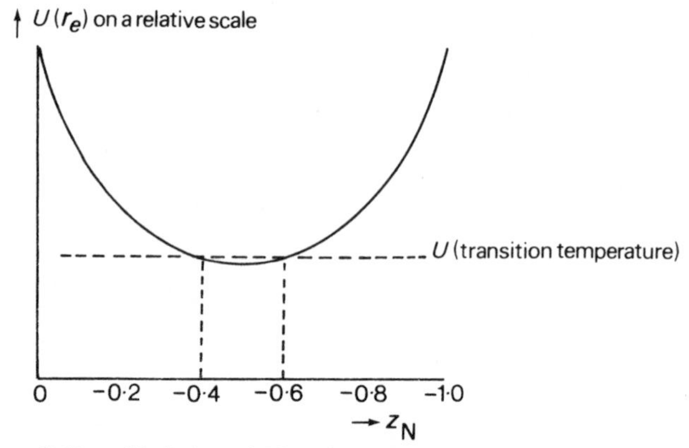


Figure 2.10 – Variation of $U(r_e)$ for KCN as a function of $z_N$ in the orthorhombic structure; the symmetry of the structure is reflected in the shape of the curve.

Thus, we can determine $U(r_e)$ for the orthorhombic structures as a function of $z_N$. By plotting $U(r_e)$ for the orthorhombic from against $z_N$ (Figure 2.10), we find that the equivalent value of $U$ from the cubic structure corresponds to $z_N$ equal to −0.4 or −0.6. The two values of $z_N$ occur symmetrically with respect to −0.5 because the structure is symmetrical with respect to spatial interchange of the C and N species. This situation does not occur with LiCN, and the following sequence of results may be obtained:

| | KCN | NaCN | LiCN |
|---|---|---|---|
| $U(z_N = -0.4)$ | −669 | −732 | −682 |
| $U(z_N = -0.6)$ | −669 | −732 | −791 |

Since it is unreasonable that $U$(LiCN) should be more positive than $U$(NaCN), because of the important proportionality with $1/r_e$, the value of $z_N = -0.6$ is indicated strongly. This result is supported by theoretical calculations, which give −0.60 and −0.40 for $z_N$ and $z_C$, respectively (Ladd, 1977). A similar analysis with $CaCO_3$ polymorphs has led to the conclusion that, in the carbonate ion $z_C \approx +0.31$ and $z_O \approx -0.77$, with a similar theoretical confirmation. However, it has been reported by Nyburg and coworkers (1978) that $z_O$ in the carbonate ion is about −0.9. Further work will undoubtedly be needed to confirm these findings.

## 2.6 ASPECTS OF CRYSTAL CHEMISTRY

Crystal chemistry is concerned with the relationship between the physical and chemical properties of solids and their internal structure and bonding. It attempts to interpret properties in the light of known structures and, conversely, to associate certain structural characteristics with measured properties. A systematic approach to the study of crystal chemistry may be said to have begun in about 1920 with the publication of Bragg's measurements of the interionic distances, $r_e$, for the alkali-metal halides (Table 2.9). The values of Δ show that the differences between the $r_e$ values for the two halides of a given cation are almost independent of the nature of the cation. Similarly, the difference between the $r_e$ values for two halides of a given anion are almost independent of the nature of the anion. These features may be explained by a model in which the ions are represented by spheres, each of a characteristic radius, and the sums of the appropriate radii are equivalent to the corresponding interionic distance. Thus,

$$r_e(\mathrm{KCl}) = r(\mathrm{K}^+) + r(\mathrm{Cl}^-) \, , \tag{2.67}$$

and

$$r_e(\mathrm{NaCl}) = r(\mathrm{Na}^+) + r(\mathrm{Cl}^-) \, . \tag{2.68}$$

Hence,

$$\Delta(\mathrm{KCl} - \mathrm{NaCl}) = r(\mathrm{K}^+) - r(\mathrm{Na}^+) \, , \tag{2.69}$$

which is independent of the nature of the halogen considered.

### 2.6.1 Ionic radii

A radius sum, $r_e$, is the experimentally measurable quantity: it is necessary to consider how $r_e$ may be divided into its components, a problem which has received much attention. The earliest method is due to Landé (1920). Six compounds that have the NaCl structure type are listed in Table 2.10, together with their interionic distances. The constancy of $r_e$ for the selenides and the sulphides indicates that the anions are in a close-packed array, with the smaller cations occupying the interstices, which situation is illustrated in Figure 2.11.

**Table 2.9 – Equilbrium interionic distances (Å) in the alkali-metal halides**

| | Li | $\Delta$ | Na | $\Delta$ | K | $\Delta$ | Rb | $\Delta$ | Cs | Mean $\Delta_-$ |
|---|---|---|---|---|---|---|---|---|---|---|
| F | 2.01 | (0.30) | 2.31 | (0.36) | 2.67 | (0.15) | 2.82 | (0.18) | 3.00 | |
| $\Delta$ | [0.56] | | [0.50] | | [0.47] | | [0.46] | | [0.56] | [0.51] |
| Cl | 2.57 | (0.24) | 2.81 | (0.33) | 3.14 | (0.14) | 3.28 | (0.28) | 3.56 | |
| $\Delta$ | [0.18] | | [0.17] | | [0.15] | | [0.15] | | [0.15] | [0.16] |
| Br | 2.75 | (0.23) | 2.98 | (0.31) | 3.29 | (0.14) | 3.43 | (0.28) | 3.71 | |
| $\Delta$ | [0.25] | | [0.25] | | [0.24] | | [0.23] | | [0.24] | [0.24] |
| I | 3.00 | (0.23) | 3.23 | (0.30) | 3.53 | (0.13) | 3.66 | (0.29) | 3.95 | |
| Mean $\Delta_+$ | | (0.25) | | (0.33) | | (0.14) | | (0.26) | | |

**Table 2.10**
**Some equilibrium Interionic distances**

| | $r_e$/Å | | $r_e$/Å |
|---|---|---|---|
| MgO | 2.10 | MnO | 2.22 |
| MgS | 2.60 | MnS | 2.61 |
| MgSe | 2.73 | MnSe | 2.73 |

In a close-packed situation we see readily that $2r_e\sqrt{2} = 4r$. Hence $r(Se^{2-})$ is 1.93 and $r(S^{2-})$ 1.84 Å. If we accept the additivity of ionic radii, illustrated by Bragg's results (Table 2.9 and (2.67), then other radii can be deduced: thus, $r(Mg^{2+}) = 0.78$ and $r(Mn^{2+}) = 0.79$ Å. However, when compared with other results, the value obtained here for $r(Mg^{2+})$ is too large by about 0.1 Å, probably because the anions and cations are not in close contact.

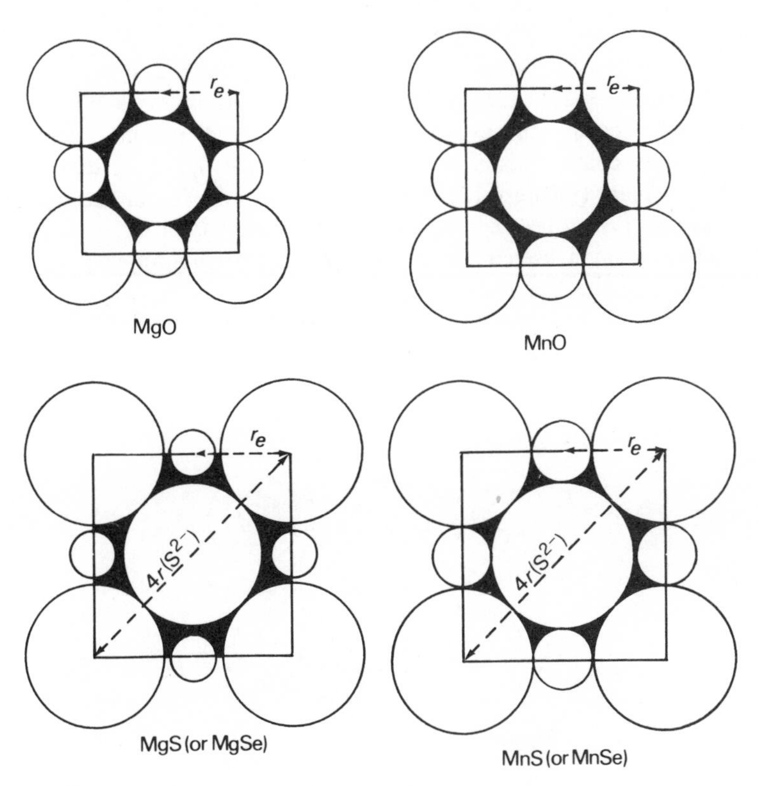


Figure 2.11 – Close-packed arrays of ions, as seen in projection; note the separation of the anions in MnO.

Pauling (1930) showed that the radius of an ion is governed mainly by the configuration of the outermost electrons in an atom. For an isoelectric series of ions, he gave the relationship

$$r_i = c/(Z_i - S) \ , \tag{2.70}$$

where $c$ is a constant for an isoelectronic series of ions, $Z_i$ is the atomic number of the $i$th ion, and $S$ is the screening constant for an electron in the outermost shell, and is obtainable from Slater's rules (Appendix 9). Considering NaF, for example, we have from (2.70)

$$r(Na^+) = c/(11 - 4.15) \ , \tag{2.71}$$

and

$$r(F^-) = c/(9 - 4.15) \ . \tag{2.72}$$

Hence

$$r(Na^+)/r(F^-) = 0.708 \ . \tag{2.73}$$

Since $r_e$(NaF) is 2.31 Å (Table 2.9), and accepting that $r_e = r_+ + r_-$, it follows that $r(Na^+)$ is 0.96 and $r(F^-)$ 1.35 Å.

Ladd (1968) showed that by applying Landé's method to LiI, for which, among the alkali-metal halides, a close-packed array of anions is most likely to exist, the same set of radii was produced as was deduced from measurements on electron density contour maps obtained by X-ray crystallographic studies on the alkali-metal halides (Figure 2.12). Table 2.11 lists Pauling's and Ladd's radii for several species: it may be seen that a small uncertainty still attaches to the value of the radius of an ion, even in the species which approach most closely spherical symmetry in their electron density.

Figure 2.12 – Diagrammatic electron density contour map for NaCl as seen in projection on to a face of the unit cell. The dashed lines are contours of zero electron density and, within experimental error, may be taken to represent the spatial limits of the ions.

Table 2.11 – Radii of some ionic species/Å

| | Pauling | Ladd | | Pauling | Ladd |
|---|---|---|---|---|---|
| $Li^+$ | 0.60 | 0.86 | $H^-$ | 2.08 | 1.39 |
| $Na^+$ | 0.95 | 1.12 | $F^-$ | 1.36 | 1.19 |
| $K^+$ | 1.33 | 1.44 | $Cl^-$ | 1.81 | 1.70 |
| $Rb^+$ | 1.48 | 1.58 | $Br^-$ | 1.95 | 1.87 |
| $Cs^+$ | 1.69 | 1.84 | $I^-$ | 2.16 | 2.12 |
| $NH_4^+$ | 1.48 | 1.66 | $O^{2-}$ | 1.40 | 1.25 |
| $Ag^+$ | 1.26 | 1.27 | $S^{2-}$ | 1.84 | 1.70 |
| $Tl^+$ | 1.40 | 1.54 | $Se^{2-}$ | 1.98 | 1.81 |
| $Be^{2+}$ | 0.31 | 0.48 | $Te^{2-}$ | 2.21 | 1.97 |
| $Mg^{2+}$ | 0.65 | 0.87 | | | |
| $Ca^{2+}$ | 0.99 | 1.18 | | | |
| $Sr^{2+}$ | 1.13 | 1.32 | | | |
| $Ba^{2+}$ | 1.35 | 1.49 | | | |

### 2.6.2 Radius ratio and *MX* structure types

The radius ratio, $R$, may be defined as the $r$(cation)/$r$(anion). It may be used as a guide to predicting structure, but it is not successful in all cases. We shall first consider this concept in relation to simple structures of the general formula $MX$.

Consider the CsCl structure type (Figure 1.8), in which the coordination pattern† is 8:8. Let the ions be of such radii that the anions at the corners of the unit cell are all simultaneously in contact with one another *and* with the central cation. Then, the unit-cell side $a$ is $2r_-$, and the body diagonal of the cubic cell is $a\sqrt{3}$, or $2(r_+ + r_-)\sqrt{3}$.

Hence
$$r_-\sqrt{3} = (r_+ + r_-) \, , \tag{2.74}$$

or
$$R_8 = r_+/r_- = 0.732 \, , \tag{2.75}$$

where $R_8$ represents the radius ratio for 8:8 coordination.

As the cation is made smaller for a constant anionic radius, the contact is lost between the anions and cations. Thus, there is a separation of charged regions, and one may ask if a more stable structure can be attained by an alteration of the coordination pattern. In fact, the NaCl structure type, with 6:6 coordination, offers an energetically more stable configuration at the smaller

†By $n$:$m$ coordination, we mean that the first species in the formula is coordinated by $n$ of the second species, and *vice versa*.

value of $R$ now under consideration. Referring to Figure 2.11, for MgS, we see readily that for the completely close-packed, or maximum contact, situation,

$$R_6 = 0.414 , \tag{2.76}$$

using the arguments developed for (2.74) and (2.75). By continuing this discussion, we obtain $R_4 = 0.225$ for the 4:4 coordination in the wurzite, α-ZnS, and blende β-ZnS, structure types (Figure 2.13). We can study these reults graphically.

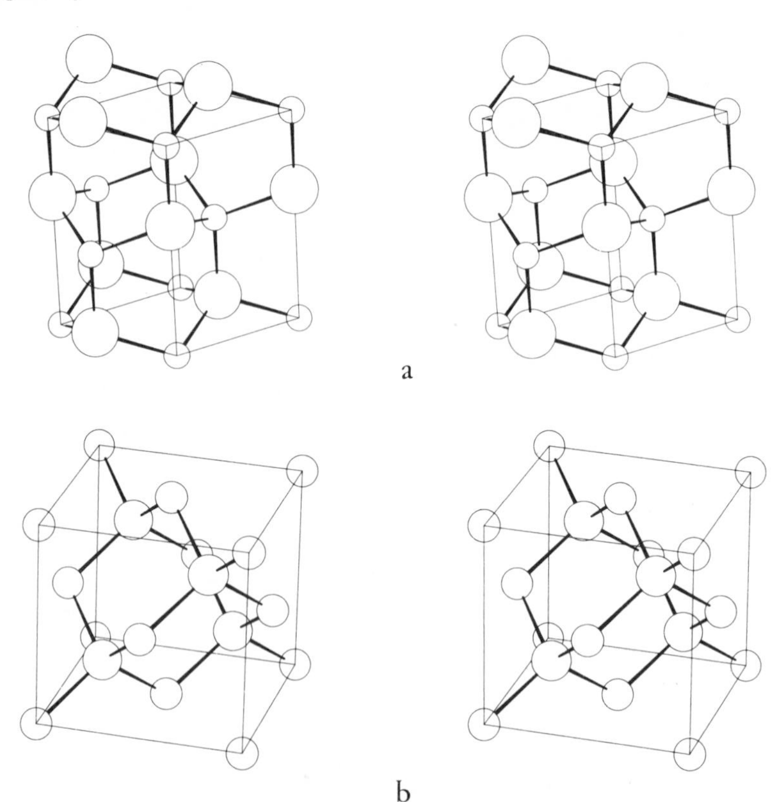


Figure 2.13 – Unit cells of the zinc sulphide structure types; circles in decreasing order of size are S and Zn. (a) Wurtzite, α-Zns, (b) Blende, β-ZnS (compare with Figure 1.14).

We write first

$$U(r_e) = \alpha/r_e , \tag{2.77}$$

where $\alpha = -10^{-3} LAz_1z_2e^2(1 - \rho/r_e)/(4\pi\epsilon_0)$ from (2.44); let $z_1 = z_2 = 1$ and let $\rho/r_e = 0.1$. Assuming that $r_e = r_+ + r_-$, and keeping a fixed value of $r_-$,

we may write
$$U(r_e) = \alpha'/(R + 1) , \tag{2.78}$$

where
$$\alpha' = \alpha/r_- .$$

The graph of $U(r_e)$ against $R$ is shown in Figure 2.14. For $R = 1$ the CsCl structure type is more stable. As $R$ is decreased, keeping $r_-$ constant, $U(r_e)$ must decrease as the graph shows. When $R = 0.732$ the ions are just in maximum contact and cannot become closer packed, even though the central cation may become smaller than its surrounding hole. Consequently $r_e$, and hence $R$, must remain constant. If, however, we change to the NaCl pattern, $U(r_e)$ can continue to decrease until $R = 0.414$. Then, by adopting the 4:4 coordination of $\beta$-ZnS, further stability can be attained. We can see from Table 2.7 that $\alpha$-ZnS is, energetically, very similar to the $\beta$ form.

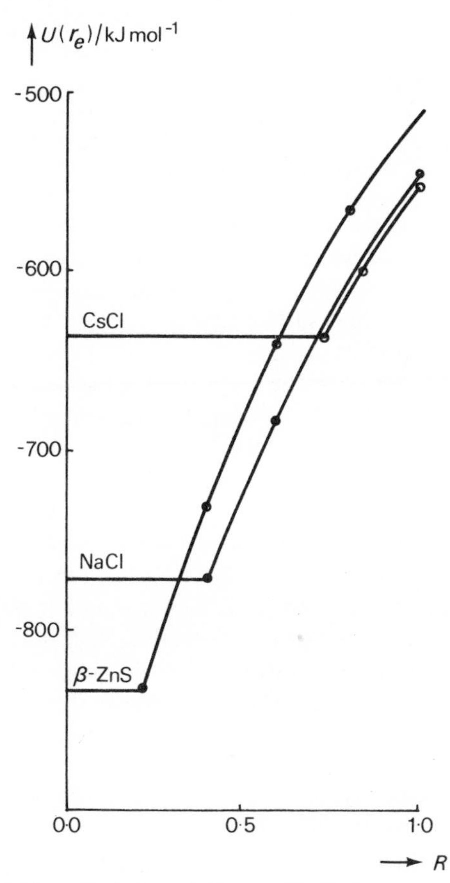


Figure 2.14 – Variation of $U(r_e)$ with $R$, for a constant value of $r$; singly-charged ions are assumed.

This analysis shows the radius ratio limits for these three important $MX$ structure types:

| | $R$ |
|---|---|
| CsCl | $\geqslant 0.732$ |
| NaCl | 0.414 – 0.732 |
| $\alpha$-ZnS, $\beta$-ZnS | 0.225 – 0.414 |

Table 2.12 lists values of $R$ for the alkali-metal halides.

**Table 2.12 – Radius ratios for alkali-metal halides**

| | Li | Na | K | Rb | Cs |
|---|---|---|---|---|---|
| F | 0.72 | 0.94 | 0.83† | 0.77† | 0.67† |
| Cl | 0.50 | 0.66 | 0.85 | 0.93 | 0.98† |
| Br | 0.46 | 0.60 | 0.77 | 0.84 | 0.98 |
| I | 0.41 | 0.53 | 0.68 | 0.74 | 0.87 |

†$r_-/r_+$: permissible for $MX$ structures with $n$:$n$ coordination.

Eight values of $R$, those enclosed by the dashed region in the Table, greater than 0.732 are found among halides which have the NaCl structure type at stp. The radius ratio is a geometrical concept based on the packing of spheres; we must not be too disturbed to find that such a simple approach has its limitations. There are three factors for us to consider.

The energy difference between the NaCl and CsCl structure types (Figure 2.14) at high values of $R$ is only about 4 kJ mol$^{-1}$. Remembering that this value reflects only the point-charge energy model of (2.44), we may expect that polarization corrections to the crystal become very important. Then we need to consider a small covalent contribution to $U$, arising from electron exchange; this effect will become clearer after a study of Chapter 3. However, it is necessary to see a possible mechanism for covalency, as well as making the suggestion that it exists. In the NaCl structure type, the p orbitals (see page 166) of adjacent ions are directed towards one another, thus facilitating orbital overlap. In the CsCl structure type, this situation does not obtain because of the different coordination pattern. Finally, we may note that RbCl at 83 K, for example, assumes the CsCl structure type, without appreciable change in its radius ratio.

Each of the ZnS structure types wurtzite or blende has, for singly-charged species, an electrostatic energy difference from that of the NaCl type at comparable values of $R$, of about 40 kJ mol$^{-1}$: it is, perhaps, significant that, at low values of the radius ratio, this structure type does not appear among the alkali-metal halides.

### 2.6.3 Radius ratio and $MX_2$ structure types

The three most important $MX_2$ structure types are fluorite $CaF_2$, rutile $TiO_2$, and β-cristobalite $SiO_2$; they are shown in Figures 1.11, 2.15 and 2.16, respectively. Since their coordination patterns are 8:4, 6:3 and 4:2, the radius ratio ($r_+/r_-$) limits for the $CaF_2$, $TiO_2$ and β-$SiO_2$ structure types are the same as for, respectively, CsCl, NaCl and ZnS. Table 2.13 lists radius ratios for some $MX_2$ halides.

**Table 2.13 – Radius ratios for some $MX_2$ halides**

| Fluorite | | Rutile | | β-Cristobalite | |
|---|---|---|---|---|---|
| $BaF_2$ | 1.25 | $CaCl_2$ | 0.69 | $BeF_2$ | 0.23 |
| $SrF_2$ | 1.11 | $CaBr_2$ | 0.63 | | |
| $BaCl_2$ | 0.88 | $MgF_2$ | 0.73 | | |
| $CaF_2$ | 0.99 | | | | |
| $SrCl_2$ | 0.78 | | | | |

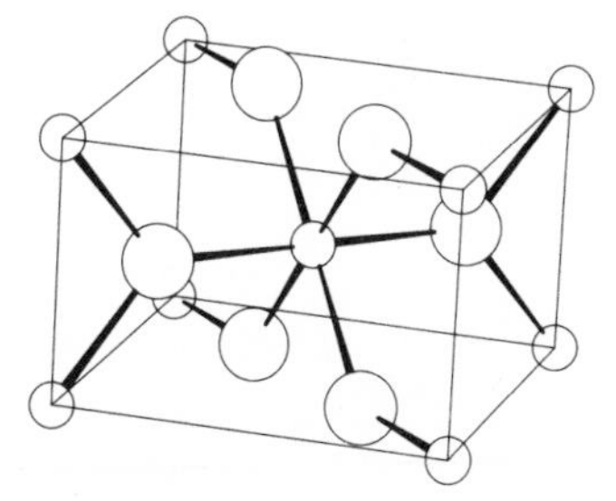
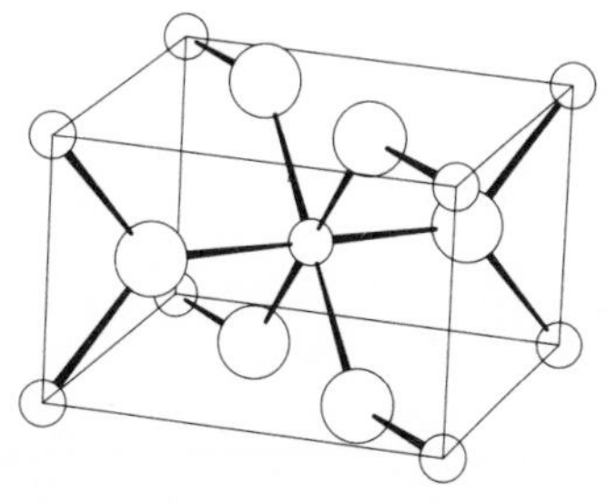

Figure 2.15 – Unit cell of the rutile, $TiO_2$, structure type; circles in decreasing order of size are O and Ti (compare with Figure 1.2).

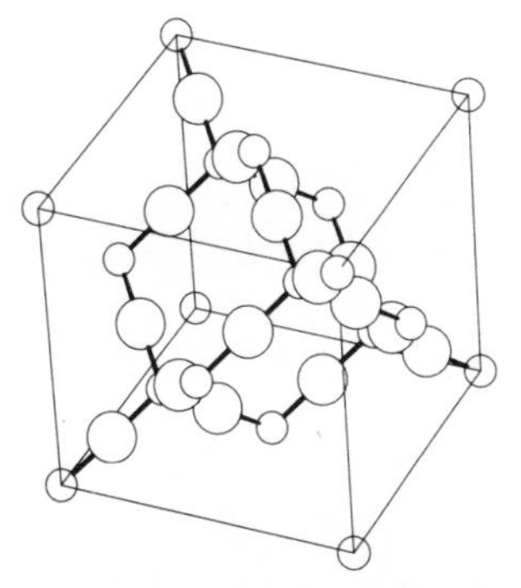
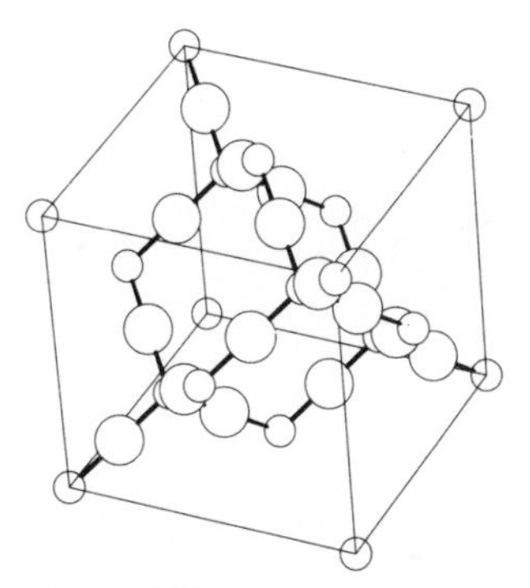

Figure 2.16 – Unit cell of the β-cristobalite, β-$SiO_2$, structure type; circles in decreasing order of size are O and Si. β-Cristobalite is related to blende (Figure 2.13b) as tridymite, another form of $SiO_2$, is to wurtzite (Figure 2.13a).

Among these halides, the radius ratio criterion is obeyed well, and Table 2.7 indicates a possible reason. From the Madelung constants, we find that the electrostatic energy differences are greater here than among the *MX* halides, for a given anion and cation. So, although polarization effects are larger with the smaller more highly-charged species, they do not determine the structure type in preference to packing considerations.

### 2.6.4 Polarization in *MX* and $MX_2$ structure types

We have likened polarization to a distortion of the electron density of ions, with a consequent increase in their energy of attraction through dipole-dipole types of forces. Its presence may be revealed by a comparison of ionic radii sums with experimental interionic distances; the silver halides provide a good illustration (Table 2.14).

**Table 2.14 – Interionic distances and radii sums in the silver halides**

| | Structure type | $r_e$/Å | $\Sigma r_i$/Å | $\Delta$/Å |
|---|---|---|---|---|
| AgF | NaCl | 2.46 | 2.46 | 0.00 |
| AgCl | NaCl | 2.77 | 2.97 | 0.20 |
| AgBr | NaCl | 2.88 | 3.14 | 0.26 |
| AgI | β-ZnS | 2.81 | 3.22† | 0.41 |

†Allowing −5% for the change of coordination from 6 to 4; from 6 to 8 coordination, the correction is about +3%.

It is known that an ion with an outermost d electron configuration (see page 115) appears to have a higher polarizability than is indicated by its radius (Tables 2.8 and 2.11), probably because of the relatively larger screening effect of d electrons. This effect is further manifested in the solubility relationships of the silver halides, as will be discussed presently.

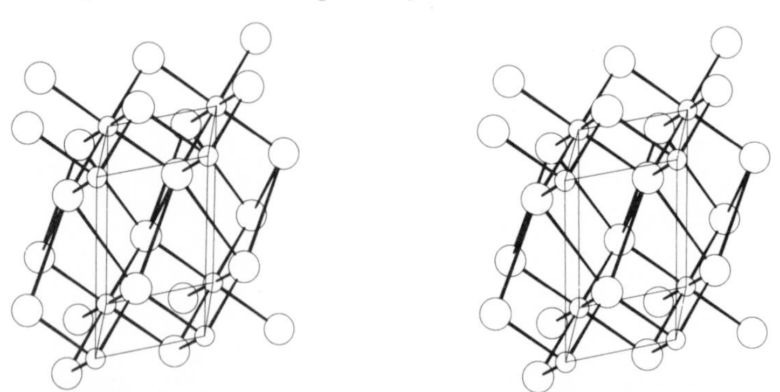

Figure 2.17 – Unit cell of the cadmium iodide $CdI_2$ structure type; circles in decreasing order of size are I and Cd. Note the asymmetry of the coordination.

Among $MX_2$ structures, increased polarization may lead to layer structures, such as $CdI_2$ (Figure 2.17). Further polarization may lead to the formation of discrete groups of atoms, and molecular (van der Waals') solids, such as $HgCl_2$, result. The sequence of changes is shown diagrammatically in Figure 2.18.

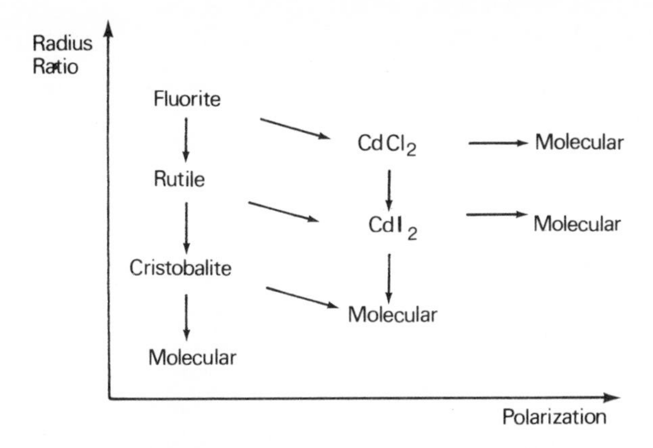


Figure 2.18 – Transformations among $MX_2$ structure types according to radius ratio and polarization.

## 2.7 STRUCTURAL AND PHYSICAL CHARACTERISTICS OF THE IONIC BOND

Ionic bonding exists between an ion and an indefinite number of neighbours. It has no particular directionality in space, and ionic structures tend to be governed, to a first approximation, by geometrical considerations, subject to electrical neutrality of the structure as a whole. Polarization and covalency enhance the crystal energy, and may modify the predictions of the radius ratio concept.

A refined electrostatic calculation gives satisfactory crystal energy values for structures which are not too complex. It should be remembered, however, that agreement between results obtained by (2.16) and (2.44) need not necessarily imply validity of the electrostatic model, but merely that it forms a satisfactory basis for the calculation of crystal energies. There are many other features, such as the crystal energies of the silver halides, or the relative stabilities of the $CdCl_2$ and $CdI_2$ structure types for the same compound; which cannot be explained without a more detailed treatment of the ionic bond.

Molecules do not exist in ionic structures, because electrons are localized in the atomic orbitals of the ions. In complex ions, such as $[NO_3]^-$ or $[CO_3]^{2-}$, covalent bonding is predominant within the complex ion, the ionic bonding taking place between these ions and the cations in the structure, as in $NaNO_3$ and $CaCO_3$; the calcite structure is shown in Figure 1.10.

Ionic solids are hard crystals of low compressibility and expansivity, but of high melting-point. They are electrical insulators in the solid state but, when molten, or in solution, where a suitable solvent can be found, conduct electricity by ion transport: this property distinguishes ionic solids most clearly from covalent and van der Waals' solids.

Trends in physical properties can be related to the crystal energy (Table 2.15) through its proportionality to $1/r_e$.

**Table 2.15 – Some physical properties of ionic solids**

| | BeO | MgO | CaO | SrO | BaO |
|---|---|---|---|---|---|
| $r_e$/Å | 1.65 | 2.10 | 2.40 | 2.57 | 2.76 |
| Hardness/Moh's scale | 9.0 | 6.5 | 5.5 | 4.1 | 3.3 |
| | NaF | NaCl | NaBr | NaI | |
| $r_e$/Å | 2.31 | 2.81 | 2.98 | 3.23 | |
| Mp/K | 1266 | 1074 | 1020 | 934 | |

## 2.8 SOLUBILITY OF IONIC COMPOUNDS

The solubility of ionic compounds in water can be treated quantitatively by thermodynamics, which shows at the same time the relation between solubility and crystal energy. Solubility is a well known property. It is often said that ionic solids are soluble in water, but that covalent solids are not. The second part of this statement is true, but the first only partly true. Again, it is stated that solubility decreases with increasing covalent character. However, consider the following series of compounds:

AgF ⟶ AgCl ⟶ AgBr ⟶ AgI
⟶ Decreasing solubility
⟶ Increasing polarization/covalent character

$CaF_2$ ⟶ $CaCl_2$ ⟶ $CaBr_2$ ⟶ $CaI_2$
⟶ Increasing solubility
⟶ Increasing polarizarion/covalent character

Evidently, a more detailed study is needed, but first we must define certain thermodynamic reference states.

### 2.8.1 Reference states for solubility

For simplicity, we shall restrict this part of the discussion to solids of the type $MX$, where $M$ and $X$ form singly-charged species; the results obtained may, however, be applied generally.

Consider the equilibrium

$$MX(\mathrm{s}) \rightleftharpoons M^{+}(\mathrm{aq}) + X^{-}(\mathrm{aq}) \ . \tag{2.79}$$

The equilibrium constant, $K$, is given, in the usual way, by

$$K = a(M^{+})a(X^{-})/a(MX) \ . \tag{2.80}$$

The activity of a pure, crystalline solid at 298.15 K and 1 atmosphere pressure is defined to be unity. Hence,

$$K = a(M^{+})a(X^{-}) \ , \tag{2.81}$$

or

$$K = c^2 f_{\pm}^2 \ . \tag{2.82}$$

where $c$ is the solubility, in mol dm$^{-1}$, and $f_{\pm}$ is the mean activity coefficient for the electrolyte in its saturated solution, $K$ is constant at a given temperature, and we shall refer it to 298 K unless otherwise stated.

The reference state for the solution is the infinitely dilute solution, in which the ratio of the activity of the solute to its molar concentration is unity. The fact that this reference state is hypothetical does not invalidate the arguments which follow from it. We may regard the reference state as a solution of mean concentration 1 mol dm$^{-3}$ and unit mean activity coefficient, and in which the partial molar heat content of the solute is the same as at infinite dilution (Appendix 10).

### 2.8.2 Solubility relationships

Let us simplify (2.79):

$$A \rightleftharpoons B \ , \tag{2.83}$$

where

$$K' = a(B)/a(A) \ . \tag{2.84}$$

The chemical potentials of the components in the system (2.83) are given by

$$\mu_A = \mu_A^{\ominus} + RT \ln a(A) \ , \tag{2.85}$$

and

$$\mu_B = \mu_B^{\ominus} + RT \ln a(B) \ . \tag{2.86}$$

At equilibrium, the solid $A$ and the saturated solution $B$ are at the same chemical potential. Hence $\mu_A = \mu_B$, and the standard free-energy change for (2.83) is given by

$$\Delta G^{\ominus} = \mu_B^{\ominus} - \mu_A^{\ominus} = -RT\ln[a(B)/a(A)] = RT\ln K' . \qquad (2.87)$$

In an analogous manner, for the process (2.79),

$$\Delta G_d^{\ominus} = -RT\ln K = -RT\ln c^2 f_{\pm}^2 . \qquad (2.88)$$

This equation represents the standard free energy of dissolution for the process *solid ⇌ ions in the reference state,* and tells us about solubility.

### 2.8.3 Two example calculations

*Silver iodide*

The solubility of silver iodide in water at 298 K is $1.02 \times 10^{-8}$ mol dm$^{-3}$. At this concentration $f_{\pm}$ is sensibly unity. Hence from (2.88) $\Delta G_d^{\ominus}$ is +91.2 kJ mol$^{-1}$.

*Lithium fluoride*

The solubility of lithium fluoride at 298 K is 0.09 mol dm$^{-1}$. Using the same procedure as for silver iodide, $\Delta G_d^{\ominus}$ is calculated as +11.9 kJ mol$^{-1}$. However, we are not justified here in taking $f_{\pm}$ as unity. From the Debye limiting equation (Appendix 11) $f_{\pm}$ is approximately 0.70. Hence a better value for $\Delta G_d^{\ominus}$ is +13.7 kJ mol$^{-1}$. If we apply the Debye limiting equation to AgI, $f_{\pm}$ is unity to within 0.01%.

Why do we need any further theory on solubility? We see that these calculations require activity coefficient data for saturated solutions. Where the solubilities are less than about 0.1 mol dm$^{-3}$, $f_{\pm}$ may be calculated with sufficient accuracy. In concentrated solutions the calculation of $f_{\pm}$ is not reliable. For example, potassium chloride is saturated at about 4.8 mol dm$^{-3}$. At this concentration $f_{\pm}$ is calculated as 0.08 from the Debye equation, whereas it is in fact 0.574. There is, however, paucity of experimental data on activity coefficents for saturated solutions. Furthermore, we need to analyse $\Delta G_d^{\ominus}$ in order to obtain a clearer understanding of solubility.

### 2.8.4 Solubility and energy

The important quantity $\Delta G_d^{\ominus}$ is given in the usual way by

$$\Delta G_d^{\ominus} = \Delta H_d^{\ominus} - T\Delta S_d^{\ominus} , \qquad (2.89)$$

where $\Delta H_d^{\ominus}$ is the standard enthalpy change for the dissolution process referred to infinite dilution and $\Delta S_d^{\ominus}$ is the corresponding change in entropy. We may write

$$\Delta S_d^{\ominus} = \Sigma \bar{S}_i^{\ominus} - S_C^{\ominus} , \qquad (2.90)$$

where $\Sigma\bar{S}_i^\ominus$ is the sum of the standard relative partial molar entropies of the hydrated ions (Appendix 12) and $S_C^\ominus$ is the standard entropy of the crystal. These data are readily available in the chemical literature.

The relationship between solubility and crystal energy is indicated by Figure 2.19, which refers to enthalpy changes.

Evidently,
$$\Delta H_C^\ominus + \Delta H_d^\ominus = \Delta H_h^\ominus \; , \qquad (2.91)$$

where $\Delta H_h^\ominus$ is the standard enthalpy change for the hydration of the gaseous ion-pair. This equation shows how, from crystal energy and enthalpy of solution data, we can obtain hydration enthalpies, which are otherwise not readily available. It is important that $\Delta H_d^\ominus$ measurements are extrapolated to infinite dilution because the process of dilution itself produces significant enthalpy changes. For example $\Delta H_d$ for $CdSO_4$ in 200 mol water is $-43.9$ kJ mol$^{-1}$, but at infinite dilution it is $-53.6$ kJ mol$^{-1}$. The difference between two such values may, in some cases, be commensurate with the value of $\Delta G_d$ itself.

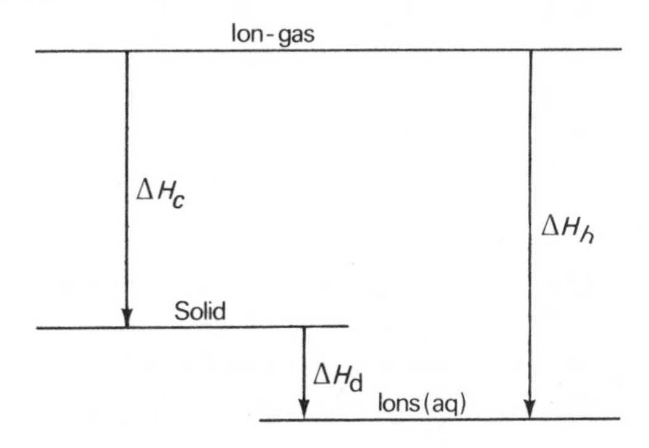


Figure 2.19 – Energy levels related to solubility

Table 2.16 lists thermodynamic data related to solubility. The interaction between ions and water molecules on the one hand, and between ions in crystal structure on the other, both increase as the ions become smaller and more highly charged, because the coulombic energy, proportional to $1/r$, predominates in both situations. Both $\Delta H_d$ and $\Delta G_d$ depend on the difference between two large quantities, one concerned with the solid, the other with the hydrated ions. A decrease in $\Delta G_h$ tends to stabilize the hydrated state, with respect to the ion-gas, and so promote solubility. However, a decrease in either $\Delta G_C$ or $\Delta S_d$ tends to decrease solubility, $\Delta G_C$ by stabilizing the crystal with respect to the ion-gas, and $\Delta S_d$ by making the hydrated state relatively less probable. The term $\Delta S_d$ becomes important with small highly-charged ions, such as in LiF and $MgF_2$.

On transferring a gaseous ion isothermally into water, two important effects occur. There is a structure-breaking effect on the water itself because of the

**Table 2.16 – Thermodynamic data relating to the solubilities of some ionic halides at 298.15 K**

| Halide | $\Delta H_d^\ominus$/kJ mol$^{-1}$ | $\Sigma \bar{S}_i^\ominus$/J mol$^{-1}$ | $S_C^\ominus$/J mol$^{-1}$ | $T\Delta S_d^\ominus$/kJ mol$^{-1}$ | $\Delta G_d^\ominus$/kJ mol$^{-1}$ |
|---|---|---|---|---|---|
| LiF | 4.6 | 4.6 | 36.0 | −9.4 | 14.0 |
| LiCl | −37.2 | 69.5 | 55.2 | 4.3 | −41.5 |
| LiBr | −49.0 | 95.0 | 69.0 | 7.8 | −56.8 |
| LiI | −63.2 | 123.4 | 75.7 | 14.2 | −77.4 |
| NaF | 0.4 | 50.6 | 58.6 | −2.4 | 2.8 |
| NaCl | 3.8 | 115.5 | 72.4 | 12.9 | −9.1 |
| NaBr | −0.8 | 141.0 | 85.8 | 16.5 | −17.3 |
| NaI | −7.5 | 169.5 | 92.5 | 23.0 | −30.5 |
| KF | −17.6 | 92.9 | 66.5 | 7.9 | −25.5 |
| KCl | 17.2 | 157.7 | 82.8 | 22.3 | −5.1 |
| KBr | 20.1 | 183.3 | 96.7 | 25.8 | −5.7 |
| KI | 20.5 | 211.7 | 104.2 | 32.1 | −11.6 |
| RbF | −26.4 | 114.6 | 72.8 | 12.5 | −38.9 |
| RbCl | 16.7 | 179.5 | 94.6 | 25.3 | −8.6 |
| RbBr | 21.8 | 205.0 | 108.4 | 28.8 | −7.0 |
| RbI | 25.9 | 233.5 | 118.0 | 34.4 | −8.5 |
| CsF | −37.7 | 123.4 | 79.9 | 13.0 | −50.7 |
| CsCl | 18.0 | 188.3 | 97.5 | 27.0 | −9.0 |
| CsBr | 25.9 | 213.8 | 121.3 | 27.6 | −1.7 |
| CsI | 33.1 | 242.3 | 129.7 | 33.6 | −0.5 |
| AgF | −20.5 | 64.4 | 83.7 | −5.8 | −14.7 |
| AgCl | 66.5 | 129.3 | 96.2 | 9.9 | 56.6 |
| AgBr | 84.1 | 154.8 | 107.1 | 14.2 | 69.9 |
| AgI | 111.7 | 183.3 | 114.2 | 20.6 | 91.1 |
| TlF | −2.5 | 111.7 | 87.9 | 7.1 | −9.6 |
| TlCl | 43.5 | 182.4 | 108.4 | 22.1 | 21.4 |
| TlBr | 57.3 | 207.9 | 119.7 | 26.3 | 31.0 |
| TlI | 74.1 | 236.4 | 123.0 | 33.8 | 40.3 |
| $MgF_2$ | −18.4 | −137.2 | 57.3 | −58.0 | 39.6 |
| $MgCl_2$ | −155.2 | −7.5 | 89.5 | −28.9 | −126.3 |
| $MgBr_2$ | −186.2 | 43.5 | 123.0 | −23.7 | −162.5 |
| $MgI_2$ | −214.2 | 100.4 | 145.6 | −13.5 | −200.7 |
| $CaF_2$ | 13.4 | −74.5 | 69.0 | −42.8 | 56.2 |
| $CaCl_2$ | −82.8 | 55.2 | 113.8 | −17.5 | −65.3 |
| $CaBr_2$ | −110.0 | 106.3 | 129.7 | −7.0 | −103.0 |
| $CaI_2$ | −120.1 | 163.2 | 142.3 | 6.2 | −126.3 |

**Table 2.16** (*continued*)

| Halide | $\Delta H_d^{\ominus}$/kJ mol$^{-1}$ | $\Sigma \bar{S}_i^{\ominus}$/J mol$^{-1}$ | $S_C^{\ominus}$/J mol$^{-1}$ | $T\Delta S_d^{\ominus}$/kJ mol$^{-1}$ | $\Delta G_d^{\ominus}$/kJ mol$^{-1}$ |
|---|---|---|---|---|---|
| $SrF_2$ | 10.5 | –58.6 | 89.5 | –44.2 | 54.7 |
| $SrCl_2$ | –51.9 | 71.1 | 117.2 | –13.7 | –38.2 |
| $SrBr_2$ | –71.5 | 122.2 | 141.4 | –5.7 | –65.8 |
| $SrI_2$ | –90.4 | 179.1 | 164.0 | 4.5 | –94.9 |
| $BaF_2$ | 3.8 | –6.7 | 96.7 | –30.8 | 34.6 |
| $BaCl_2$ | –13.0 | 123.0 | 125.5 | –0.7 | –12.3 |
| $BaBr_2$ | –25.5 | 174.1 | 148.5 | 7.6 | –33.1 |
| $BaI_2$ | –47.7 | 230.9 | 171.1 | 17.8 | –65.5 |

interaction of the ions with water molecules, which become coordinated around the ions in hydration shells; then there is a structure-making effect arrising from the formation of these hydration shells. The first of these effects is important with large ions as it tends to increase the $\bar{S}_i^{\ominus}$ values, as may be seen by comparing $SrF_2$ and $BaF_2$, for example. The second effect is prominent with small ions, since it acts so as to decrease $\bar{S}_i^{\ominus}$, as exemplified in the comparison of $CaF_2$ and $MgF_2$. Any given case involves an interplay of these factors, but, while solubility may be difficult to explain in molecular terms, except qualitatively, the thermodynamic analysis is quantitative, and the results are true within the limits of experimental error.

We can now explain the trends in solubility with which we introduced this section. In the series of calcium halides, $\Delta G_h$ increases† less rapidly than $\Delta G_C$ from fluoride to iodide so that $\Delta G_d$ becomes more negative in this same direction. This trend occurs despite some large negative $\Delta S_d$ terms. For the series of silver halides an opposite tendency exists: $\Delta G_C$ decreases from fluoride to iodide much more abruptly and rapidly than would be expected for halides containing an ion of the size of $Ag^+$ (1.27 Å). The partial covalent character of these compounds, which increases markedly from fluoride to chloride, is responsible for the greater stability of the crystal, with respect to the ion-gas and hence for the decrease in solubility along this series.

Substances like MgO and CaO, for example, are highly ionic but effectively insoluble in water. The crystal energies are numerically large, because the Madelung energy terms include the product of the charges on the ions (Table 2.7). On the other hand, the $O^{2-}$ ion hydrates‡ as a pair of singly-charged $[OH]^-$ ions, which with the cations, do not produce as low an energy as in the crystal. Hence, these ionic solids are insoluble in water.

†Becomes less negative.
‡For example, $2\Delta H_h$ ($OH^-$,g) ≈ – 1000 kJ mol$^{-1}$; $\Delta H_h$ ($CO_3^{2-}$,g) ≈ –1250 kJ mol$^{-1}$.

## 2.9 SPECTRA OF IONIC COMPOUNDS

In studying the electrostatic model, we treated an ionic solid as a regular array of point charges. We considered modifications of this picture, by polarization and covalency, but have left the impression of a static system of particles. Now ions possess vibrational motion, even at 0 K, and this motion contributes to the heat capacity of a solid.

### 2.9.1 Absorption and colour

All ionic solids absorb in the ultraviolet region of the spectrum; some compounds absorb also in the visible region. The absorption does not produce opacity, unlike metals: often, apparently opaque ionic crystals are simply intensely coloured, and transmission of light can be observed in thin section. Ionic solids are coloured for two main reasons. Some may contain ions which give rise to a characteristic colour through transitions among their d electrons, as with $[Fe(H_2O)_6]^{2+}$ and $[Co(NH_3)_6]^{3+}$. However, in the $[MnO_4]^-$ ion, for example, the colour arises through a charge transfer of electrons from the ligand atoms into the d orbitals of the metal atoms. Other compounds may contain ions which, although colourless in solution, undergo polarization in the solid with the appearance of colour arising from transitions among partially delocalized electrons, as in AgI, $Ag_3PO_4$ or $Pb_3O_4$ for example. If the absorption moves from the ultraviolet region just into the visible, the wavelengths absorbed would be in the blue region of the spectrum. Consequently, ionic solids whose colour arises in this manner are often yellow to red.

### 2.9.2 Heat capacity

At constant volume, the molar heat capacity $C_V$ is defined by

$$C_V = (\partial U/\partial T)_V \ . \qquad (2.92)$$

As the temperature of a solid is increased, its vibrational energy increases. If, as in the case of CsCl, for example, there are no translational energy modes, and since rotation of a species about its own axis does not constitute a degree of freedom, all of the energy imparted to a solid in the form of heat enhances the vibrational degrees of freedom. They are the number of independent square terms, that is, those dependent on the square of a coordinate of a velocity parameter, needed to specify the total energy of the species in the solid. As CsCl contains $2L$ ions per mol, and each ion can vibrate independently in three dimensions, there are $6L$ vibrational degrees of freedom per mol. Since, from the principle of equipartition of energy,† each vibrational degree of freedom contributes an amount of energy $kT$, the total vibrational thermal energy, ignoring the zero energy, is $6LkT$, or $6RT$ per mole. Thus from (2.92) $C_V = 6R$, which is approximately 50 kJ mol$^{-1}$. For monatomic species $C_V = 25$ kJ mol$^{-1}$, which embodies the historic law of Dulong and Petit.

†See also Appendix 21.

The principal molar heat capacities are related by

$$C_P - C_V = \beta^2 TV/\kappa \ , \tag{2.93}$$

where $\beta$ is the volume expansivity and $\kappa$ is given by (2.47). For NaCl at 298 K, $\beta = 1.1 \times 10^{-4}\ \text{K}^{-1}$, $\kappa = 4.1 \times 10^{-11} \text{N}^{-1}\ \text{m}^2$ and $V = 2.7 \times 10^{-5}\ \text{m}^3\ \text{mol}^{-1}$. Hence $C_P - C_V = 2.4\ \text{J mol}^{-1}\ \text{K}^{-1}$. For an ideal gas $C_P - C_V = R$; the smaller value for the solid arises from the fact that its heat capacity is determined by the vibrational energy, and this quantity does not vary appreciably in constant pressure and constant volume conditions.

Planck postulated in 1901 that a vibrating atom could acquire energy in quanta of $h\nu$, the probability that a vibrating atom can acquire this amount of energy being proportional to $\exp(-h\nu/kT)$. At room temperature, almost all vibrational degress of freedom are active and $C_V$ tends to its limiting value (Figure 2.20). As the temperature is decreased vibrations of increasingly lower frequency cease to be excited. Thus $C_V$ decreases with decreasing temperature because the average energy of the substance decreases. In the limit as $T \longrightarrow 0$ $C_V \longrightarrow 0$, from the third law of thermodynamics.

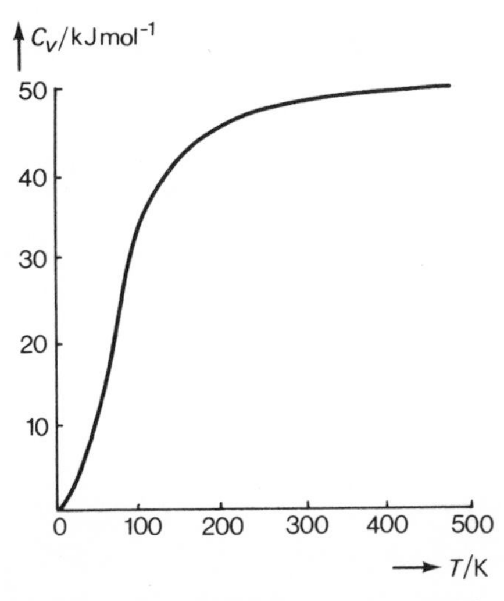


Figure 2.20 – Variation of $C_V$ with $T$ illustrated by NaCl.

The effect of change in temperature of a compound on its atomic vibrations can be indicated by the following example. The HCl molecule has a vibration 'frequency' of 2989 $\text{cm}^{-1}$. The ratio of the probabilities that this oscillator will acquire the corresponding energy at 500 K and 50 K is about $4 \times 10^{33}$ from the exponential law given above. The energy absorbed by a vibrating system of atoms is used mainly to increase the amplitudes of the vibrations.

Many molecules can be raised to excited energy states by the absorption of radiation. The requirement for a vibration to be active in the infrared spectral region is that it should contain an oscillating dipole. The rate of change of dipole moment with time determines the ability of a species to absorb infrared radiation. If as a first approximation we treat a vibrating diatomic molecule as a harmonic oscillator, then the frequency of the oscillation is given by the well known expression

$$\nu = (1/2\pi)\sqrt{f/\mu} \; ; \tag{2.94}$$

$\mu$ is the reduced mass of the vibrating system which, for two species of masses $m_1$ and $m_2$, is given by

$$1/\mu = 1/m_1 + 1/m_2 \; , \tag{2.95}$$

as described in Appendix 13; $f$ is the bond force constant (restoring force per unit displacement) between the two species.

For the HCl molecule $\mu = 0.981 \times 1.660 \times 10^{-27}$ kg; hence $f = 516$ N m$^{-1}$. Force constants are often reported in the units of mdyn Å$^{-1}$; $f$ (HCl) = 5.16 mdyn Å$^{-1}$. Force constants can be useful properties for characterising bond energies, as the following results show:

| | HF | HCl | HBr | HI |
|---|---|---|---|---|
| $f$/N m$^{-1}$ | 989 | 516 | 405 | 308 |
| $10^{10}r_e$/m | 0.92 | 1.27 | 1.41 | 1.61 |
| $D_0$(H$X$)/eV | 5.83 | 4.43 | 3.75 | 3.06 |

In ionic solids, each pair of oppositely charged ions is equivalent to an oscillating dipole, and its vibrations are excited by infrared radiation: NaCl gives a single absorption band at about 164 cm$^{-1}$; the heavier the species, the smaller are the wavenumbers of its vibrations.

## 2.10 DEFECTS IN CRYSTALS

From perfect crystals of vibrating atoms, we consider finally in this chapter the even more real situation of such crystals containing imperfections. The subject of crystal defects is extensive, and we shall investigate here only that class of imperfections known as point defects.

The simplest point defect is a vacancy at a lattice site, the atom having been transferred from the interior of the crystal to its surface; this imperfection is known as a Schottky defect (Figure 2.21a). In a crystal at thermal equilibrium with its surrounding, a certain number of lattice vacancies always exist and contribute to the entropy of a crystal at any temperature above 0 K. If we take a constant volume of a crystal, we can write for the creation of defects

$$\Delta A = \Delta U - T\Delta S \;, \tag{2.96}$$

where $A$ is the Helmholtz free energy.

Let $n$ defects be distributed over $N$ lattice sites. Then the first defect can be arranged in $N$ ways, the second defect in $(N - 1)$ ways, and the $n$th in $[N - (n-1)]$ ways. Thus the total number of arrangements of $n$ defects is given by

$$w = N(N - 1)(N - 2) \ldots\ldots [N - (n - 1)] = N!/(N - n)! \,. \tag{2.97}$$

However, the defects are indistinguishable from one another; we cannot differentiate between defects $i$ and $j$ on site $k$ for example. There are $n$ indistinguishable ways of obtaining the first defect $(n - 1)$ ways of obtaining the second, $(n - 2)$ for the third, and so on, and in all $n!$ Hence, the total number of arrangements that we require is given by

$$W = N!/[(N - n)!\, n!] \;. \tag{2.98}$$

Let $\Delta u$ represent the energy change for the formation of a single defect. Then,

$$\Delta A = n\Delta u - T\Delta S \;, \tag{2.99}$$

for $n$ defects. Using the Boltzmann equation (1.7), taking $W_1$ equal to unity as a reference state of perfect order, and $W_2$ from (2.98),

$$\Delta S = k\ln\{N!/[(N - n)!n!]\} \tag{2.100}$$

and

$$\Delta A = n\Delta u - kT\ln\{N!/[(N - n)!n!]\} \;. \tag{2.101}$$

At equilibrium

$$[\partial(\Delta A)/\partial n]_T = 0 \;, \tag{2.102}$$

and using Stirling's approximation for factorials,

$$\ln X! = X\ln X - X \;, \tag{2.103}$$

we obtain

$$\Delta u - kT\ln[(N - n)/n] = 0 \;. \tag{2.104}$$

If we assume that $n \ll N$, then

$$n = N\exp(-\Delta u/kT) \;. \tag{2.105}$$

In ionic crystals, it is most favourable energetically to form pairs of Schottky defects, in order to maintain the best electrical balance in the crystal. For pairs of vacancies, $W$ is squared; hence,

$$n_{\pm} = N\exp(-\Delta u/2kT) \ . \tag{2.106}$$

The energy† of a nearest neighbour bond in a solid is approximately 1 eV, so this amount of work has to be done in order to create a vacancy. On the other hand the structure can then relax around the vacancy, and approximately two-thirds of the energy expended is regained in this process. The entropy change is positive and so contributes to the driving force for the creation of defects, according to (2.99).

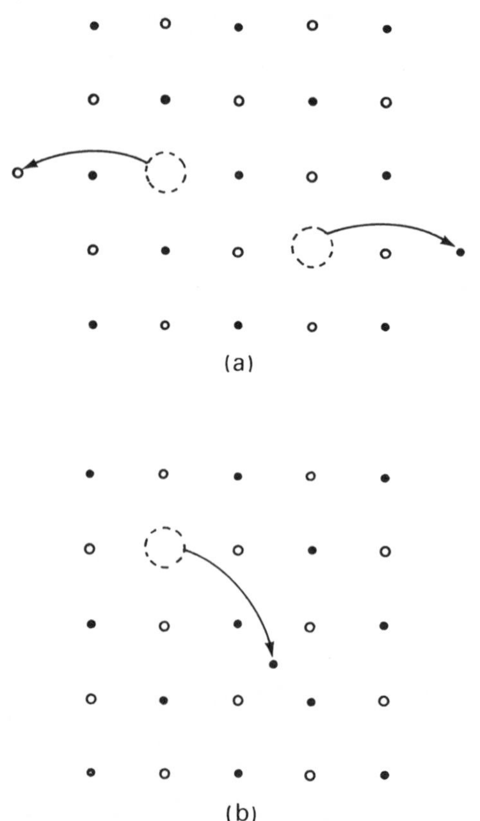


Figure 2.21 – Point defects: (a) Pair of Schottky defects in the NaCl structure type; (b) Frenkel defect in AgCl.

As an example calculation, consider KCl at 300 K. The value of $\Delta u$ for a pair of defects is approximately 1 eV, or $\Delta U = 96$ kJ mol$^{-1}$.

†Average value.

Hence, using (2.106),

$$n_{\pm}/N = \exp(-64000/600R) = 2.68 \times 10^{-6} \quad . \tag{2.107}$$

If $N$ is set equal to $L$, the Avogadro constant, $n_{\pm}$ becomes $1.6 \times 10^{18}$, or approximately 1 defect pair per 370000 sites. At 600 K, this ratio increases to 1 in about 600.

Another type of defect, found in AgCl for example, is called a Frenkel defect. It implies that a vacant cation site exists, the $Ag^+$ ion having been transferred to an interstitial position in the structure (Figure 2.21). If we consider $N$ lattice sites and $N'$ interstices, $W$ (Frenkel) is $N!N'!/[(N-n)!\ n!\ (N'-n)!\ n!]$, where $n$ has the same meaning as before. Carrying out an analysis similar to that for the Schottky defect leads to

$$n = \sqrt{NN'} \exp(-\Delta u/2kT) \quad , \tag{2.108}$$

for the creation of $n$ Frenkel defects consisting of the interstitial ion *and* its hole. The reason that AgCl exhibits, preferentially, Frenkel defects is not a size effect because $r(Na^+) < r(Ag^+) < r(K^+)$: more probably it is connected with its greater tendency towards covalent character† than is the case with NaCl and KCl. Furthermore, the silver cation is under enhanced compressive stress in an interstitial position and this situation will increase the tendency for partial covalency.

### 2.10.1 Defects and ion mobility

The electrical conductivity of an ionic solid is dependent upon the nature and concentration of defects. This concentration may be increased by heating the solid to a high temperature. If the material is then quenched rapidly, the defects become locked into the structure and, because ionic mobility is related to defect concentration, measurements of electrical conductivity reflect changes in the defect concentration. It turns out that in the alkali-metal halides for example, the electrical conductivity is proportional to the ionic mobilities and to the exponential factor in (2.105).

If a crystal structure were perfect, it would be very difficult to envisage a mechanism for ion transport. Studies on the movement of ions in ionic solids have been carried out by a radioactive tracer technique. In a typical experiment, a thin slice of $^{24}NaCl$ was sandwiched between two crystal plates of NaCl, and the sample was maintained at a constant temperature. The defect concentrations through the crystal were determined by cutting thin sections of the sandwich after a given time and then counting the radioactivity in each section. The defect concentration, $d$, at a distance $r$ from the surface after a time $t$, was found to obey the equation

$$\ln d = a - r^2/(Dt) \quad , \tag{2.109}$$

†Calculate the effective atomic numbers for $Na^+$, $K^+$ and $Ag^+$ from Slater's rules.

where $a$ is a constant and $D$ is the diffusion coefficient for the substance under examination. Further experiments showed that the variation of $D$ with temperature followed a Boltzmann distribution (1.10):

$$D = D_0 \exp(-U_d/RT) \ . \tag{2.110}$$

The activation energy $U_d$ for the diffusion process in NaCl is approximately 173 kJ mol$^{-1}$, which includes the energy needed both to create a vacancy and to move an ion into the vacant site. Thus, for a single ion, and using our earlier datum (see page 114), we have

$$U_d \approx 96 + U_m \ . \tag{2.111}$$

Thus $U_m$, the energy needed to induce migration, is about 77 kJ mol$^{-1}$ in NaCl; the activation energy is approximately equally divided between the energies needed to create a vacancy and to move an ion into the hole.

### 2.10.2 Doping and colour centres

Defects can be created by introducing foreign ions into a structure. Thus NaCl can be doped by crystallization from a melt containing up to ½% $CaCl_2$. Because $Ca^{2+}$ has about the same radius as $Na^+$, it can fit into the NaCl structure: electrical neutrality is maintained by leaving one $Na^+$ site vacant for each $Ca^{2+}$ introduced during the crystallization process.

If solid NaCl is heated in sodium vapour, it acquires a yellow colour: the crystal contains an excess of occupied $Na^+$ sites, so that some $Cl^-$ vacancies can be expected. A single electron from $Cl^-$ may remain delocalized at the vacant Schottky site, surrounded by 6 $Na^+$ ions. Such an electron constitutes an $F$ centre. The $F$-centre electron has hydrogen-like character, and undergoes 1s $\longrightarrow$ 2p transitions, which are responsible for absorption in the visible region of the spectrum.

The change of colour is often associated with a change in the refractive index of the material and $F$-centre materials can be used for storage and read out of information. In some materials, such as $LiNbO_3$ for example, the refractive index variation associated with $F$-centre defects are very large, and the material can be used to store holographic information and to reproduce it with good efficiency. The intrinsic value of these applications is decreased to some extent by poor erase characteristics, but it has been suggested that in their absence of grain characteristics they may prove to be superior to photographic film in certain applications.

## APPENDIX 5 – LEAST-SQUARES LINE

If it is desired to fit a straight-line relationship to a number of observations in excess of two, it is often appropriate to use the method of least squares. Let the equation be of the form

$$y = ax + b \ , \tag{A5.1}$$

where $a$ and $b$ are constants which have to be determined. For any observation $i$,

$$ax_i + b - y_i = e_i \ , \tag{A5.2}$$

where $e_i$ is an error which will be assumed both to be random and to reside in the value of the dependent variable $y_i$, the error in the independent variable $x_i$ being relatively negligible. According to the principle of least squares, the best values of $a$ and $b$ are chosen such that the sum of the squares of the errors $e_i$ is a minimum.

Thus $$\text{Min}\,\{\sum_i e_i^2\} = \text{Min}\{\sum_i (ax_i + b - y_i)^2\} \ . \tag{A5.3}$$

The required minimum value may be found by differentiating the right-hand side of (A5.3) partially with respect to both $a$ and $b$, and setting each of the derivatives equal to zero.

Hence $$\partial\{\sum_i e_i^2\}/\partial a = 2\sum_i (ax_i^2 + bx_i - x_i y_i) = 0 \tag{A5.4}$$

and $$\partial\{\sum_i e_i^2\}/\partial b = 2\sum_i (ax_i + b - y_i) = 0 \ . \tag{A5.5}$$

Thus, we may derive $$a[x^2] + b[x] - [xy] = 0 \ , \tag{A5.6}$$

and $$a[x] + bN - [y] = 0 \ . \tag{A5.7}$$

Equations (A5.6) and (A5.7) are known as the normal equations and $[x]$, for example, means $\sum_i x_i$ over the number $N$ of observations. If each observation has a weight $w$ then the normal equations become

$$a[wx^2] + b[wx] - [wxy] = 0 \ , \tag{A5.8}$$

and $$a[wx] + b[w] - [wy] = 0 \ . \tag{A5.9}$$

Solving for $a$ and $b$,

$$a = ([w]\ [wxy] - [wx]\ [wy])/\Delta\ , \tag{A5.10}$$

and

$$b = ([wx^2]\ [wy] - [wx]\ [wxy])/\Delta\ , \tag{A5.11}$$

where $\Delta$ is given by

$$\Delta = [w]\ [wx^2] - [wx]\ [wx]\ . \tag{A5.12}$$

If all of the weights are unity, $[w] = N$.

The standard deviations in $a$ and $b$ may be estimated by the following procedure. From (A5.2),

$$[e^2] = \sum_i w_i\,(ax_i + b - y_i)^2\ . \tag{A5.13}$$

Then, we write, without proof here,

$$\sigma^2(a) = \{[e^2]/(N-2)\}[w]/\Delta \tag{A5.14}$$

and

$$\sigma^2(b) = \{[e^2]/(N-2)\}[wx^2]/\Delta\ , \tag{A5.15}$$

$\sigma$ being an estimated standard deviation and $\sigma^2$ the corresponding variance.

It is recommended that the least-squares line be compared, where feasible, with a plot of the experimental $x,y$ values. In the light of this inspection, certain observations may be reasoned to be unreliable. It must be remembered that a least-squares procedure will always give the best fit to the observations, including the bad ones.

The least-squares technique may be extended to functions of higher degree. Thus, the quadratic function

$$y = ax^2 + bx + c \tag{A5.16}$$

may be fitted to experimental data in excess of three by this method, starting from an equation similar to (A5.3).

## APPENDIX 6 NUMERICAL INTEGRATION

A numerical integration procedure computes the value of a definite integral from a set of values of the integrand. When the integrands are ordinates of a curve, the integration defines the area under the curve; we meet numerical integration often in this form.

It may happen that a curve can be simulated by a least-squares fit to an appropriate function. Then there is no problem, because we can obtain an analytical solution from the fitted function. Suppose that the curve in Figure A6.1 can be fitted by the function

$$y = 2x^2 + 3x + 2\ . \tag{A6.1}$$

Then

$$\int_0^{0.8} y\mathrm{d}x = 2x^3/3 + 3x^2/2 + 2x\Big]_0^{0.8} = 17.408/6 = 2.90133\ldots\ . \tag{A6.2}$$

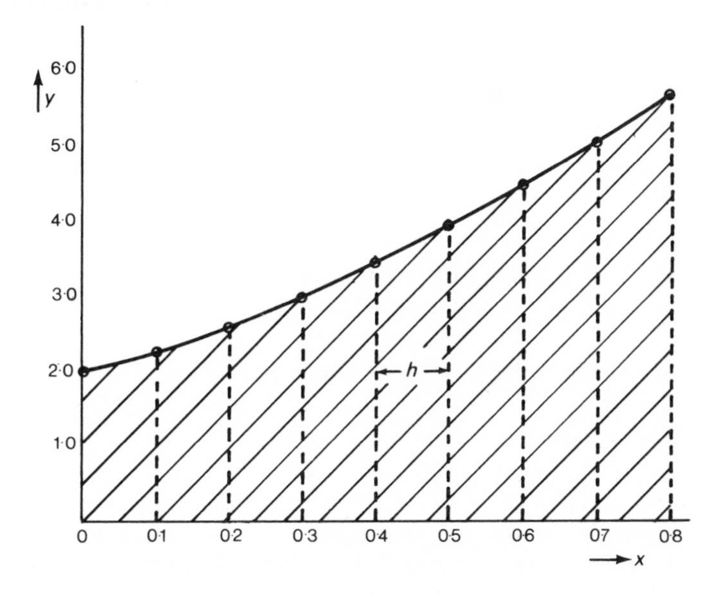


Figure A6.1 – A function, $y = f(x)$.

## A6.1 Numerical and other methods

It may not be possible to deduce a function which will represent satisfactorily a set of experimental measurements, and it becomes necessary to adopt another procedure. Two rather different methods will be described here.

Some numerical integration techniques depend on the relationship

$$\int_a^c f(x)\mathrm{d}x = \int_a^b f(x)\mathrm{d}x + \int_b^c f(x)\mathrm{d}x\ . \tag{A6.3}$$

One such method is embodied in Simpson's rule.

### A6.1.1 *Simpson's Rule*

In this procedure a curve such as that in Figure A6.1 is divided into an EVEN number $n$ of intervals of equal width $h$. Then Simpson's rule, given without proof here, states that for any function $y = f(x)$

$$\int_{x_0}^{x_n} y\mathrm{d}x = \int_{x_0}^{x_0+nh} y\mathrm{d}x = (h/3)[y_0 + y_n + 2(y_2 + y_4 + y_6 + \ldots + y_{n-2}) + 4(y_1 + y_3 + y_5 + \ldots + y_{n-1})] \qquad \text{(A6.4)}$$

*Example*

Let the curve in Figure A6.1 be divided into 8 intervals of width 0.1 units. Then, from the curve, we have the following data:

| $x$ | 0.0 | 0.1 | 0.2 | 0.3 | 0.4 | 0.5 | 0.6 | 0.7 | 0.8 |
|---|---|---|---|---|---|---|---|---|---|
| $y$ | 2.0 | 2.3 | 2.7 | 3.1 | 3.5 | 4.0 | 4.5 | 5.1 | 5.7 |

Using (A6.4),

$$\int_0^{0.8} y\mathrm{d}x = (0.1/3)\ [2.0 + 5.7 + 2(2.7 + 3.5 + 4.5) + 4(2.3 + 3.1 + 4.0 + 5.1)] = 2.903\ , \qquad \text{(A6.5)}$$

which agrees with the analytical result to within 0.07%.

A6.1.2 *Direct weighing method*

The curve in Figure A6.1 is drawn on good quality graph paper. The shaded area is cut out carefully and weighed; let its mass be 0.2759 g. A certain known number of squares are cut out and weighed so as to provide a calibration relationship between mass and area. Let 500 squares have a mass of 4.7563 g and, using the graph scales, let 1 square be equal to 0.1 units of area. Then

$$\text{Mass of curve/Area under curve} = \text{Mass of 500 squares/(500} \times \text{scale)}\ . \qquad \text{(A6.6)}$$

Hence, $\quad \text{Area under curve} = 50 \times 0.2759/4.7563 = 2.90\ , \qquad \text{(A6.7)}$

which is within 0.05% of the analytical result.

Problems requiring numerical integration arise, for example, in evaluating thermodynamic functions, such as $\Delta H$ or $S$.

## APPENDIX 7 CRYSTAL STRUCTURE ANALYSES OF KCl AND NaCl

The first crystal structure analyses, those of KCl and NaCl, were reported in 1912, and confirmed by Bragg in 1913; we shall consider briefly the latter work. It may be desirable for the reader to make reference to Appendices 2 and 3 from which results will be drawn.

Potassium and sodium chlorides crystallize as cubes. In order to study the reflexion of X-rays from planes other than the cube faces, which are equivalent

planes to be investigated. Bragg's work was based on data for the planes of the types ($h00$), ($hh0$) and ($hhh$), parallel to (100), (110) and (111) faces respectively, as shown in Figure A7.1.

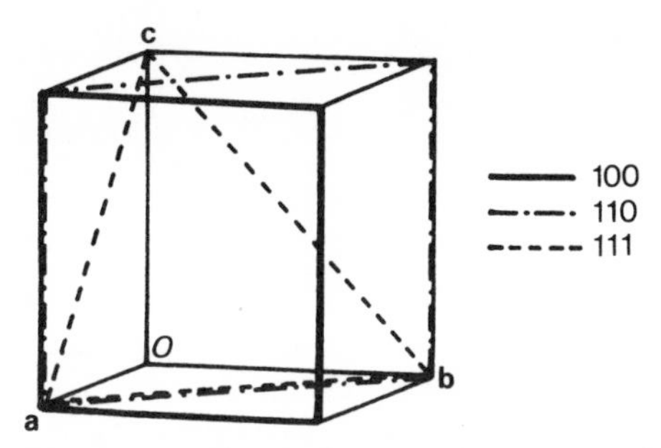


Figure A7.1 – Cube, showing the orientations of the (100), (010) and (001) faces.

X-rays were allowed to fall on these planes in turn, and the intensity, $I$, of the reflected (diffracted) radiation was measured as a function of the Bragg angle, $\theta$. The six sets of results are presented in the form of bar graphs in Figure A7.2.

We shall use the Bragg equation in the form

$$n\lambda = 2d\sin\theta \quad . \tag{A7.1}$$

Any of the sets of spectra obey the Bragg equation. Consider the $h00$ spectra for KCl. The three reflexions occur at $\theta$-values of 5.36, 10.76 and 16.26°. The ratios of the corresponding values of $\sin\theta$ are 1:2.00:3.00, which are integral within experimental error.

Consider next the first reflexion for KCl in the $h00$, $hh0$ and $hhh$ sets, their $\theta$-values being 5.36. 7.59 and 4.65°. Since $d \propto 1/\sin\theta$, the ratios of these $d$-values are $1 : 1/\sqrt{2} : 2/\sqrt{3}$. In the original work, the $hhh$ reflexion at $\theta = 4.65°$ was not recorded: this fact led to the $d$-ratios being taken as $1 : 1/\sqrt{2} : 1/\sqrt{3}$, which indicated a primitive cubic arrangement of atoms. Such a structure could be based on the cubic unit cell in Figure A7.1: we can see from the figure that the $d$-ratios for the (100), (110) and (111) planes are $1 : 1/\sqrt{2} : 1/\sqrt{3}$.

In the NaCl results, the true $d$-ratios are clearly established, and these ratios are consistent with a face-centred, $F$, arrangement of atoms (Figure 1.2). We can now explain the intensity data for both NaCl and KCl more fully in terms of the $F$ unit cell. The $h00$ and $hh0$ series of spectra must be labelled 200, 400, 600 and 220, 440, 660, respectively ($n = 2,4,6$). Reflexions such as 100 or 110 ($n = 1$) are not possible, because successive planes are interleaved by planes of equivalent scattering power, at exactly $d/2$ apart, and complete interference takes place.

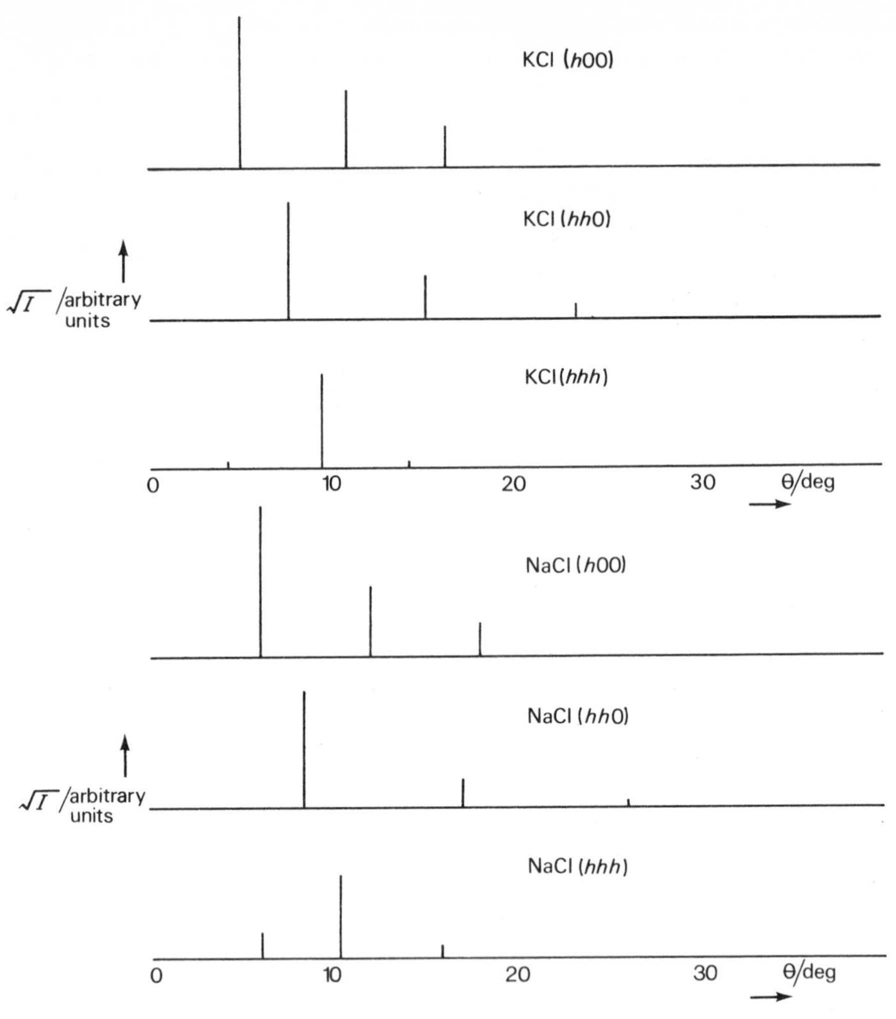


Figure A7.2 – Bar graph of $\sqrt{I}$ for KCl and NaCl up to $\theta \approx 26°$. For each compound the ordinates are normalised to the value of $\sqrt{I}$ for the first $h00$ reflexion.

The (*hhh*) planes, however, are populated by either all cations or all anions, and can be labelled, in order, (111), (222) and (333). Consider the (111) family of planes. They are interleaved by planes at a spacing $d_{111}/2$: from the Bragg equation, it is clear that when successive (111) planes are in the reflexion position, the interleaving plane scatters exactly out of phase, because the path distance between the planes of anions and cations is from (A7.1) $2(d_{111}/2)\sin\theta$. Eliminating $2\sin\theta$, the path difference becomes $n\lambda/2$, because $d$ in (A7.1) is $d_{111}$. Thus when $n$ is an odd number, the reflexion intensity will be weaker than when $n$ is even, and this is clearly indicated in Figure A7.2. We can see further how $111_{KCl}$ is much weaker in intensity than $111_{NaCl}$. X-rays are scattered by electrons, and thus to a first approximation, the intensity of scattering will be governed by the atomic number of a species. Now $K^+$ and $Cl^-$ are isoelectronic, with 18 electrons each, while $Na^+$ has 10 electrons: hence the out-of-phase scattering has a lesser effect on the intensity in NaCl than in KCl, which is again completely in accord with experiment.

We have shown that the unit cell of the NaCl structure type (Figure 1.2) contains 4 cations and 4 anions (see page ). The density of NaCl is 2165 kg m$^{-3}$ and the relative atomic masses of Na and Cl are 22.990 and 35.453 respectively. Using the relationship density = mass/volume in terms of the unit cell, we may write

$$2165 = 4 \times 58.443 \times 1.6606 \times 10^{-27}/a^3 \ , \qquad \text{(A7.2)}$$

where $a$ is the length of the unit cell edge. Hence $a = 5.64 \times 10^{-10}$ m, or 5.64 Å.

We have not used the wavelength of the X-radiation involved in the experimental measurements of KCl and NaCl. Indeed, X-ray wavelengths were unknown when these measurement were made. Taking the $\theta$-value for the 200 reflexion from NaCl as 6.00° we have, from (A7.1), with $n = 2$ and $d$, the fundamental $h00$ spacing, set equal to $a$, $\lambda = 0.589$ Å. This wavelength corresponds to Pd $K\alpha$ X-radiation used by Bragg in 1913.

It should be noted that crystal structures are not determined today by this simple analysis, although several of its principles have their counterparts in modern X-ray crystallography.

## APPENDIX 8 EQUATION OF STATE FOR A SOLID

The Helmhotz free energy $A$ is defined thermodynamically by

$$A = U - TS \ . \qquad \text{(A8.1)}$$

Hence, since $A$ is an extensive property,

$$\mathrm{d}A = \mathrm{d}U - T\mathrm{d}S - S\mathrm{d}T \ . \qquad \text{(A8.2)}$$

At constant temperature, and since

$$T\mathrm{d}S = \mathrm{d}U + P\mathrm{d}V \ , \qquad \text{(A8.3)}$$

$$\mathrm{d}A = \mathrm{d}U - (\mathrm{d}U + P\mathrm{d}V) \ . \qquad \text{(A8.4)}$$

Hence,

$$\mathrm{d}A = -P\mathrm{d}V \ , \qquad \text{(A8.5)}$$

and $$(\partial A/\partial V)_T = -P \tag{A8.6}$$

From (A8.2), we obtain $$\left(\frac{\partial A}{\partial V}\right)_T = \left(\frac{\partial U}{\partial V}\right)_T - T\left(\frac{\partial S}{\partial V}\right)_T \tag{A8.7}$$

Using Maxwell's relations†, $$\left(\frac{\partial S}{\partial V}\right)_T = \left(\frac{\partial P}{\partial T}\right)_V \tag{A8.8}$$

Furthermore $$\left(\frac{\partial P}{\partial T}\right)_V \left(\frac{\partial T}{\partial V}\right)_P \left(\frac{\partial V}{\partial P}\right)_T = -1 , \tag{A8.9}$$

or $$\left(\frac{\partial P}{\partial T}\right)_V = -\frac{(\partial V/\partial T)_P}{(\partial V/\partial P)_T} . \tag{A8.10}$$

From the thermodynamic definitions of expansivity ($\beta$) and compressibility ($\kappa$), it follows that the right-hand side of (A8.10) is equal to $\beta/\kappa$. We obtain now,

$$\left(\frac{\partial A}{\partial V}\right)_T = \left(\frac{\partial U}{\partial V}\right)_T = T\beta/\kappa , \tag{A8.11}$$

and, with (A8.6), $$\left(\frac{\partial U}{\partial V}\right)_T = -P + T\beta/\kappa . \tag{A8.12}$$

At 0 K, $\left(\frac{\partial U}{\partial V}\right)_T = -P$: $T\beta/\kappa$ is not apparently negligible at $T = 298.15$ K, the temperature at which we need to calculate $U(r_e)$ in order to compare it conveniently with the corresponding thermodynamic value. A detailed analysis‡ shows that for NaCl the error in taking $(\partial U/\partial V)_T = -P$ at 298.15 K is about 1%, $-765$ kJ mol$^{-1}$ instead of $-774$ kJ mol$^{-1}$ (Table 6.1). Among the alkali-metal halides, this error ranges from 0.5% to 1.5%. The full analysis incorporates the expansivity, the temperature coefficient of compressibility and the pressure coefficient of compressibility. For our present purposes, we shall avoid these complexities, and tolerate the small error involved in setting $(\partial U/\partial V)_T$ equal to $-P$ at 298.15 K.

†Suggested Reading: *Thermodynamics* (Denbigh).

‡Suggested Reading: *Solid-State Physics* (Tosi).

## APPENDIX 9 SLATER'S RULES

In wave-mechanical calculations it is often sufficiently accurate, for principal quantum numbers $n$ up to 4, to use the approximate analytical wave functions which have been derived by Slater. These functions apply to a single electron in a central field, that is, in a field in which the potential energy is a function of only a radial parameter $r$ and provided by an effective charge $\zeta e$. The difference between $\zeta$ and the atomic number $Z$ is the screening constant $S$ for the particular electron. It measures the degree to which the other electrons in the atom screen the electron under consideration from the nuclear charge.

Atomic orbitals are assumed to be of the following forms, for s and p electrons:

$$\psi(1\text{s}) = N_{1\text{s}}\exp(-\zeta r),\ \psi(2\text{s}) = N_{2\text{s}}r\exp(-\zeta r/2),\ \psi(3\text{s}) = N_{3\text{s}}r^2\exp(-\zeta r/3),$$

$$\psi(2\text{p}_x) = N_{2\text{p}}x\exp(-\zeta r/2),\ \psi(3\text{p}_x) = N_{3\text{p}}xr\exp(-\zeta r/3)\ . \qquad \text{(A9.1)}$$

The atomic orbitals $\psi(2\text{p}_y)$ or $\psi(3\text{p}_z)$, for example, can be written down by analogy with $\psi(2\text{p}_x)$ or $\psi(3\text{p}_x)$, respectively. The normalizing constants $N$ are chosen such that $\int\psi^2 \text{d}\tau = 1$, and lengths such as $r$ are defined relative to the Bohr radius $a_0$.

$$a_0 = 4\pi\epsilon_0 h^2/me^2 = 5.29466 \times 10^{-11}\,\text{m}\ , \qquad \text{(A9.2)}$$

where $\epsilon_0$ is the permittivity of a vacuum; $r$ is an absolute value $R$ divided by $a_0$.

By integration (see later) we find the following values:

$$N_{1\text{s}} = (\zeta^3/\pi)^{1/2},\ N_{2\text{s}} = (\zeta^5/96\pi)^{1/2},\ N_{3\text{s}} = (0.4\zeta^7/3^9\pi)^{1/2},$$

$$N_{2\text{p}} = (\zeta^5/32\pi)^{1/2},\ N_{3\text{p}} = (0.4\zeta^7/3^8\pi)^{1/2}\ . \qquad \text{(A9.3)}$$

The exponent $\zeta$ is given by

$$\zeta = Z'\,(\text{effective}) = Z - S\ , \qquad \text{(A9.4)}$$

and the values of $S$ are found by the following rules.

First, the atomic orbitals which are occupied are divided into the groups 1s / 2s,2p / 3s,3p / 3d / 4s,4p / 4d / 4f / and so on. Then $S$ is formed by summing the following contributions:

(a) From any orbital of energy *higher* than that of the group considered, zero;

(b) From each other electron in the group considered, 0.35 per electron, or 0.30 per electron if the group considered is 1s;

(c) From the electron group of next lowest energy to that of the group considered, 0.85 per electron if the electron being considered is s or p, and 1.00 per electron for all lower energy groups. If the electron being considered is d (or f), 1.00 per electron for all *lower* energy electron groups.

*Examples*

| Atom | $Z$ | Electron considered | $S$ | | $\zeta$ |
|---|---|---|---|---|---|
| He | 2 | 1s | (1×0.30) | = 0.30 | 1.70 |
| Be | 4 | 2s | (1×0.35) + (2×0.85) | = 2.05 | 1.90 |
| C | 6 | 1s | (1×0.30) | = 0.30 | 5.70 |
| C | 6 | 2s,2p | (3×0.35) + (2×0.85) | = 2.75 | 3.25 |
| Na | 11 | 3s | (8×0.85) + (2×1.00) | = 8.30 | 2.20 |
| $Na^+$ | 11 | 2s,2p | (7×0.35) + (2×0.85) | = 4.15 | 6.85 |
| Ni | 28 | 3d | (7×0.35) + (8×1.00) + (8×1.00) + (2×1.00) | = 20.45 | 7.55 |

Alternative sets of analytical wave functions are now available; some of them are discussed by Slater.†

### A9.1 Calculation of $N_{1s}$

As an example from (A9.3), we shall calculate the normalizing constant, $N_{1s}$. It is defined by the equation

$$\int\psi^2(1s)d\tau = 1 \; ; \qquad (A9.5)$$

$\psi$ is here taken to be real, and $\psi^2$ measures the probability of finding an electron in a spherical shell of radius $r$ and thickness $r + dr$. Hence, $d\tau$ is the volume of the spherical shell in Figure A9.1, and may be seen to be $4\pi r^2 dr$. Hence from (A9.1)

$$4\pi\int_0^\infty N_{1s}^2 \exp(-2\zeta r) r^2 dr = 1 \; , \qquad (A9.6)$$

since the range of variable $r$ is from 0, the centre of the atom, to infinity. To evaluate this integral let $t = 2\zeta r$. Then

$$dt = 2\zeta dr \; . \qquad (A9.7)$$

The gamma function, $\Gamma(n)$, is defined by (see also Appendix 16)

$$\Gamma(n) = \int_0^\infty t^{n-1}\exp(-t)dt \; , \qquad (A9.9)$$

† Suggested Reading: *Quantum Mechanics* (Slater).

and, for integral values of the argument, $n$,

$$\Gamma(n) = (n - 1)! \ . \tag{A9.10}$$

Hence, the integral in (A9.8) is equal to $\Gamma(3)$, and, therefore,

$$N_{1s} = (\zeta^3/\pi)^{1/2} \ . \tag{A9.11}$$

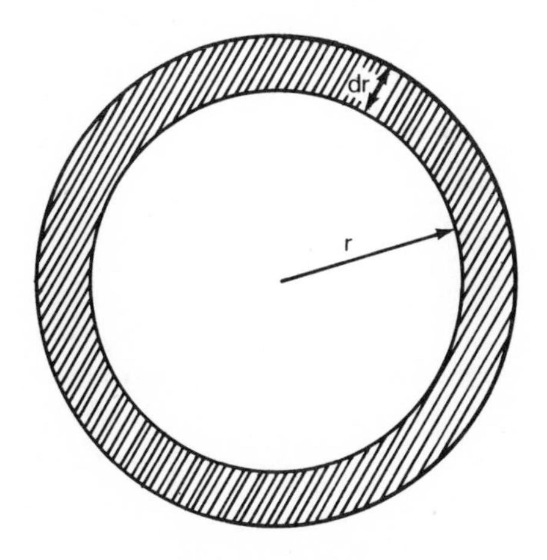


Figure A9.1 – Central section of a spherical shell of radius $r$ and thickness d$r$.

## APPENDIX 10 STANDARD STATE FOR ELECTROLYTE SOLUTIONS

The student who is not familiar with partial molar quantities may need to study first Appendix 12.

In an equilibrium between a solute and its solution, the solid and the solute in the saturated solution are at the same chemical potential. Hence, for any species $i$,

$$\mu_i = \mu_i^{\ominus} + RT\ln a_i \ , \tag{A10.1}$$

or

$$R\ln a_i = \mu_i/T - \mu_i^{\ominus}/T \ . \tag{A10.2}$$

We should remember that quantities such as $a_i$, the activity of a single ionic species, are not measurable experimentally; nevertheless, it is convenient to discuss them as though they were. Under the conditions of constant pressure and composition

$$R\{\partial(\ln a_i)/\partial T\}_{P,n_j} = \{\partial(\mu_i/T)/\partial T\}_{P,n_j} - \{\partial(u_i^{\ominus}/T)/\partial T\}_{P,n_j} \ . \tag{A10.3}$$

From (12.16), dividing by $T^2$,

$$\mu_i/T^2 - (1/T)(\partial u_i/\partial T)_{P,n_j} = \bar{H}_i/T^2 \ , \qquad \text{(A10.4)}$$

or

$$-[\partial(\mu_i/T)/\partial T]_{P,nj} = -\bar{H}_i/T^2 \ . \qquad \text{(A10.5)}$$

Hence from (A10.3)

$$\{\partial(\ln a_i)/\partial T\}_{P,n_j} = (H_i^{\ominus} - \bar{H}_i)/RT^2 \ ; \qquad \text{(A10.6)}$$

$\bar{H}_i$ is the partial molar heat content of the $i$th constituent in the solution, and $H_i^{\ominus}$, equivalent here to $\bar{H}_i^{\ominus}$, is the corresponding value in the pure state of the $i$th constituent. We know that

$$a_i = c_i f_i \ , \qquad \text{(A10.7)}$$

but concentrations in mol $dm^{-3}$, the usual units of solubility, are not independent of temperature. So we may write, from (A10.6),

$$\{\partial(\ln f_i)/\partial T\}_{P,n_j} = (H_i^{\ominus} - \bar{H}_i)/RT^2 + \{\partial[\ln(\rho_0/\rho)]/\partial T\}_{P,n_j} \ , \qquad \text{(A10.8)}$$

where $\rho_0$ and $\rho$ are the densities of the solvent and solution, respectively; $c_i$ is constant under differentiation with respect to $T$.

### A10.1 Standard state

The standard state of a solution is based on the concept of unit activity, and there is freedom in its choice. The standard state should be convenient for the given application, and capable of forming a basis for comparison of different electrolyte solutions. It is permissible to choose different standard states for a given substance in two phases at equilibrium with each other. The equilibrium constant will then be altered in value, but not in its constancy, at a given temperature. However, when different standard states are chosen for two such substances, their activities are not equal, although their chemical potentials must be identical because they are in equilibrium. Thus, for any species in phase I,

$$\mu_{\mathrm{I}} = \mu_{\mathrm{I}}^{\ominus} + RT\ln a_{\mathrm{I}} \ , \qquad \text{(A10.9)}$$

and for phase II in equilibrium with phase I

$$\mu_{\mathrm{II}} = \mu_{\mathrm{II}}^{\ominus} + RT\ln a_{\mathrm{II}} \ . \qquad \text{(A10.10)}$$

Now $\mu_I = \mu_{II}$; hence

$$\mu_I^{\ominus} + RT\ln a_I = \mu_{II}^{\ominus} + RT\ln a_{II} . \quad (A10.11)$$

If the standard states are one and the same, $\mu_I^{\ominus} = \mu_{II}^{\ominus}$ and $a_I = a_{II}$.

The standard state for solutions is the infinitely dilute solution, the activity, $a$ of the solute in solution being defined such that the ratio $a/c \longrightarrow 1$ as $c \longrightarrow 0$, $c$ being the concentration of the solution in mol dm$^{-3}$ (see also Appendix 11). The standard state is hypothetical: it corresponds to a solution of concentration one mol dm$^{-3}$, in which, from (A10.8) since $\rho \longrightarrow \rho_0$ as $c \longrightarrow 0$ and $f_i \longrightarrow 1$, the partial molar heat content of the solute in the standard state, $\bar{H}_i$, has the same value as in the infinitely dilute solution $H_i^{\ominus}$. Thus, in discussing solubility, it is convenient to refer $\Delta H_d^{\ominus}$ to an infinitely dilute solution, and $\Delta G_d^{\ominus}$ to a hypothetical solution of unit activity, knowing that these descriptions apply, in our context, to one and the same standard state.

In practice, we are concerned with the measurable quantity, the mean activity $a_{\pm}$, but we may use a similar definition of the standard state. The reader may wish to refer to Appendix 11 for the definition of mean activity properties.

Consider next the equilibrium

$$\underset{\phantom{(saturated solution)}}{NaCl(s)} \rightleftharpoons \underset{\text{(saturated solution)}}{Na^+(aq) + Cl^-(aq)} . \quad (A10.12)$$

The standard state is a hypothetical solution of unit mean concentration $c_{\pm}$ and unit mean activity coefficient $f_{\pm}$. This choice has the required property that

$$a_{\pm}(NaCl) = 1 = c^2 f_{\pm}^2 , \quad (A10.13)$$

where $c$ the stoichiometric concentration has, in this example, the same value as $c_{\pm}$. The concentration of each ionic species is also unity. Hence, the condition

$$f_{\pm} = f_+ = f_- \quad (A10.14)$$

holds, and permits us to write

$$\mu^{\ominus}(NaCl) = \mu^{\ominus}(Na^+) + \mu^{\ominus}(Cl^-) . \quad (A10.15)$$

Now consider an unsymmetrical electrolyte, such as $MgCl_2$, in saturated solution of stoichiometric concentration $c$:

$$MgCl_2(s) \rightleftharpoons \underset{\text{(saturated solution)}}{Mg^{2+}(aq) + 2Cl^-(aq)} . \quad (A10.16)$$

The equation $\quad \mu(MgCl_2) = \mu^{\ominus}(MgCl_2) + RT\ln a(MgCl_2) \quad$ (A10.17) requires that, in standard state,

$$a_{\pm}(MgCl_2) = 1 = 4c^3 f_{\pm}^3 \ . \tag{A10.18}$$

The standard state refers to unit mean concentration $4c^3$ and unit mean activity coefficient. The concentration of $MgCl_2$ in the standard state is $4^{-\frac{1}{3}}$, that of $Mg^{2+}$ being $4^{-\frac{1}{3}}$ and that of $Cl^-$ $2 \times 4^{-\frac{1}{3}}$. It is clear that $c_{\pm}^3 = (4^{-\frac{1}{3}}) \times (2 \times 4^{-\frac{1}{3}})^2 = 1$. Apparently the standard state for $Cl^-$ is different in the NaCl and $MgCl_2$ solutions in the same concentration terms. The following argument may be used in order to combat this apparent inconsistency.

Let 1 mol of $Mg^{2+}$ be concentrated from the hypothetical solution of concentration $4^{-\frac{1}{3}}$ to a new solution of unit concentration, while 2 mol of $Cl^-$ are diluted from the hypothetical solution of concentration $(2 \times 4^{-\frac{1}{3}})$ to a new solution of unit concentration. Both of the new solutions will be deemed to obey the requirement that $f(Mg^{2+}) = f(Cl^-) = 1$. Hence,

$$\Delta G(Mg^{2+}) = RT\ln(4^{-\frac{1}{3}}) = \tfrac{1}{3}\, RT\ln(4) \ , \tag{A10.19}$$

and

$$\Delta G(Cl^-) = -2RT\ln(2 \times 4^{-\frac{1}{3}}) = \tfrac{2}{3}RT\ln(4) - 2RT\ln(2) \ . \tag{A10.20}$$

The total free energy change is zero, and we can write

$$\mu^{\ominus}(MgCl_2) = \mu^{\ominus}(Mg^{2+}) + 2\mu^{\ominus}(Cl^-) \ . \tag{A10.21}$$

to compare with (A10.15). As long as we refer to the standard state of unit concentration *and* unit activity coefficient, we can compare electrolyte solutions on a common thermodynamic basis.

## APPENDIX 11 DEBYE-HÜCKEL LIMITING LAW

A strong electrolyte in aqueous solution, while fully dissociated, may not behave as though the concentration of free ions is equal to the corresponding stoichiometric concentration. On dissolution in water the ions in an electrolyte become hydrated: they become attached, albeit loosely, to a number of water molecules in a hydration sphere, and the ionic charge is, to some extent, distributed over this sphere. Positive and negative hydrated ions attract one another electrostatically, and every hydrated ion may be regarded as being surrounded by oppositely-charged species, so forming an ionic atmosphere. The hydrated ions cluster and disperse dynamically, but over a period of time which is long in comparison with the lifetime of any cluster, there will be a certain fraction of the total stoichiometric ionic concentration which is unavailable as free ions. This effect is expressed by the activity $a$ of a species, defined such that

$$a_i = c_i f_i \ , \tag{A11.1}$$

where $c_i$ and $f_i$ are the concentration and activity coefficient of the $i$th species. It is further defined that

$$\underset{c_i \to 0}{\text{Limit}} \; f_i = 1 \; . \qquad \text{(A11.2)}$$

In many thermodynamic arguments, particularly those involving strong electrolytes, it is necessary to know the value of $a$ or $f$. Although single-ion activities cannot be measured, the Debye-Hückel theory of strong electrolytes leads to an approximate equation for the calculation of the activity coefficient of a single ion. Without proof here, we write

$$\ln f_i = -Az_i^2\sqrt{I} \; ; \qquad \text{(A11.3)}$$

$A$ is given by

$$A = 1.8247 \times 10^6 / (\epsilon_r T)^{\frac{3}{2}} \; , \qquad \text{(A11.4)}$$

where $\epsilon_r$ is the relative permittivity of the solvent, $T$ is the absolute temperature, $z_i$ is the charge on the ion, and $I$ is the total ionic strength of the solution given by

$$I = \tfrac{1}{2} \sum_i c_i z_i^2 \; . \qquad \text{(A11.5)}$$

Equation (A11.3) provides satisfactory values of $f_i$ provided that $I \leqslant 0.02$, for which reason it is referred to as the Debye-Hückel limiting law.

The measurable activity properties are the mean activity $a_\pm$ and the mean activity coefficient $f_\pm$. For a generalised electrolyte $A_{\nu+}^{z+}B_{\nu-}^{z-}$ $f_\pm$ is given by

$$f_\pm^\nu = (f_+^{\nu+} f_-^{\nu-}) \; , \qquad \text{(A11.6)}$$

where

$$\nu = \nu_+ + \nu_- \; ; \qquad \text{(A11.7)}$$

$a_\pm$ and $c_\pm$ are defined in a similar manner. Then (A11.3) becomes

$$\ln f_\pm = -Az_+z_-\sqrt{I} \; . \qquad \text{(A11.8)}$$

Extensions of the Debye-Hückel equation for higher ionic strengths have been proposed, one of the most satisfactory being the Davies equation:

$$\ln f_\pm = -Az_+z_- \left\{ \frac{\sqrt{I}}{(1 + \sqrt{I})} - 0.3I \right\} . \qquad \text{(A11.9)}$$

The following data for aqueous $K_2SO_4$ at 25 °C exemplifies these equations:

| $c$ | $I$ | $A$ | $f_\pm$(A11.8) | $f_\pm$(A11.9) | $f_\pm$(Experimental) |
|---|---|---|---|---|---|
| 0.001 | 0.003 | 1.172 | 0.88 | 0.88 | 0.88 |
| 0.005 | 0.015 | 1.172 | 0.75 | 0.77 | 0.77 |
| 0.01 | 0.030 | 1.172 | 0.67 | 0.69 | 0.69 |
| 0.05 | 0.15 | 1.172 | 0.40 | 0.47 | 0.51 |
| 0.10 | 0.30 | 1.172 | 0.28 | 0.35 | 0.42 |
| 0.50 | 1.5 | 1.172 | 0.06 | 0.28 | 0.37 |

## APPENDIX 12 PARTIAL MOLAR QUANTITIES

In describing the state of a solution, or indeed that of any other body, two kinds of property are used. There are intensive properties such as density, viscosity and refractive index, which are independent of the amount of substance considered, and there are extensive properties like volume, energy and entropy, which do depend on the amount of substance under consideration. It may be noted that while volume, for example, is an extensive property, volume per unit mass (density) is an intensive property. The extensive properties of a given amount of solution can be measured, often directly, and in our thermodynamic investigation of solubility we need to know how some of these properties vary with composition.

### A12.1 Partial molar volume

We wish to study open system, those in which composition is a variable. Typical undergraduate introductory courses in thermodynamics are concerned largely with closed systems, those in which the composition of the system is constant, so that the thermodynamics of the present context is perhaps the less familiar.

Consider a solution consisting of $n_A$ mol of a solvent species $A$ and $n_B$ mol of a solute species $B$. On addition of an amount $\mathrm{d}n_A$ mol of $A$ the volume increase will be $\mathrm{d}V$; then we may write

$$\mathrm{d}V/\mathrm{d}n_A = \bar{V}_A \,, \tag{A12.1}$$

and $\bar{V}_A$ is the partial molar volume of the solvent in the solution.

Similarly, we write

$$\mathrm{d}V/\mathrm{d}n_B = \bar{V}_B \,, \tag{A12.2}$$

for the partial molar volume of the solute in the solution.

A partial molar quantity is a property of the solution as a whole and not solely of the component in question. Thus $\bar{V}_B$ is the change in total volume when solute molecules are added. An important part of this change is concerned with the packing of adjacent solvent molecules and therefore with the forces between the species in solution.

In order to assist in the appreciation of a partial molar property, Figure A12.1 is given as an illustration of the partial molar volume of $CaCl_2$ in water, $\bar{V}(CaCl_2)$ as a function of the number of moles of $CaCl_2$, $n(CaCl_2)$ at 25°C.

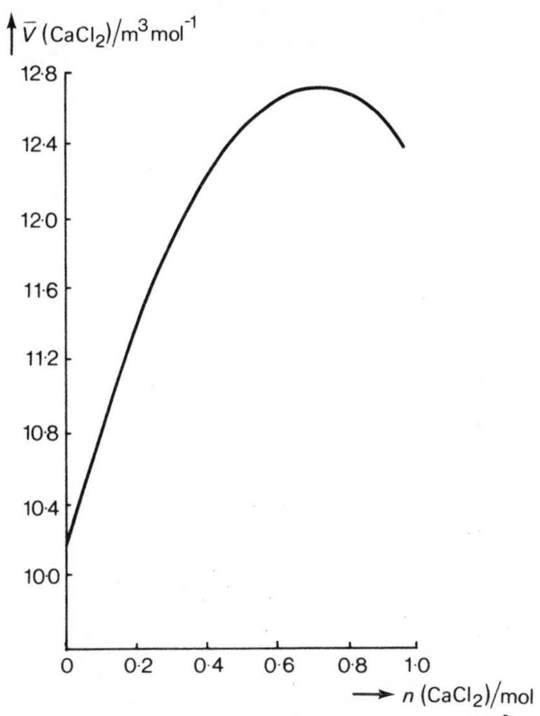


Figure A12.1 – Partial molar volume of $CaCl_2$ in water at 25 °C as a function of moles $CaCl_2$.

## A12.2 Partial molar entropy

In studying solubility in Chapter 2, we were concerned with the change in free energy per mol of solute $\Delta G_d^\ominus$ in establishing an equilibrium state, starting with the solid and hydrated ions in their standard states. We saw how $\Delta G_d^\ominus$ could be divided between the enthalpy of solution to infinite dilution $\Delta H_d^\ominus$ and the corresponding entropy change, which included the sum of the partial molar entropies of the hydrated ions; the reader may wish to refer again to Figure 2.19 and (2.90). The process of dissolution involves changes in the solute, the solvent and the solution: we shall find that the term $\sum_i \bar{S}_i^\ominus$ may be either positive or negative for different pairs of ions.

## A12.3 Measurement of the partial molar entropy of an ion

The partial molar entropy of hydrated ions may be determined from measurements of the temperature variation of emf in a suitable galvanic cell. We shall consider $ZnCl_2$ as a solute, and the following cell could be set up:

$$\text{Zn} \mid \text{ZnCl}_2(c) \,\vdots\, \text{HCl}(a_\pm = 1) \mid \text{H}_2(1\text{atm}),\text{Pt} \; , \tag{A12.3}$$

for which the spontaneous cell reaction is

$$\text{Zn} + 2\text{H}^+ \longrightarrow \text{Zn}^{2+} + \text{H}_2 \; . \tag{A12.4}$$

The emf is measured at several values of the concentration $c$ and extrapolated, conveniently against $\sqrt{c}$, to zero concentration (infinite dilution). These measurements are repeated at three four temperatures between about 15 to 35 °C so as to obtain $\mathrm{d}E^{\ominus}/\mathrm{d}T$; since

$$\Delta S^{\ominus} = nF\mathrm{d}E^{\ominus}/\mathrm{d}T \; , \tag{A12.5}$$

we can obtain $\Delta S^{\ominus}$ for reaction (A12.4): $n$ is the number of electrons involved in the reaction and $F$ is the Faraday constant. But $\Delta S^{\ominus}$ is given also by

$$\Delta S^{\ominus} = \{\bar{S}^{\ominus}(\text{Zn}^{2+}) + S^{\ominus}(\text{H}_2)\} - \{S^{\ominus}(\text{H}^+) + \bar{S}^{\ominus}(\text{Zn})\} \; . \tag{A12.6}$$

By convention, we put $\bar{S}^{\ominus}(\text{H}^+) = 0$. Hence,

$$\bar{S}^{\ominus}(\text{Zn}^{2+}) = \Delta S^{\ominus} + S^{\ominus}(\text{Zn}) - S^{\ominus}(\text{H}_2) \; . \tag{A12.7}$$

In a typical experiment, $\mathrm{d}E^{\ominus}/\mathrm{d}T$ (at 25 °C) was found to be $-1.00 \times 10^{-4}$ V K$^{-1}$; $S^{\ominus}(\text{Zn}) = 41.6$ J mol$^{-1}$ K$^{-1}$ and $S^{\ominus}(\text{H}_2) = 130.6$ J mol$^{-1}$ K$^{-1}$. From (A12.5). $\Delta S^{\ominus} = -19.30$ J mol$^{-1}$ K$^{-1}$, and, thus, $\bar{S}^{\ominus}(\text{Zn}^{2+}) = 108.3$ J mol$^{-1}$ K$^{-1}$. This result is, of course, relative to the chosen value of $\bar{S}^{\ominus}(\text{H}^+)$ as zero.

**A12.4 Generalised description of partial molar quantities**

Any extensive property, $X$, is determined by both the state of the system, and the amounts of substances present. Thus

$$X = f(T, P, n_i) \quad (i = 1, 2, \ldots, N) \; . \tag{A12.8}$$

Hence,

$$\mathrm{d}X = (\partial X/\partial T)_{P,n_i}\,\mathrm{d}T + (\partial X/\partial P)_{T,n_i}\,\mathrm{d}P + \sum_{i=1}^{N} (\partial X/\partial n_i)_{T,P,n_j}$$

$$(j = 1, 2, \ldots, N; j \neq i) \; . \tag{A12.9}$$

The derivative $(\partial X/\partial n_i)_{T,P,n_j}$ is the partial molar property $\bar{X}_i$ for the component $i$ in the whole system.

If the extensive property is the Gibbs free energy, the partial molar property is identical with the chemical potential. Thus

$$(\partial G/\partial n_i)_{T,P,n_j} = \bar{G}_i = \mu_i \ . \tag{A12.10}$$

Differentiating (A12.10) with respect to $T$, we write

$$(\partial^2 G/\partial n_i \partial T)_{P,n_j} = (\partial \mu_i/\partial T)_{P,n_j} \ . \tag{A12.11}$$

Since we have
$$(\partial G/\partial T)_P = -S \ , \tag{A12.12}$$

differentiating with respect to $n_i$ gives

$$(\partial^2 G/\partial T \partial n_i)_{P,n_j} = -(\partial S/\partial n_i)_{P,n_j} = -\bar{S}_i \ . \tag{A12.13}$$

Since the order of differentiation in (A12.11) and (A12.13) is immaterial,

$$(\partial \mu_i/\partial T)_{P,n_j} = \bar{S}_i \ . \tag{A12.14}$$

Using $G = H - TS$ and differentiating with respect to $n_i$ we have

$$\bar{G}_i = \bar{H}_i - T\bar{S}_i = \mu_i \ . \tag{A12.15}$$

Using (A12.14) and rearranging, we obtain

$$\mu_i - T(\partial \mu_i/\partial T)_{P,n_j} = \bar{H}_i \ . \tag{A12.16}$$

We have now developed general expressions for the partial molar entropy, partial molar free energy and partial molar enthalpy.

## APPENDIX 13 REDUCED MASS

The calculation of a reduced mass is a central force problem in mechanics. We shall consider a two-particle problem in which the potential energy of the system is determined by the distance between the two particles.

Let two particles of masses $m_1$ and $m_2$ be vibrating along a line joining their centres. The speeds of the particles at any instant are $v_1$ and $v_2$ respectively, and the distance between them is $r$ (Figure A13.1); $C$ is the centre of mass of the system.

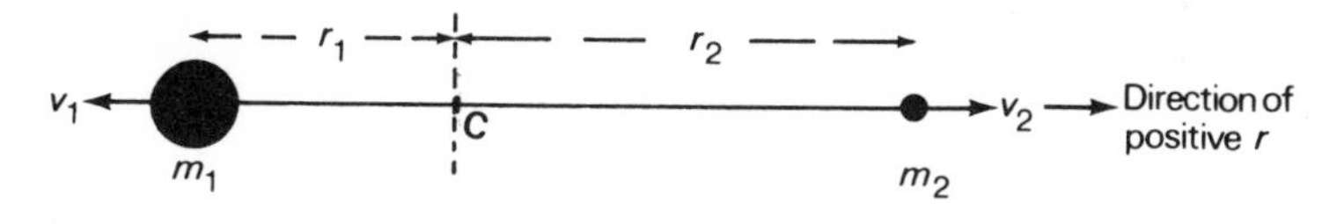


Figure A13.1 – Two-particle system vibrating about a centre of mass, $C$.

The kinetic energy $T$ of the system is given by

$$T = \tfrac{1}{2} m_1 v_1^2 + \tfrac{1}{2} m_2 v_2^2 \tag{A13.1}$$

and the centre of mass is defined by

$$m_1 r_1 + m_2 r_2 = 0 \ . \tag{A13.2}$$

From Figure A13.1, noting the positive direction of $r$,

$$r = r_2 - r_1 \ . \tag{A13.3}$$

Therefore,

$$m_1(r_2 - r) = -m_2 r_2 \ , \tag{A13.4}$$

or

$$r_2 = m_1 r/(m_1 + m_2) \ . \tag{A13.5}$$

Hence from (A13.3) and (A13.5),

$$r_1 = -m_2 r/(m_1 + m_2) \ . \tag{A13.6}$$

Since

$$v_i^2 = \mathrm{d}r_i/\mathrm{d}t = r_i \ , \tag{A13.7}$$

$$T = \tfrac{1}{2} m_1 \dot{r}_1^2 + \tfrac{1}{2} m_2 \dot{r}_2^2 \ , \tag{A13.8}$$

or

$$T = \tfrac{1}{2} m_1 m_2 \dot{r}/(m_1 + m_2) + \tfrac{1}{2} m_1 m_2 \dot{r}/(m_1 + m_2) \ . \tag{A13.9}$$

Equation (A13.9) may be written

$$T = \mu \dot{r}^2 \ , \tag{A13.10}$$

where $\mu$, given by,

$$\mu = m_1 m_2/(m_1 + m_2) \ , \tag{A13.11}$$

is the reduced mass of the system. Evidently, (A13.10) represents the kinetic energy of two particles of equal mass, $\mu$, separated by a distance $r$, and vibrating about their centre of mass mid-way between them. Thus,

$$T = \tfrac{1}{2} \mu(\dot{r}/2)^2 + \tfrac{1}{2} \mu(\dot{r}/2)^2 = \mu \dot{r}^2 \ . \tag{A13.12}$$

**PROBLEMS TO CHAPTER 2**

1. The convergence limit for the process

$$\text{Tl(g, ground state)} \longrightarrow \text{Tl}^+\text{(g)} + \text{e}^- \tag{P2.1}$$

is 49250 $\text{cm}^{-1}$. Calculate $I(\text{Tl})$ in kJ $\text{mol}^{-1}$.

2. The following values for the vapour pressure of molten lead were derived from measurements of the rate of effusion of the vapour into a vacuum through a small hole, at different temperatures:

| $T/K$ | 895.4 | 922.1 | 964.5 | 1009.7 | 1045.5 |
|---|---|---|---|---|---|
| $P/\text{N m}^{-2}$ | 0.0783 | 0.205 | 0.539 | 1.40 | 3.40 |

Plot a graph of $\ln P$ against $1/T$. Set up an appropriate linear equation by the method of least squares to represent the dependnence of the vapour pressure of lead on temperature over the given range. Hence determine the enthalpy of evaporation $\Delta H_e$ of lead for the experimental temperature range. Estimate the standard deviation in $\Delta H_e$ arising from the least-squares fit to the given data (see Appendix 23).

3. Refer to Problem 2. The molar heat capacity of lead may be represented by the equations

$$C_P = 23.56 + 0.00975T \text{ in J mol}^{-1}\text{ K}^{-1} \text{ between 298 and 600 K} \quad \text{(P.2.2)}$$

and

$$C_P = 32.43 - 0.00310T \text{ in J mol}^{-1}\text{ K}^{-1} \text{ between 600 and 1200 K} \quad \text{(P2.3)}$$

The enthalpy of fusion of lead at the melting point, 600 K is 4.81 $\text{mol}^{-1}$. If the result already obtained for $\Delta H_e$ be taken to apply at the average temperature of 970 K, extrapolate the result to 298 K to obtain $S_M$(Pb): it may be assumed that the vapour behaves ideally. It can be helpful first to construct a thermodynamic cycle to show the various processes involved in the calculation.

4. Given the following data:
$I(\text{Mg}) = 6.09$ eV, $I(\text{Mg}^+) = 11.82$ eV, $S_M(\text{Mg}) = 149.0$ kJ $\text{mol}^{-1}$,
$D_0(\text{Cl}_2) = 243.0$ kJ $\text{mol}^{-1}$, $E(\text{Cl}) = -348.6$ kJ $\text{mol}^{-1}$,
$\Delta H_f(\text{MgCl,s}) = -221.8$ kJ $\text{mol}^{-1}$ and $\Delta H_f(\text{MgCl}_2\text{,s}) = -641.8$ kJ $\text{mol}^{-1}$,
show how, on reaction of Mg(s) with $Cl_2$(g), the formation of $MgCl_2$(s) is preferred to that of MgCl(s).

5. Use the following data to construct a thermochemical cycle for the formation of $NH_4Cl$(s), and then find the standard value of this quantity.

| | | $\Delta H^\ominus/\text{kJ mol}^{-1}$ | |
|---|---|---|---|
| $\tfrac{1}{2}N_2(g) + \tfrac{3}{2}H_2(g)$ | $\longrightarrow NH_3(g)$ | −46.0 | (P2.4) |
| $NH_3(g) + aq$ | $\longrightarrow NH_4^+(aq) + OH^-(aq)$ | −34.7 | (P2.5) |
| $\tfrac{1}{2}H_2(g) + \tfrac{1}{2}Cl_2(g)$ | $\longrightarrow HCl(g)$ | −92.5 | (P2.6) |
| $HCl(g) + aq$ | $\longrightarrow H^+(aq) + Cl^-(aq)$ | −74.9 | (P2.7) |
| $NH_4^+(aq) + OH^-(aq) + H^+(aq) + Cl^-(aq)$ | $\longrightarrow NH_4^+(aq) + Cl^-(aq)$ | −52.3 | (P2.8) |
| $NH_4Cl(s) + aq$ | $\longrightarrow NH_4^+(aq) + Cl^-(aq)$ | +15.1 | (P.29) |

6. Use the spherical conducting shell model to calculate an approximate value for the Madelung constant of the NaCl structure type.

*7. From the following data on CaO calculate the affinity of oxygen for two electrons.

| | |
|---|---|
| Structure type of CaO | NaCl |
| Unit-cell side, $a$ | 4.811 Å |
| Madelung constant, $A$ | 1.7476 – remember the ionic charges |
| Compressibility, $\kappa$ | $0.70 \times 10^{-11}$ $N^{-1}$ $m^2$ |
| $I(Ca)$ | 589.5 kJ $mol^{-1}$ |
| $I(Ca^+)$ | 1145. kJ $mol^{-1}$ |
| $S_M(Ca)$ | 176.6 kJ $mol^{-1}$ |
| $D_0(O_2)$ | 489.9 kJ $mol^{-1}$ |
| $\Delta H_f(CaO,s)$ | −635.5 kJ $mol^{-1}$ |

8. From the following data, together with data on the dissociation energies and electron affinities of the halogens and the enthalpies of solution of the strontium halides given in the text, determine an average value for the standard enthalpy of hydration $\Delta H_h$, of the $Sr^{2+}$ ion:

| | $\Delta H$/kJ $mol^{-1}$ |
|---|---|
| $I(Sr)$ | 549.4 |
| $I(Sr^+)$ | 1064. |
| $S_M(Sr)$ | 163.6 |
| $\Delta H_f^\ominus(SrF_2)$ | −1209. |
| $\Delta H_f^\ominus(SrCl_2)$ | − 828.0 |
| $\Delta H_f^\ominus(SrBr_2)$ | − 715.5 |
| $\Delta H_f^\ominus(SrI_2)$ | − 569.4 |
| $\Delta H_h^\ominus(F^-,g)$ | − 513.0 |
| $\Delta H_h^\ominus(Cl^-,g)$ | − 371.1 |
| $\Delta H_h^\ominus(Br^-,g)$ | − 340.6 |
| $\Delta H_h^\ominus(I^-,g)$ | − 301.7 |

9. The sulphate ion may be considered as a regular tetrahedral arrangement of oxygen atoms around sulphur. If the charge on oxygen is $-0.7e$ and the S–O distance 1.30 Å, calculate the electrostatic self-energy of the sulphate ion.

10. If $r(Cs^+)$ and $r(I^-)$ are 1.69 and 2.16 Å, respectively, and the departure of $r_e$ from additivity is +0.10 Å, calculate the density of CsI.

*11. In rutile $TiO_2$, the coordination of O around Ti is 6-fold but it is not regular: the Ti atoms around each O form an isosceles triangle. Use the following data to calculate the Ti–O bond lengths and the O–Ti–O bond angles.

Tetragonal crystal system: $a = b = 4.593$ Å, $c = 2.959$ Å.
Two formula-entities per unit cell at fractional coordinates

2 Ti 0,0,0; ½, ½, ½

4 O $x,x,0$; $\bar{x},\bar{x},0$; $½+x,½-x,½$; $½-x,½+x,½$

with $x = 0.3056$.

It may be found helpful first to make sketches of the structure in one unit cell, in three dimensions and in plan, from which it may be determined how the tetragonal symmetry reduces the number of independent calculations that are needed.

12. Show that the radius ratio for the wurtzite structure with the atoms in maximum contact is 0.225.
13. In an experiment, precipitated silver iodide was dissolved in aqueous solutions of potassium iodide and sodium iodide; the heats evolved were 9.45 kJ mol$^{-1}$ and 7.46 kJ mol$^{-1}$, respectively. Next, finely divided silver was suspended in similar solutions of the two alkali-metal iodides. On adding iodine, the silver dissolved rapidly to form silver iodide: the heats evolved were 72.0 kJ mol$^{-1}$ (of AgI) in the potassium iodide solution, and 70.3 kJ mol$^{-1}$ in the sodium iodide solution.

    Set up equations to represent the chemical reactions taking place, and calculate an average value for the heat of formation of silver iodide; all processes may be considered to have taken place at 25 °C.
14. Determine the standard free energy of solution of $MgF_2$, given that $\Delta H_d^{\ominus} = -18.4$ kJ mol$^{-1}$, $S_C = 57.3$ J mol$^{-1}$ K$^{-1}$ and $\Sigma \bar{S}_i^{\ominus} = -137.2$ J mol$^{-1}$ K$^{-1}$ with tespect to the appropriate standard states. The solubility of $MgF_2$ is 0.075 g dm$^{-3}$; calculate the mean activity coefficient $f_{\pm}$ for this salt.
15. Derive an expression for the number $n$ of Frenkel defects in a crystal of AgCl containing $N$ lattice sites and $N'$ interstices. If the energy needed to create a single Frenkel defect in AgCl is 1 eV, calculate the fraction of Frenkel defects in this substance at 500 K.

*16. The probability of Schottky defects in KCl may be represented by

$$n/N = \exp(-\Delta H/2kT) \ , \qquad \text{(P2.10)}$$

per formula-entity. What is the corresponding equation for $CaCl_2$.

17. Radioactive silver was allowed to diffuse through a silver/indium alloy at 1000 K. The penetration depth, $x$, was determined after time intervals, $t$, of $6 \times 10^4$s by measuring the radioactivity, $\beta_t$. Show that the diffusion process follows the equation

$$\beta_t = A\exp(-x^2/Dt) \ , \qquad \text{(P2.11)}$$

and find the constants $A$ and $D$ of the equation.

| $x$/mm | $\beta_t$/arbitrary units | $x$/mm | $\beta_t$/arbitrary units |
|---|---|---|---|
| 0.000 | 600 | 0.329 | 88 |
| 0.084 | 540 | 0.376 | 50 |
| 0.132 | 450 | 0.425 | 25 |
| 0.183 | 360 | 0.470 | 12 |
| 0.230 | 250 | 0.520 | 5 |
| 0.279 | 160 | 0.568 | 2 |

18. In a further radioactive experiment, the values of the diffusion coefficient $D$ were obtained as a function of temperature. Given that $D$ is related to $T$ by a Boltzmann equation,

$$D = D_0 \exp(-U/RT) \;, \qquad \text{(P2.12)}$$

determine the value of the molar activation energy $U$ for the diffusion process.

| $T$/K | 878 | 1007 | 1176 | 1253 | 1322 |
|---|---|---|---|---|---|
| $D$/m$^{-2}$ s | $1.6 \times 10^{-18}$ | $4.0 \times 10^{-17}$ | $1.1 \times 10^{-15}$ | $4.0 \times 10^{-15}$ | $1.0 \times 10^{-14}$ |

What is the value of $D_0$ and what is its significance?

19. Calculate the standard molar entropy of nickel from the following data:

| T/K | 15.05 | 25.20 | 47.10 | 67.13 | 82.11 | 133.4 | 204.1 | 256.5 | 283.0 |
|---|---|---|---|---|---|---|---|---|---|
| $C_P$/J mol$^{-1}$ K$^{-1}$ | 0.1945 | 0.5994 | 3.532 | 7.639 | 10.10 | 17.88 | 22.72 | 24.81 | 26.09 |

20. Calculate the Bragg angle $\theta$ for the 300 reflexion from CsI of Cu $K\alpha$ X-radiation ($\lambda$ = 1.542 Å), given that $a$ = 4.562 Å. Would the relative intensity of this reflexion be expected to be weak, strong or something in between? Give reasons for your answer.

Chapter 3

# Covalent Compounds

## 3.1 INTRODUCTION

In studying ionic compounds we were concerned with forces between discrete, charged species. It is well known that oppositely charged particles attract one another and we have little difficulty in accepting this situation for, say, $Na^+$ and $Cl^-$ ions. It is not so obvious how two neutral hydrogen atoms attract each other to form a hydrogen molecule $H_2$. Yet we expect that the forces involved will be electrical, because such is the nature of matter.

Most students have, at some time, used the dot diagrams of Lewis (1916). He was aware of the special stability of the inert gases and postulated that, in compound formation, atoms could achieve an effective inert-gas configuration by sharing a pair of electrons, each atom providing one electron to the bond pair. Thus, methane was written as

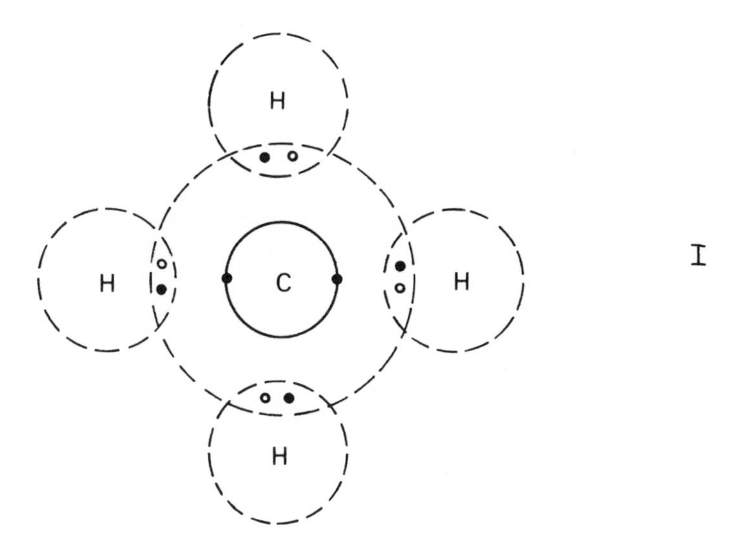

and carbon and hydrogen obtain, effectively, the configurations of neon and helium respectively. Lewis explained multiple bonds by the sharing of two or more pairs of electrons, and so ethene (ethylene) was written as

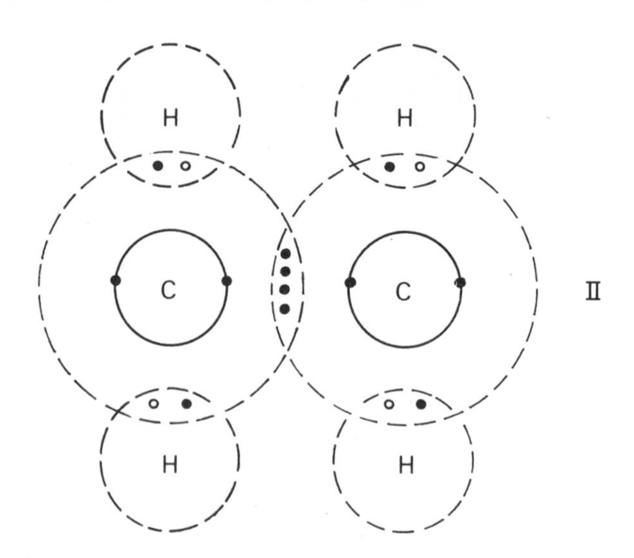


The basic idea of the shared pair of electrons is still acceptable, qulitatively, but it is necessary to be able to show how it is in methane, for example, that carbon forms four equivalent C–H bonds of length 1.09 Å with H–C–H angles of 109.5°, whereas in ethene there are apparently only three bonds from carbon, with H–C–H angles of 120.0°; for these requirements are part of a theory of valence. We need to be able to understand also how molecules form at all, and why they do so in definite atomic proportions and with well defined stereochemistry. The Lewis description was unable to supply these details, and we shall find that wave mechanics provides a more acceptable valence theory. Because we shall be concerned intimately with electrons in the wave-mechanical theory, we shall consider first some of the important properties of the electron.

## 3.2 WAVE-PARTICLE DUALITY

At the turn of the 20th century there were two, apparently distinct, patterns of physical behaviour. One of them concerned particles of definite mass, moving according to Newton's laws of motion. The other pattern referred to waves such as electromagnetic radiations, moving in a continuous medium and subject to the laws governing wave motion. It is interesting to review some properties of light as examples of these behavioural patterns and to compare them with similar properties of matter (Figure 3.1).

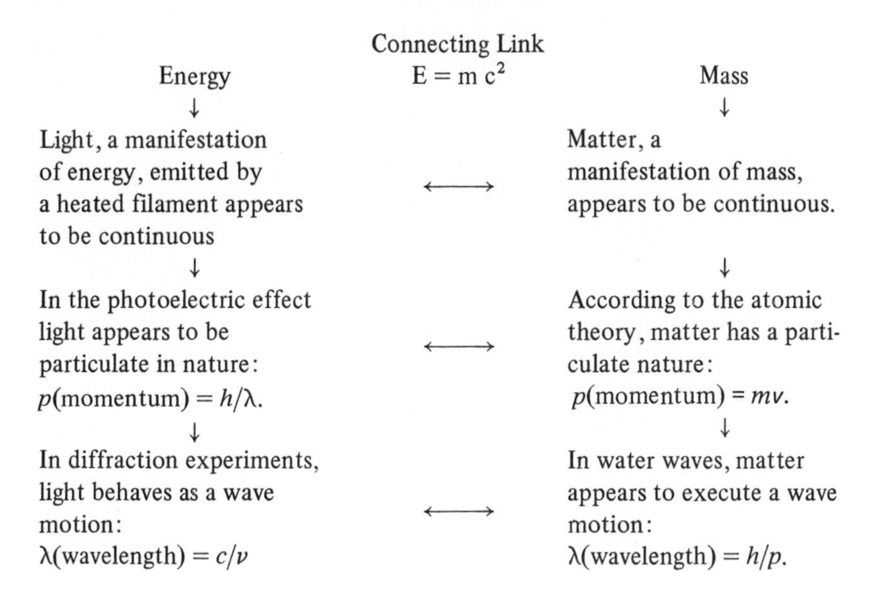


Figure 3.1 – Wave-particle duality in light and matter.

### 3.2.1 Black-body radiation

When objects are heated they emit radiation. As the temperature of the radiator is increased, the frequency of the emitted radiation increases from the infrared end of the spectrum, through the visible and into the ultraviolet region; an indication of the temperature of the radiator is given by its colour. The phenomenon is called black-body radiation, and Figure 3.2 shows the energy distribution $E(\nu)$ as a function of frequency, $\nu$. The maximum $\nu_{max}$ in $E(\nu)$ moves to lower frequencies as the temperature of the radiator is decreased, and Wien's experiments (1894) showed that

$$T/\nu_{max} = \text{constant} . \tag{3.1}$$

Rayleigh considered that radiation was emitted by molecular oscillators inside the radiation cavity. He determined first the number $N(\nu)$ of oscillators in a cubical enclosure of side $c/\nu$, where $c$ is the speed of light in a vacuum. The result, as amended by Jeans, is

$$N(\nu) = 8\pi \nu^2/c^3 . \tag{3.2}$$

We show in Appendix 21 that the mean energy of a classical oscillator at a temperature $T$ is $kT$, where $k$ is the Boltzmann constant. Hence the energy distribution† is represented by the Rayleigh-Jeans equation,

†Strictly, we mean the energy density in the interval between $\nu$ and $(\nu + d\nu)$, that is $E(\nu)d\nu$.

$$E(\nu) = N(\nu)kT = 8\pi\nu^2kT/c^3 \quad . \tag{3.3}$$

Evidently this equation does not accord with Wien's observation (3.1), and this fact is illustrated also from Figure 3.2. Furthermore (3.3) predicts a very large accumulation of energy in the very high-frequency region, leading to the so-called ultraviolate catastrophe.

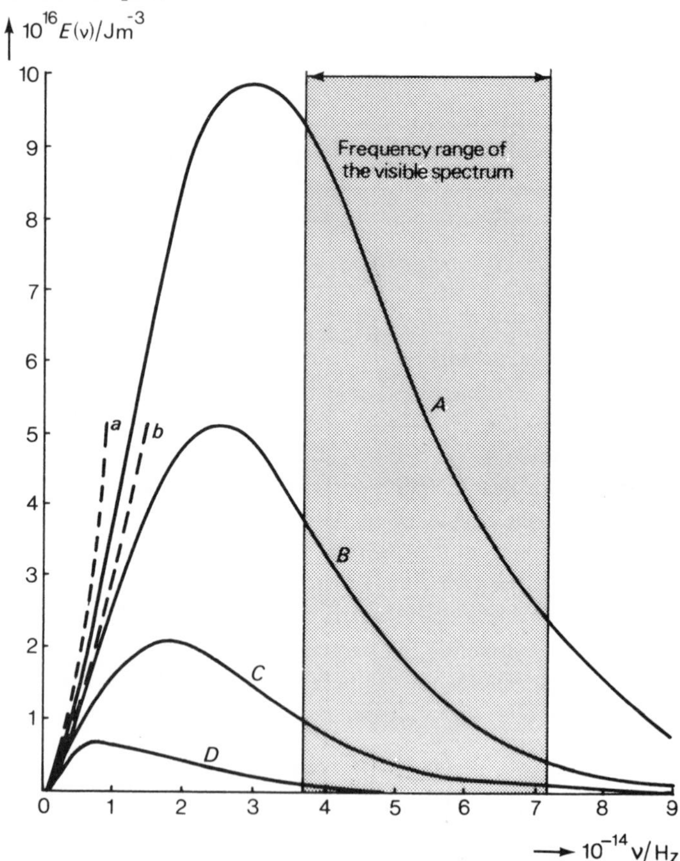


Figure 3.2 – Black-body radiation curves: Rayleigh-Jeans equation *a* 5000 K
*b* 4000 K
Planck equation *A* 5000 K
*B* 4000 K
*C* 3000 K
*D* 2000 K

Planck (1900) made a major revision to the Rayleigh-Jeans equation. He postulated that an oscillator could radiate energy only in certain amounts called quanta, and that the energy $E$ of a single quantum of radiation (the minimum quantity of energy) is given by

$$E = h\nu \ , \qquad (3.4)$$

where $h$ is the Planck constant, and $\nu$ is the frequency of the oscillator. Planck reconciled the Wien and Rayleigh-Jeans equations, which now emerge as special cases of the Planck equation:

$$E(\nu) = (8\pi h\nu^3/c^3)\,\{\exp(h\nu/kT) - 1\}^{-1} \ . \qquad (3.5)$$

Classical conditions correspond to no quantization, that is to the limit of $E(\nu)$ as $h \rightarrow 0$. From (3.5), we have

$$E(\nu) = (8\pi h\nu^3/c^3)\,\{(h\nu/kT) + (h\nu/kT)^2/2 + (h\nu/kT)^3/6 + \ldots\}^{-1} \ , \qquad (3.6)$$

or

$$E(\nu) = (8\pi h\nu^3/c^3)\,\{(h\nu/kT)\,[1 + (h\nu/kT)/2 + (h\nu/kT)^2/6 + \ldots]\}^{-1} \ . \qquad (3.7)$$

Hence, dividing the numerator and denominator by $h$,

$$\underset{h\rightarrow 0}{\text{Limit}}\ E(\nu) = 8\pi\nu^2 kT/c^3 \ , \qquad (3.8)$$

which is the Rayleigh-Jeans equation. Since $h$ is really a constant, it is preferable to consider that (3.8) is reached as the dimensionless quantity $(h\nu/kT)$ tends to a very small value. Thus, the classical condition is attained for $(h\nu/kT) \ll 1$, that is at low frequencies and high temperatures, typically $\nu = 10^{11}$ Hz and $T = 3000$ K. If we consider high frequencies or low temperatures, then $h\nu \gg kT$, and (3.5) becomes

$$E(\nu) = (8\pi h\nu^3/c^3)\exp(-h\nu/kT) \ ; \qquad (3.9)$$

by differentiating $E(\nu)$ with respect to $\nu$, and setting the derivative equal to zero, we can show that

$$T/\nu_{max} = h/(3k) = \text{constant} \ . \qquad (3.10)$$

So Wien had observed quantum behaviour, whereas the Rayleigh-Jeans equation was derived for black-body radiation under classical conditions. The value of the constant in (3.10) has been determined experimentally as $1.6\times10^{-11}$ K s; hence, the Planck constant is approximately $6.6\times10^{-34}$ J s. If we consider a mechanical oscillator, such as a simple pendulum, then for a length of 248.4 mm its frequency will be 1 Hz. Thus $h\nu$ is about $6.6\times10^{-34}$ J and, from (3.4), the separation of the energy levels of this mechanical oscillator would be of this magnitude, and so present a continuum with respect to methods for their detection. For optical frequencies in the region of $5\times10^{14}$ Hz the energy quantum is about $3.3\times10^{-19}$ J, or 200 kJ $mol^{-1}$, which value is significant.

At any temperature $T$ each oscillator can be excited by the absorption of one quantum of energy $h\nu$. In classical terms, when the average thermal energy is $kT$, oscillators could be excited continuously, and even the highest-frequency oscillators would contribute to the black-body radiation. One of the successes of Planck's theory is that it provides the necessary damping of the high-frequencies, through the exponential term in (3.9), so that the ultraviolet catastrophe cannot occur.

### 3.2.2 Photoelectric effect

In Figure 3.3 a monochromatic light source incident upon a metal cathode results in electron flow from the photocathode, provided that the external emf source $E$ is suitably adjusted. Experimentally it was found that

(a) the magnitude of the current flowing is proportional to the intensity of the light source;

(b) the kinetic energy of the electrons is independent of the light intensity;

(c) the number of electrons emitted is proportional to the light intensity;

(d) the mean kinetic energy of the electrons increases with an increase in the frequency of the light source;

(e) no photoelectrons are emitted unless the frequency of the incident light exceeds a certain minimum value;

(f) there is no delay in emission on irradiation of the photocathode, provided that $h\nu$ is sufficiently large to initiate electron flow.

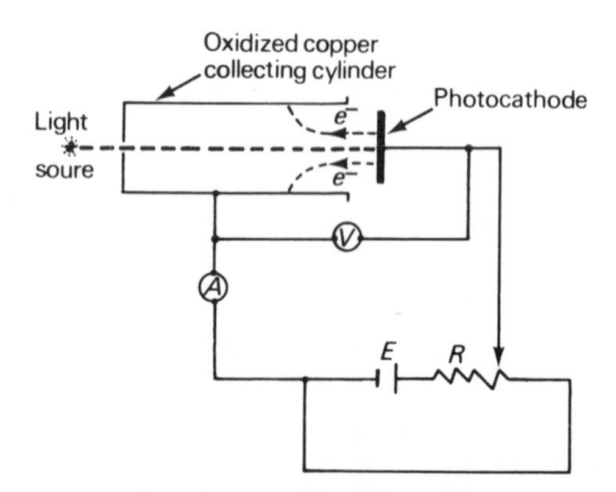


Figure 3.3 – Schematic diagram of apparatus for demonstrating the photoelectric effect. A contact potential difference exists between the dissimilar photocathode and cylinder metals: it is assumed here that this potential difference makes the cathode positive with respect to the cylinder ($A$ ammeter or electrometer; $V$ voltmeter; $R$ variable resistor; $E$ external battery).

Einstein summarized these findings, in terms of Planck's quantum hypothesis by the equation

$$mv^2/2 = h\nu - \phi_M \ , \tag{3.11}$$

where $m$ and $v$ are respectively the mass and speed of an electron; (3.11) may be written in terms of the momentum, $p$, of the electron as

$$p^2/(2m) = h\nu - \phi_M \ . \tag{3.12}$$

The term $\phi_M$ represents an energy barrier which an electron must overcome in order to escape from the binding forces in the metal. It is called the electronic work function of the metal. When the incident energy $h\nu$ exceeds $\phi_M$, an electron is ejected from the metal with a kinetic energy $m\,v^2/2$ equal to the excess of $h\nu$ over $\phi_M$. The reader is invited to consider how the experimental findings (a) to (f) are explained by (3.11).

This elegant theory makes a demand on our interpretation of the nature of light. It implies that a quantum of incident energy is transferred to a single electron; in other words the light beam behaves as though it were particulate. The light particles, or corpuscles, are called photons. However if light is particulate, the photons should have a momentum, $p$. This momentum is given by

$$p = mc \ , \tag{3.13}$$

where $m$ is the mass associated with the photon. Using the Einstein equation which relates mass and energy,

$$E = mc^2 \ , \tag{3.14}$$

together with (3.4), for a single quantum, we obtain

$$p = E/c = h\nu/c \tag{3.15}$$

It is useful to have an idea of the magnitudes of some of these quantities in practical situations. If we raise the temperature of 1 mol of water, without heat loss, from 293 to 373 K, the heat energy needed would be $0.01802 \times 4184 \times 80$, or 6030 J. If the heating apparatus were an infrared lamp of frequency $3 \times 10^{14}$ Hz, the number of quanta needed would be $6030/(6.6262 \times 10^{-34} \times 3 \times 10^{14})$, or $3.03 \times 10^{22}$, or 0.05 mol of quanta of photons. This amount of energy is closely equivalent to that consumed by a 100 watt electric lamp burning for 1 minute. If we regarded the radiation as particulate, the momentum of each infrared photon would be $6.6262 \times 10^{-34} \times 3 \times 10^{14}/(2.9979 \times 10^{8})$, or $6.63 \times 10^{-28}$ N s. The total momentum is $2.00 \times 10^{-5}$ N s, which is approximately that attained by a 10 mg weight at the end of a 20 cm fall, under gravity, from rest.

### 3.2.3 Light and electrons

The topics in Figure 3.1 other than the photoelectric effect are generally familiar and need not be elaborated here. We shall accept that light can behave with a wave-like nature in some experiments and with a particle-like nature in other experiments. We also see that there is an equivalence between mass and energy, of which light is one form, which are linked through (3.14), by the speed of light.

In 1924, de Broglie suggested that a particle travelling with momentum $p$ should have an associated wavelength $\lambda$ given by

$$p = h/\lambda \ . \tag{3.16}$$

Experiments with electrons show that they behave somewhat akin to light photons. Davisson and Germer (1925) showed that electrons could be diffracted by metallic nickel, whereas Thompson's earlier experiments on cathode rays (electrons) demonstrated their particulate character. The diffraction experiment confirmed de Broglie's prediction of one year earlier. Equation (3.16) may be regarded as a reconciliation of opposites: momentum is a property which we associate with particles, whereas wavelength is clearly a feature of waves.

### 3.2.4 Heisenberg's uncertainty principle

Suppose that we wish to determine simultaneously and precisely both the momentum and position of an electron. We shall find that it is not a feasible proposition and the following argument, based on the diffraction of electrons at a slit, will show why this is so.

Let a beam of electrons be passed through a slit (Figure 3.4) and fall on a photographic plate detector: the intensity of the pattern has the form shown. If the width of the slit is $\Delta x$, then every electron recorded on the plate must have passed through the slit, and so has a positional uncertainty of $\Delta x$ in the direction normal to that of the incident beam. If the slit width is decreased, the value of $\Delta x$ becomes smaller. However with a narrow slit, a diffraction pattern is obtained: it may be interpreted to mean that the electron receives a certain momentum $\Delta p_x$, as it passes through the slit.

If the momentum of an electron is $p$ and its angle of scatter is $\alpha$, the component of the momentum along $x$ is

$$\Delta p_x = p \sin\alpha \ . \tag{3.17}$$

The angular width of the diffraction pattern is given by

$$\Delta x = \lambda/\sin\beta \ . \tag{3.18}$$

Using the de Broglie equation (3.16), we obtain

$$\Delta p_x \Delta x = h\sin\alpha/\sin\beta \ . \tag{3.19}$$

Since generally, $\alpha \approx \beta$,

$$\Delta p_x \Delta x \approx h \ . \tag{3.20}$$

Equation (3.20) is one representation of the Heisenberg uncertainty principle and it expresses a limit to the simultaneous, precise knowledge of both the momentum and the position of an electron.

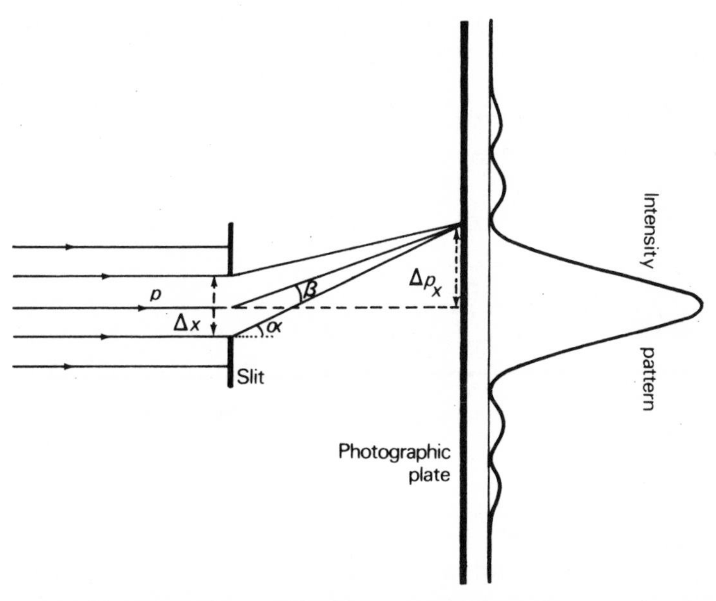


Figure 3.4 – Diffraction of electrons at a slit.

## 3.3 THE BOHR ATOM

When hydrogen atoms are excited by an electric discharge their emission spectrum includes light at four frequencies in the visible region (Figure 3.5). Balmer (1895) showed that the frequencies fitted an equation of the type

$$\tilde{\nu} = R_{\mathrm{H}}\{(1/n_2^2) - (1/n_1^2)\} \ , \tag{3.21}$$

where $\tilde{\nu}$ is the wavenumber $(1/\lambda)$ of the spectral line; $R_{\mathrm{H}}$, the Rydberg constant for hydrogen† is $1.0967758 \times 10^7$ m$^{-1}$, $n_2$ is 2, and $n_1$ is another integer, greater than 2. It was a feature of Bohr's atomic theory that it explained the atomic spectrum of hydrogen in terms of transitions between two energy levels. An electron, only when moving from one energy level $E_2$ to a level of lower energy $E_1$, was stated to emit radiation of frequency given by

$$\nu = (E_2 - E_1)/h \ . \qquad (3.22)$$

The theory predicted the energy levels correctly and gave a value for the Rydberg constant in excellent agreement with experiment. The energy levels in atomic hydrogen are given by the equation

$$E_n = -\mu e^4/(8n^2h^2\epsilon_0^2) \ , \qquad (n = 1, 2, 3 \ldots) \ , \qquad (3.23)$$

where $\mu$ is the reduced mass of the system of one proton and one electron (see Appendix 13). Evaluating $\{\mu e^4/(8h^2\epsilon_0^2)\}/(hc)$ gives $1.096773 \times 10^7$ m$^{-1}$ for $R_H$.

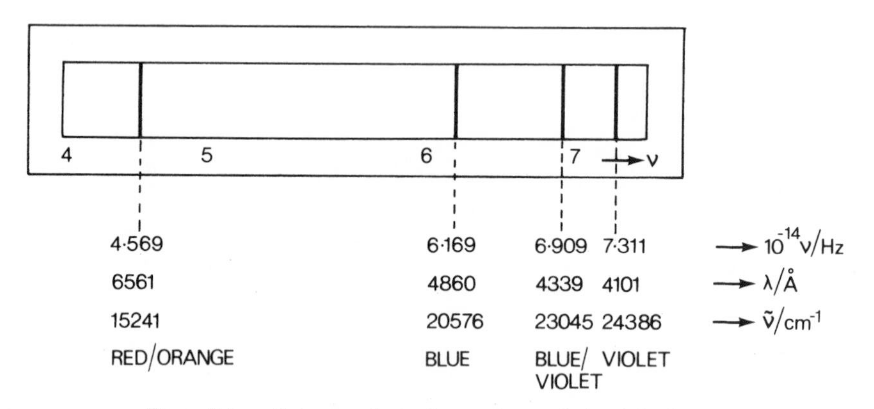


Figure 3.5 – Balmer series in the spectrum of atomic hydrogen.

Bohr's theory was based on classical mechanics, but with quantum conditions imposed in order both to fit experimental results and to explain how the electrons in Rutherford's atomic model did not follow a spiral path into the nucleus. It was therefore a patchwork theory and, more seriously, could not explain spectra other than that of the simple example of atomic hydrogen. The theory supposed that electrons moved around the nucleus in orbits, rather like the planets in our solar system; indeed, the term planetary electrons was used in this context. However, two fundamental objections arise when applying planetary theory to electrons; we need to define both the position and the momentum of an electron, and to follow its orbit. The uncertainty

†$R_H = R_\infty/[1 + (m_e/m_p)]$.

principle tells us that these objectives cannot be achieved simultaneously. That it can be done with planets and not with electrons depends on the different sizes of these bodies in relation to the methods used for their observation.

If, then, we are not to obtain the required success by following descriptions based on the particulate nature of the electron, it seems not unreasonable to consider next theories based upon its wave properties, and so we turn our attention to wave mechanics.

## 3.4 WAVE EQUATION

Since an electron has wave properties, its behaviour can be represented by a wave equation. This equation cannot be derived, but may be presented through certain basic ideas, and justified *a posteriori* by the success of the results obtained with it.

The classical wave equations have certain special solutions: they are known as standing (time-independent) waves, which have positions of zero amplitude, or nodes. In the case of a stretched string, these standing waves are the fundamental and overtone vibrations of the string and can be labelled 0, 1, 2, . . . depending on the number of nodes in the string between its fixed ends. The integers in (3.21), which represent Bohr's principal quantum number, arise from the standing-wave solution of a wave equation for electrons.

A harmonic standing wave in one dimension, $x$, can be written in the form

$$\psi = \mathrm{A}\sin 2\pi\,\{(x/\lambda) - \nu t\} \cdot \sin 2\pi\nu t \ . \tag{3.24}$$

Differentiating (3.24) twice with respect to $x$ shows that

$$\mathrm{d}^2\psi/\mathrm{d}x^2 = -(4\pi/\lambda^2)\psi \ , \tag{3.25}$$

and using de Broglie's equation, (3.16), we obtain

$$\mathrm{d}^2\psi/\mathrm{d}x^2 + -(4\pi p^2/h^2)\psi \ ; \tag{3.26}$$

we shall not be concerned with the time-dependency of $\psi$.

The classical law of conservation of energy is

$$E = T + V \ , \tag{3.27}$$

where $T$ and $V$ are, respectively, the kinetic and potential energies of a system of total energy $E$. For a particle of mass $m$ moving with speed $v$ in the $x$ direction, in a field of potential energy $V$.

$$E = mv^2/2 + V \ , \tag{3.28}$$

or
$$E = p^2/(2m) + V . \tag{3.29}$$

If we replace $p^2$ in (3.26) by its value from (3.29) we have

$$d^2\psi/dx^2 + (8\pi^2 m/h^2)(E - V)\psi = 0 , \tag{3.30}$$

which is the time-independent Schrödinger equation in one dimension. In three dimensions we may write

$$\nabla^2\psi/dx^2 + (8\pi^2 m/h^2)(E - V)\psi = 0 , \tag{3.31}$$

where $\nabla^2$ (del squared) is the laplacian operator

$$\partial^2/\partial x^2 + \partial^2/\partial y^2 + \partial^2/\partial z^2 . \tag{3.32}$$

More concisely, we can write $\mathcal{H}\psi = E\psi$ (3.33)

where $\mathcal{H}$ is the Hamiltonian operator; in (3.30), it takes the form

$$\mathcal{H} = -\{h^2/(8\pi^2 m)\}\partial^2/\partial x^2 + V \tag{3.34}$$

Thus we may say that the classical equation of motion, (3.29), has been converted into the corresponding wave equation by the replacement

$$p \longrightarrow -[ih/(2\pi)]\, d/dx . \tag{3.35}$$

In (3.31) $\psi$ is called a wave function, and it may be used to represent the amplitude of an electron wave. We can introduce a physical meaning through $|\psi|^2$. Let us consider that an electron is spread out to form a density distribution, or charge cloud: then $|\psi|^2$ may be said to measure† the electron density at any point. Rather more precisely, we may say that $|\psi|^2 d\tau$ measures the probability that an electron may be found in the elemental volume of space, $d\tau$.

In order that $\psi$ shall represent a physically meaningful solution of the wave equation, two or more conditions must be imposed. For an electron which is bound to an atom, $\psi$ must be finite, single-valued (having only one value at any point $x,y,z$) and continuous. The wave equation will then be capable of being normalized, that is

$$\int|\psi|^2 d\tau = 1 , \tag{3.36}$$

†Since $\psi$ is, in general, a complex quantity, $|\psi|^2 = \psi\psi^*$, where $\psi^*$ is the complex conjugate of $\psi$. The quantity with physical meaning is $|\psi|^2$, and in many cases we can obtain this result from $\psi\psi$ as well as from $\psi\psi^*$, implying that $\psi$ is mathematically real.

where the integral, taken over the range of the variable, indicates that the total probability of finding an electron somewhere in space must be unity.

These conditions are significant. Equations (3.31) and (3.33) contain the total energy $E$ of the electron and thus the solutions must depend on $E$. The important solutions of (3.33) are called stationary states: $E$ is invariant with time (a conservative system), and hence $|\psi|^2$ is also independent of time. The stationary states are our main interest: their wave functions are called eigenfunctions, and the corresponding energies are termed eigenvalues. An important feature of the wave equation is that the existence of discrete energies is introduced explicitly, whereas Bohr had needed to introduce them empirically in his theory.

### 3.4.1 Electron in a box

Let an electron be confined to motion in a one-dimensional box† of length $a$, and let the box be terminated by a potential barrier of infinite height, such that the potential energy $V$ is zero for $0 \leqslant x \leqslant a$, but infinite for $0 > x > a$ (Figure 3.6a). The appropriate wave equation is (3.30), with $V = 0$, and the solution (see Appendix 14) is

$$\psi = A\exp(ikx) + B\exp(-ikx) \ , \tag{3.37}$$

where $A$ and $B$ are constants: from (3.30), it follows that $k = (8\pi^2 mE/h^2)^{1/2}$. Using de Moivre's theorem,

$$\exp(i\theta) = \cos\theta + i\sin\theta \ , \tag{3.38}$$

and

$$\psi = C\cos kx + D\sin kx \ . \tag{3.39}$$

One boundary condition is that $\psi = 0$ at $x = 0$; this is attained for $C = 0$. The function must vanish also at $x = a$: since $\psi$ must be finite, this condition may be achieved by allowing $k$ to take the form

$$k = n\pi/a \ . \tag{3.40}$$

Thus,

$$\psi_n = D\sin(n\pi x/a) \ , \tag{3.41}$$

and the allowed energies are given, from (3.30), by

$$E_n = n^2h^2/(8ma^2) \ . \tag{3.42}$$

†Something like a single bead on a bead-frame wire.

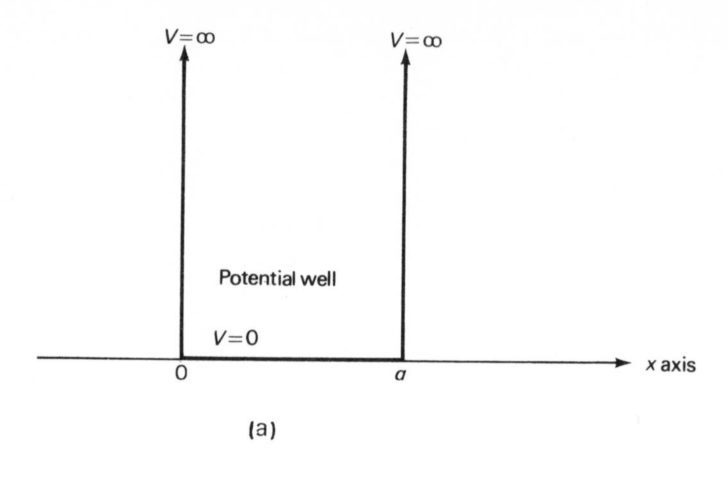


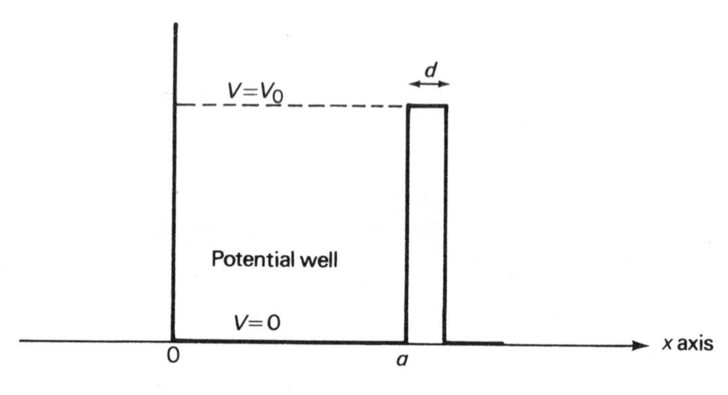


Figure 3.6 – Electron in a box: (a) Infinite potential well; (b) Finite potential well.

Equation (3.42) introduces quantization through $n$. The lowest state, $\psi_1$, has an energy, $E_1$, the zero energy. It is kinetic energy, since $V = 0$: even in the lowest energy state, the electron is in motion. This property is entirely wave-mechanical, because if $h$ were zero the minimum energy would be zero, and energy would not be quantized. The existence of the zero energy is in accord with Heisenberg's uncertainty principle: if the kinetic energy of an electron were zero, its momentum would be exactly known (zero) and $\Delta p \Delta x$ would be zero too.

The probability of finding the electron lying in the interval $x$ to $(x + dx)$, somewhere between $x = 0$ and $x = a$, is $|\psi|^2 dx$. From (3.36) and (3.41),

$$\int_0^a D^2 \sin^2(n\pi x/a) = 1 \; . \qquad (3.43)$$

Integration of this expression leads to the result

$$D = (2/a)^{1/2} \ . \tag{3.44}$$

Figure 3.7 shows the solutions of (3.41) for $n = 1$ to $n = 5$, taking $D = 1$: they are similar to the amplitudes of the fundamental ($\psi_1$) and first four overtone ($\psi_2,\psi_3,\psi_4,\psi_5$) vibrations of a stretched string.

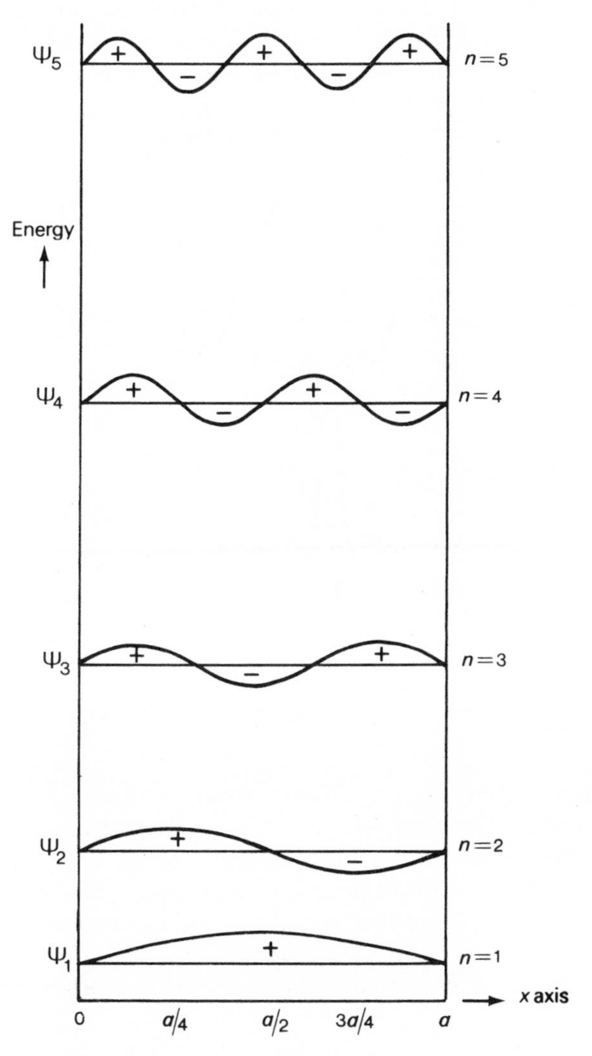


Figure 3.7 – Eigenfunctions of the first five solutions of the electron in a box problem; the number of nodes is $n$–1.

As $a$ is made smaller, any given energy difference, $(E_{n-1} - E_n)$, becomes larger. Conversely, as $a$ becomes larger, energy differences become smaller: in the limit as $a \longrightarrow \infty$, all values of $E$ are allowed, which is the case where there is no box at all; in other words, we have a free electron. In this situation, the state $\psi$, of the electron is described by a superposition of eigenfunctions of the type $c\exp(\pm ikx)$, and given by the expression $\psi = \sum_n c_n \exp(\pm ikx)$. Then, interference occurs and a wave packet results that progresses in time with a centre of gravity corresponding to the centre of mass of the equivalent particle. The electron can be said to be localised, provided that we do not attempt to define precisely both its position and its momentum. The superposition is carried out by a Fourier transformation, but we shall not discuss this topic. Figure 3.8 illustrates wave packets: since all energy states are allowed, free elections produce a continuuum in an atomic spectrum (see Figure 2.3).

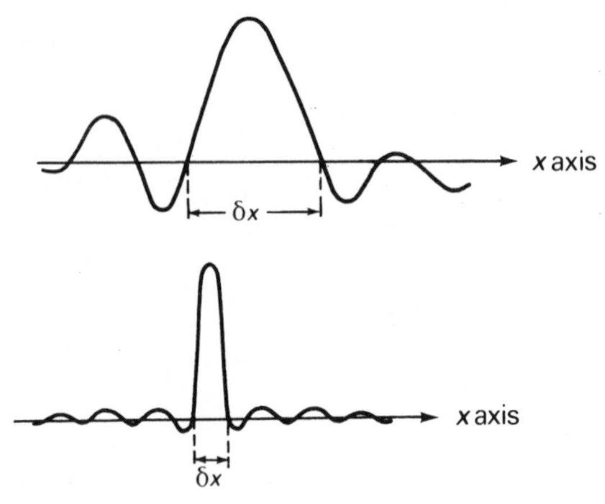


Figure 3.8 – Wave packets: (a) superposition of a small number of waves; (b) superposition of a large number of waves. The most probable location of the particle is given by the range $\delta x$. Similar curves are encountered in the location of atoms by electron-density synthesis. (Suggested Reading: *Crystallography* Ladd and Palmer).

*Tunnelling*

Let us take a tennis ball and lock it in a well constructed metal safe. Newtonian mechanics tell us that the probability of the tennis ball being found outside the safe, unless someone takes it out, is zero. In the case of the electron-in-a-box problem, quantum mechanics reveals a different picture.

Let an electron be confined to a one-dimensional box by a potential-energy wall of height $V_o$ and thickness $d$ (Figure 3.6b). Within the box the term $(E - V)$ in (3.30) is positive, and the solution of wave equation is (3.41): the right-hand side of this equation may be written as $D\sin(2\pi/h)\sqrt{2m(E-V_o)}x$, or more generally as $D\exp\text{-}[(i2\pi/h)\sqrt{2m(E-V_o)}x]$-, since $C$ in (3.39) is zero.

In the region *within* the potential barrier $V_0$ is greater than E, and the general solution of the wave equation for this region is $D\exp\{-(2\pi/h)\sqrt{2m(V_0-E)}x\}$. Thus the probability of finding an electron in the region of negative kinetic energy is not zero, but falls of exponentially with the distance $x$ of penetration within the barrier. As long as the barrier is neither infinitely high nor infinitely wide, there is a finite probability that electrons will *tunnel* through the barrier.

*Boxes of higher dimensions*

We can extend the box to two dimensions (Figure 3.9), and the appropriate wave equation is

$$\partial^2\psi/\partial x^2 + \partial^2\psi/\partial y^2 + (8\pi^2 m/h^2)(E - V)\psi \ . \tag{3.45}$$

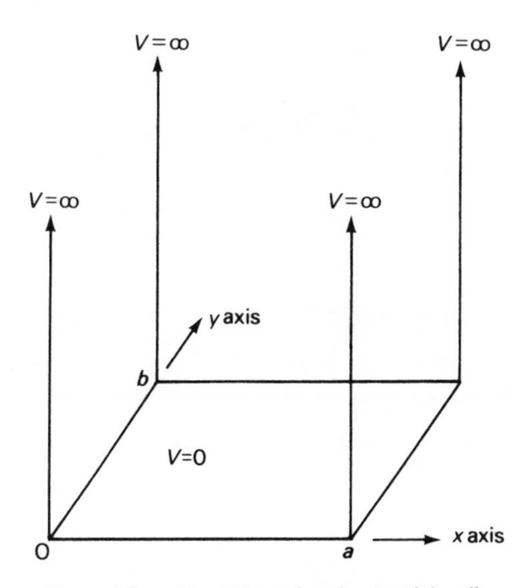


Figure 3.9 – Two-dimensional potential well.

This differential equation is said to be seperable, that is the eigenfunctions can be written as products in $x$ and $y$. By analogy with (3.41), including the boundary conditions as before, we may write

$$\psi_{n,l} = D'\sin(n\pi x/a)\sin(l\pi y/b) \ , \tag{3.46}$$

and for the total energy

$$E_{n,l} = E_n + E_l \ . \tag{3.47}$$

From a consideration of (3.42) and (3.44),

$$E_{n,l} = (h^2/8m)\{(n^2/a^2) + (l^2/b^2)\}, \quad (3.48)$$

and

$$D' = 2/(ab)^{1/2} \quad (3.49)$$

The lowest energy state is given by

$$E_{1,1} = (h^2/8m)\{1/a^2) + (1/b^2)\}, \quad (3.50)$$

and the corresponding eigenfunction is

$$\psi_{1,1} = D'\sin(\pi x/a)\sin(\pi y/b). \quad (3.51)$$

The next highest states are $\psi_{1,2}$ and $\psi_{2,1}$, with energies

$$E_{1,2} = (h^2/8m)\{(1/a^2) + (4/b^2)\}, \quad (3.52)$$

and

$$E_{2,1} = (h^2/8m)\{(4/a^2) + (1/b^2)\} \quad (3.53)$$

For a square box, $a = b$ and $E_{1,2} = E_{2,1}$; the energies are said to be degenerate (Figure 3.10).

In three dimensions, the energy levels are characterised by three integers (quantum numbers): it is not difficult to show that a cubic box of side $a$ has energies given by

$$E_{n,l,m_l} = \{h^2/(8m\,a^2)\}\{n^2 + l^2 + m_l^2\}. \quad (3.54)$$

We shall illustrate a three-dimensional solution by reference to the hydrogen atom.

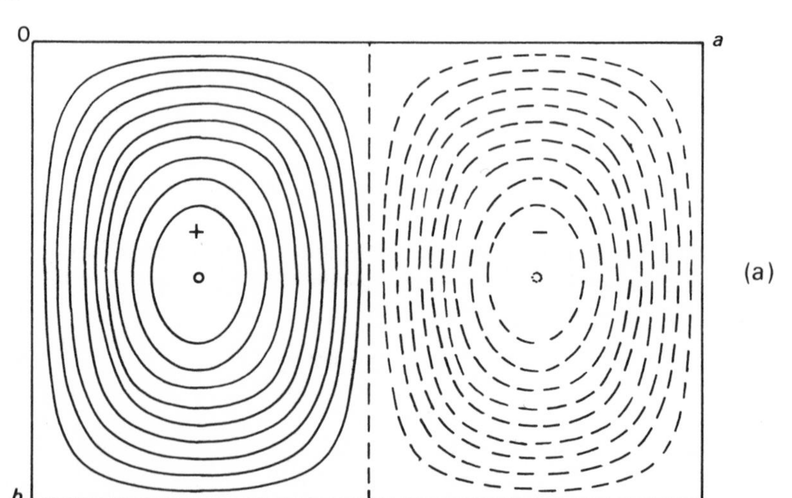

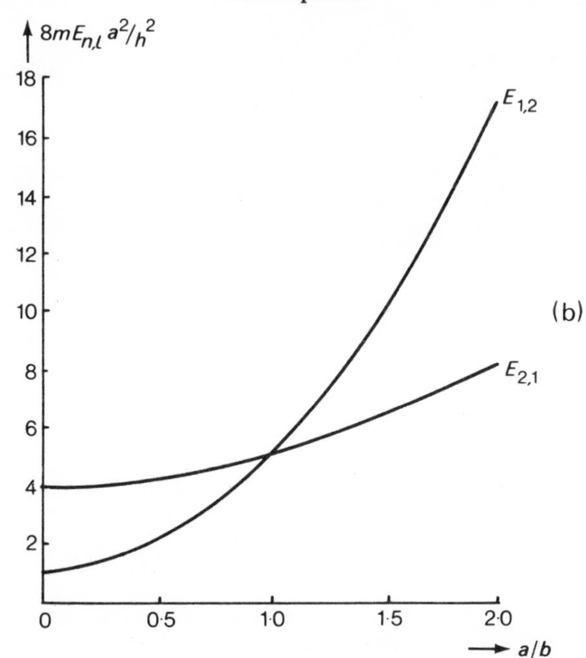


Figure 3.10 – Two-dimensional electron in a box solutions:
(a) $\psi_{2,1}$ for $a/b = 1.5$; (b) $E_{2,1}$ and $E_{1,2}$ as a function of $a/b$.

### 3.4.2 Hydrogen atom

For the hydrogen atom, the Hamiltonian has the form

$$\mathcal{H} = -[h^2/(8\pi^2\mu)]\cdot\nabla^2 - e^2/(4\pi\epsilon_0 r) \ , \tag{3.55}$$

where $\mu$ is the reduced mass of the system of one proton and one electron and $r$ is the distance between the proton and the electron.

It is convenient to transform $\nabla^2$ from Cartesian coordinates to spherical polar coordinates. We introduce the colatitude $\theta$ the azimuth $\phi$ and the radius vector $r$ (Figure 3.11); the relationships between $x$, $y$, $z$ and $r$, $\theta$, $\phi$ follow readily from the diagram:

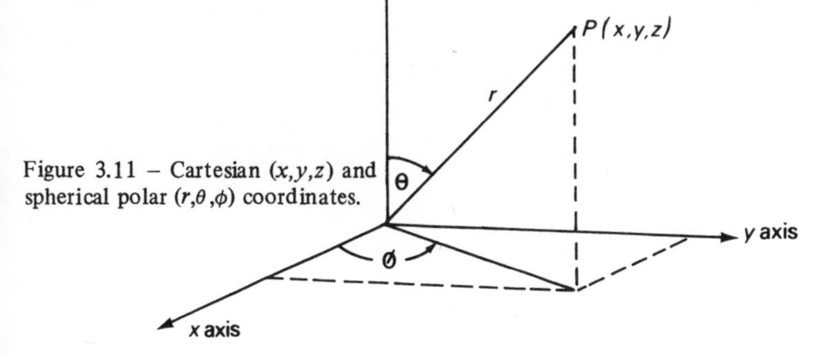


Figure 3.11 – Cartesian $(x,y,z)$ and spherical polar $(r,\theta,\phi)$ coordinates.

$$\begin{aligned} z &= r\cos\theta \\ y &= r\sin\theta\sin\phi \\ x &= r\sin\theta\cos\phi \\ x^2 + y^2 + z^2 &= r^2 \\ z/r &= \cos\theta \\ y/x &= \tan\phi \end{aligned} \tag{3.56}$$

It is shown in Appendix 15 that in polar coordinates

$$\nabla^2 = r^{-2}\left\{\frac{\partial}{\partial r}\left[r^2\frac{\partial}{\partial r}\right] + \frac{1}{\sin^2\theta}\frac{\partial^2}{\partial\phi^2} + \frac{1}{\sin\theta}\frac{\partial}{\partial\theta}\left[\sin\theta\frac{\partial}{\partial\theta}\right]\right\}. \tag{3.57}$$

Hence, the wave equation for the hydrogen atom may be written as

$$\frac{\partial}{\partial r}\left[r^2\frac{\partial\psi}{\partial r}\right] + \frac{1}{\sin^2\theta}\frac{\partial^2\psi}{\partial\phi^2} + \frac{1}{\sin\theta}\frac{\partial}{\partial\theta}\left[\sin\theta\frac{\partial\psi}{\partial\theta}\right] + (8\pi^2\mu r^2/h^2)\{E + e^2/(4\pi\epsilon_o r)\} = 0\ . \tag{3.58}$$

This differential equation is separable into terms, each term depending on only one of the variables $r$, $\theta$ and $\phi$. We shall write first

$$\psi = R(r)Y(\theta,\phi)\ . \tag{3.59}$$

Substituting (3.59) in (3.58), and remembering, for example, that

$$\frac{\partial\psi}{\partial r} = Y\,\partial R/\partial r = (\psi/R)\partial R/\partial r\ , \tag{3.60}$$

we obtain

$$\frac{1}{R}\frac{\partial}{\partial r}\left[r^2\frac{\partial R}{\partial r}\right] + (8\pi^2\mu r^2/h^2)\,[E + e^2/(4\pi\epsilon_o r)] = -\frac{1}{Y}\left[\frac{1}{\sin^2\theta}\frac{\partial^2 Y}{\partial\phi^2} + \frac{1}{\sin\theta}\frac{\partial}{\partial\theta}\left\{\sin\theta\frac{\partial Y}{\partial\theta}\right\}\right] \tag{3.61}$$

Since $r$, $\theta$ and $\phi$ are independent variables, both sides of (3.61) must be equal to a constant, $K$. Thus, with rearrangements,

$$\frac{1}{r^2}\frac{\partial}{\partial r}\left[r^2\frac{\partial R}{\partial r}\right] + \{(8\pi^2\mu/h^2)\,[E + e^2/(4\pi\epsilon_o r)] - (K/r^2)\}\,R = 0 \tag{3.62}$$

and

$$\frac{1}{\sin^2\theta}\frac{\partial^2 Y}{\partial\phi^2} + \frac{1}{\sin\theta}\frac{\partial}{\partial\theta}\left[\sin\theta\frac{\partial Y}{\partial\theta}\right] + KY = 0\ ; \tag{3.63}$$

(3.62) is the radial-dependent part and (3.63) the angular-dependent part of (3.61). The angular-dependent equations can be further separated:

$$Y(\theta,\phi) = \Theta(\theta)\,\Phi(\phi)\ . \tag{3.64}$$

Thus, following the procedure as before, we obtain

$$-\frac{1}{\Phi}\frac{\partial^2\Phi}{\partial\phi^2} = \frac{\sin\theta}{\Theta}\frac{\partial}{\partial\theta}\left[\sin\theta\frac{\partial\Theta}{\partial\theta}\right] + K\sin^2\theta\ , \tag{3.65}$$

and both sides of (3.65) may be equated to a constant, say $m_l^2$.

3.4.2.1 *Solution of the Φ-equation*

We have

$$\partial^2\Phi/\partial\phi^2 + m_l^2\Phi = 0 \tag{3.66}$$

From Appendix 14, we have the solution

$$\Phi = A\exp(im_l\phi)\ . \tag{3.67}$$

In order that $\Phi$ may be single-valued at $\phi = \phi_0$ and $\phi = \phi_0 + 2\pi$, $m_l$ must be integral; it is a quantum number, and may take the values 0, ±1, ±2, . . . We know also that $A\exp(-im_l\phi)$ is an equivalent solution, and following (3.34), $A = 1/(2\pi)^{1/2}$.

3.4.2.2 *Solution of the Θ-equation*

Let $\cos\theta = z$, and let $P(z)$ replace $\Theta(\theta)$. Then, the limits $\pi$ and 0 in $\theta$ become −1 and +1 in $z$, and

$$\partial\Theta/\partial\theta = (\partial P/\partial z)(\partial z/\partial\theta) = -(\sin\theta)\partial P/\partial z\ . \tag{3.68}$$

From the right-hand side of (3.65), we have

$$\frac{\partial}{\partial z}\left[(1-z^2)\frac{\partial P}{\partial z}\right] + \left[K - \frac{m_l}{(1-z^2)}\right]P = 0\ , \tag{3.69}$$

which is known as the associated Legendre differential equation†; $K = l(l+1)$, where $l$ is another quantum number. Some solutions of (3.69) are listed in Table 3.1. The normalization constant, $N$, is obtained from

$$N^2\int_{-1}^{1} P(z)P(z)\mathrm{d}z = 1\ . \tag{3.70}$$

†Suggested reading: *Mathematics* (Margenau and Murphy).

**Table 3.1** – One electron $\Theta_{l,m_l}$ functions

| $l$ | $m_l$ | $P_{l,m_l}(z)$ | $\Theta_{l,m_l}(\theta)$ | $\Theta_{l,m_l}(\theta)$ (normalized) |
|---|---|---|---|---|
| 0 | 0 | 1 | 1 | $1/(2)^{½}$ |
| 1 | 0 | $z$ | $\cos\theta$ | $(3/2)^{½}\cos\theta$ |
| 1 | +1 } | $(1 - z^2)^{½}$ | $\sin\theta$ | $(3/4)^{½}\sin\theta$ |
| 1 | −1 } | | | |

It may be checked readily that $\Theta_{l,m_l}(\theta)$ is invariant for an increment of $2\pi$ in $\theta$, that is, it is single-valued.

We may now combine the $\Theta_{l,m_l}$ and the $\Phi_{m_l}$ functions to obtain the complete angular dependent set of functions $Y_{l,m_l}$, or spherical harmonics, listed in Table 3.2.

**Table 3.2** – One-electron $Y_{l,m_l}(\theta,\phi)$ functions, or spherical harmonics; the quantum number $m_l$ occurs in the exponential term of $Y$.

| $l$ | $m_l$ | $Y_{l,m_l}(\theta,\phi)$ (normalized) | Orbital type |
|---|---|---|---|
| 0 | 0 | $1/(4\pi)^{½}$ | s |
| 1 | 0 | $(3/4\pi)^{½}\cos\theta$ | $p_z$ |
| 1 | +1 | $(3/8\pi)^{½}\sin\theta\exp(i\phi)$ | p |
| 1 | −1 | $(3/8\pi)^{½}\sin\theta\exp(-i\phi)$ | p |

From 3.4.2.2 we may write

$$\Phi^{+} = 1/(2\pi)^{½}(\cos m_l\phi + i\sin m_l\phi) \quad (3.71)$$

and

$$\Phi^{-} = 1/(2\pi)^{½}(\cos m_l\phi - i\sin m_l\phi)\ . \quad (3.72)$$

Now, if two or more functions are eigenfunctions of a linear operator†, then linear combinations of two or more functions on one centre are also eigenfunctions of the operator provided that they are degenerate; the appropriate normalizing constants must be applied. Hence, we have

$$(2)^{-½}(\Phi^{+} + \Phi^{-}) = (\pi)^{-½}\cos\phi \quad (3.73)$$

$$\frac{(2)^{-½}}{i}(\Phi^{+} - \Phi^{-}) = (\pi)^{-½}\sin\phi\ . \quad (3.74)$$

†If $\alpha(f + g) = \alpha f + \alpha g$, where $f$ and $g$ are any two functions, then the operator $\alpha$ is linear: $d/dx$ is an example of a linear operator.

Equations (3.73) and (3.74) relate to the $p_x$ and $p_y$ orbitals respectively, and may be added to Table 3.2 in place of $(1/2\pi)^{1/2}\exp(\pm i\phi)$ in the last two entries therein.

3.4.2.3 *Solution of the R-equation*

Equation (3.62) may be written as

$$\frac{\partial^2 R}{\partial r^2} + \frac{2}{r}\left[\frac{\partial R}{\partial r}\right] + (8\pi^2\mu/h)\left\{E + e^2/(4\pi\epsilon_0 r) - \frac{l(l+1)h^2}{8\pi^2\mu r^2}\right\} R = 0 \ , \qquad (3.75)$$

since we have taken $K$ to be $l(l+1)$. We make the substitutions

$$\left.\begin{aligned} x &= r\zeta \\ (\zeta/2)^2 &= (8\pi^2\mu E)/h^2 \end{aligned}\right\} \qquad (3.76)$$

Then using

$$\partial R/dx = \frac{\partial R/\partial r}{\partial x/\partial r} \qquad (3.77)$$

and

$$\partial^2 R/\partial x^2 = \left[\frac{1}{\partial x/\partial r}\right]\frac{\partial}{\partial r}\left[\frac{\partial R/\partial r}{\partial x/\partial r}\right] = \frac{(\partial^2 R/\partial r^2)(\partial x/\partial r) - (\partial^2 x/\partial r^2)(\partial R/\partial r)}{(\partial x/\partial r)^3} \qquad (3.78)$$

we obtain

$$x\frac{\partial^2 R}{\partial x^2} + 2\frac{\partial R}{\partial x} + [8\pi^2\mu e^2/(4\pi\epsilon_0 h^2\zeta) - x/4 - l(l+1)/x]R = 0 \,. \qquad (3.79)$$

The differential equation which satisfies the associated Laguerre polynomials†,

$$x\frac{\partial^2 R}{\partial x^2} + 2\frac{\partial R}{\partial x} + [n' - (k-1)/2 - x/4 - (k^2-1)/(4x)]R = 0 \ , \qquad (3.80)$$

is clearly of the same form. Form (3.79) and (3.80), it is clear

that

$$k = 2l + 1 \qquad (3.81)$$

and

$$n' - l = n = 8\pi^2\mu e^2/(4\pi\epsilon_0 h^2\zeta) \ , \qquad (3.82)$$

where $n$, the principal quantum number, is integral and greater than zero. Thus $R$ depends on both $n$ and $l$, and may be normalized according to

†Suggested Reading: *Mathematics* (Margenau and Murphy).

$$N^2 \int_0^\infty R_{n,l}^2(r) r^2 \mathrm{d}r = 1 \; ; \tag{3.83}$$

some results are listed in Table 3.3. It may be noted that any one-electron atom will have the same function of $R$, provided that the potential energy function is modified to $Z\,e^2/(4\pi\epsilon_0 r)$, where $Z$ is the atomic number.

**Table 3.3** – One-electron radial functions, $R_{n,l}(r)$

| $n$ | $l$ | $R_{n,l}(r)$ | $R_{n,l}(r)$ (normalized) | Orbital type |
|---|---|---|---|---|
| 1 | 0 | $2\exp(-\zeta r/2)$ | $(Z/a_0)^{\frac{3}{2}}\,2\exp(-\zeta r/2)$ | 1s |
| 2 | 0 | $(2\zeta r - 4)\exp(-\zeta r^2)$ | $\frac{1}{2\sqrt{2}}(Z/a_0)^{\frac{3}{2}}(2-\zeta r)\exp(-\zeta r/2)$ | 2s |
| 2 | 1 | $6\zeta r\exp(-\zeta r/2)$ | $\frac{1}{2\sqrt{6}}(Z/a_0)^{\frac{3}{2}}\zeta r\exp(-\zeta r/2)$ | 2p |

In this Table, $a_0$ is the Bohr radius, the most probable distance of a 1s electron from the nucleus of a hydrogen atom. It is given by

$$a_0 = (4\pi\epsilon_0)h^2/\mu e^2) \; , \tag{3.84}$$

and its value is 0.5295 Å, or $0.5295 \times 10^{-10}$ m. From (3.76) and (3.82), it follows that

$$\mathrm{E}_{n,l,m_l} = -\mu e^4/(8n^2h^2\epsilon_0^2) \; . \tag{3.85}$$

This equation should be compared with the Bohr equation (3.23): the negative sign indicates that we are concerned with bound states for the electron, for which energy is conventionally a negative quantity, being zero at infinite separation. We may note that the energy does not here depend on $l$ and $m_l$. This result is a property of the Coulomb type of potential energy, and each energy level above the lowest will have a $(2l + 1)$ degeneracy.

Next, we may make the substitution

$$\rho = n\zeta r/2 \; , \tag{3.86}$$

and list in Table 3.4 the complete hydrogenic wave functions which we have discussed; the quantum number, $n$, is now apparent in the exponential part of $\psi$.

**Table 3.4 – Hydrogenic (one-electron) wave functions, $\psi_{n,l,m_l}$**

| | | |
|---|---|---|
| $\psi_{1,0,0}$ | $(\pi)^{-\frac{1}{2}}(Z/a_o)^{\frac{3}{2}}\exp(-\rho)$ | 1s |
| $\psi_{2,0,0}$ | $(32\pi)^{-\frac{1}{2}}(Z/a_o)^{\frac{3}{2}}(2-\rho)\exp(-2\rho/2)$ | 2s |
| $\psi_{2,1,0}$ | $(32\pi)^{-\frac{1}{2}}(Z/a_o)^{\frac{3}{2}}\rho\exp(-\rho/2)\cos\theta$ | $2p_z$ |
| $\psi_{2,1,1}$ | $(32\pi)^{-\frac{1}{2}}(Z/a_o)^{\frac{3}{2}}\rho\exp(-\rho/2)\sin\theta\cos\phi$ | $2p_x$ |
| $\psi_{2,1,-1}$ | $(32\pi)^{-\frac{1}{2}}(Z/a_o)^{\frac{3}{2}}\rho\exp(-\rho/2)\sin\theta\sin\phi$ | $2p_y$ |

Finally, we consider the normalization of the wave functions in Table 3.3, taking $R_{21}$ as an example. From (3.76) and (3.85), and, for completeness, introducing $Z$ into the potential energy expression, we find

$$\zeta = 2Z/na_o \ . \tag{3.87}$$

From (3.83)

$$36N^2\zeta^2 \int_0^\infty r^4\exp(-\zeta r)\mathrm{d}r = 1 \ . \tag{3.88}$$

Let $\zeta r = t$, so that $\mathrm{d}r = \mathrm{d}t/\zeta$. Then, we have

$$36N^2/\zeta^3 \int_0^\infty t^4\exp(-t)\mathrm{d}t = 1 \ . \tag{3.89}$$

The integral is the gamma function $\Gamma(5)$, for which the solution is 4! (see Appendix 16). Hence, $N = \zeta^{\frac{3}{2}}/(12\sqrt{6})$, and using (3.87), the result in Table 3.3 follows.

### 3.4.3 Pictorial orbitals

The solutions of the wave equation in terms of $n$, $l$, $m_l$ and $m_s$ (see page 170) are, in general, complex (see Table 3.2). However, by taking linear combinations, as exemplified by (3.71) to (3.74), we can obtain real functions which can be represented in a pictorial manner.

In Figure 3.7, we showed some wave functions for the electron in a one-dimensional box; $\psi$ was plotted against $x$. Clearly, in order to represent $\psi_{n,l,m_l}$ it would be necessary to plot in four dimensions; more conveneintly, we can separate $R_{n,l}$ for constant $\theta$ and $\phi$ and $Y_{l,m_l}$ for constant $r$.

The 1s orbital is spherically symmetrical; it has no angular dependence, and $\psi_{1,0,0}$ decreases exponentially with the distance $r$ from the nucleus. The probability of finding an electron between distances $r$ and $(r + \mathrm{d}r)$ from the nucleus is given by (3.36) with $\mathrm{d}\tau$ equal to $4\pi r^2\mathrm{d}r$, the volume enclosed by spherical shells of radii $r$ and $(r + \mathrm{d}r)$. Figure 3.12 shows the functions $R$, $R^2$ and $4\pi r^2R^2$, each as a function of $r$, for some hydrogenic atomic orbitals. Figure 3.13 illustrates the angular-dependent part, $Y_{l,m_l}$ for the same wave functions.

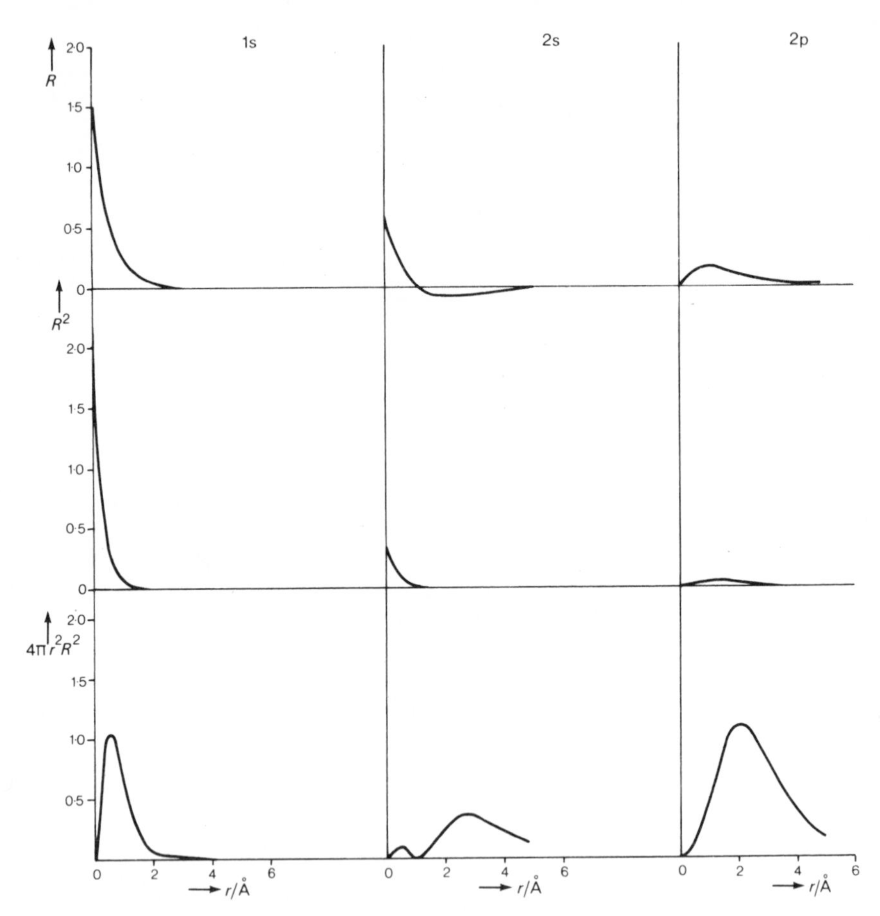


Figure 3.12 – One-electron radial eigenfunction, $R_{n,l}(r)$ (normalized) from Table 3.3 ($Z = 1$); all diagrams are on the same scale.

From Figure 3.12, we may note some important features of orbitals:

(a) Orbitals of the same $l$ value increase in size as $n$ increases;

(b) The s orbitals have a finite density, $\psi^2$, at the nucleus, whereas p orbitals (and d) have zero density at the nucleus;

(c) The number of nodes, or zeros of $\psi$, is $n-l-1$;

Figure 3.13 shows that whereas s orbitals are spherically symmetrical, p orbitals are directional, lying along the Cartesian $x$, $y$ and $z$ axes; the sign of the angular function is indicated in each lobe. The three p orbitals have the same energy; this fact is evident from (3.85). A more descriptive illustration of an orbital is the

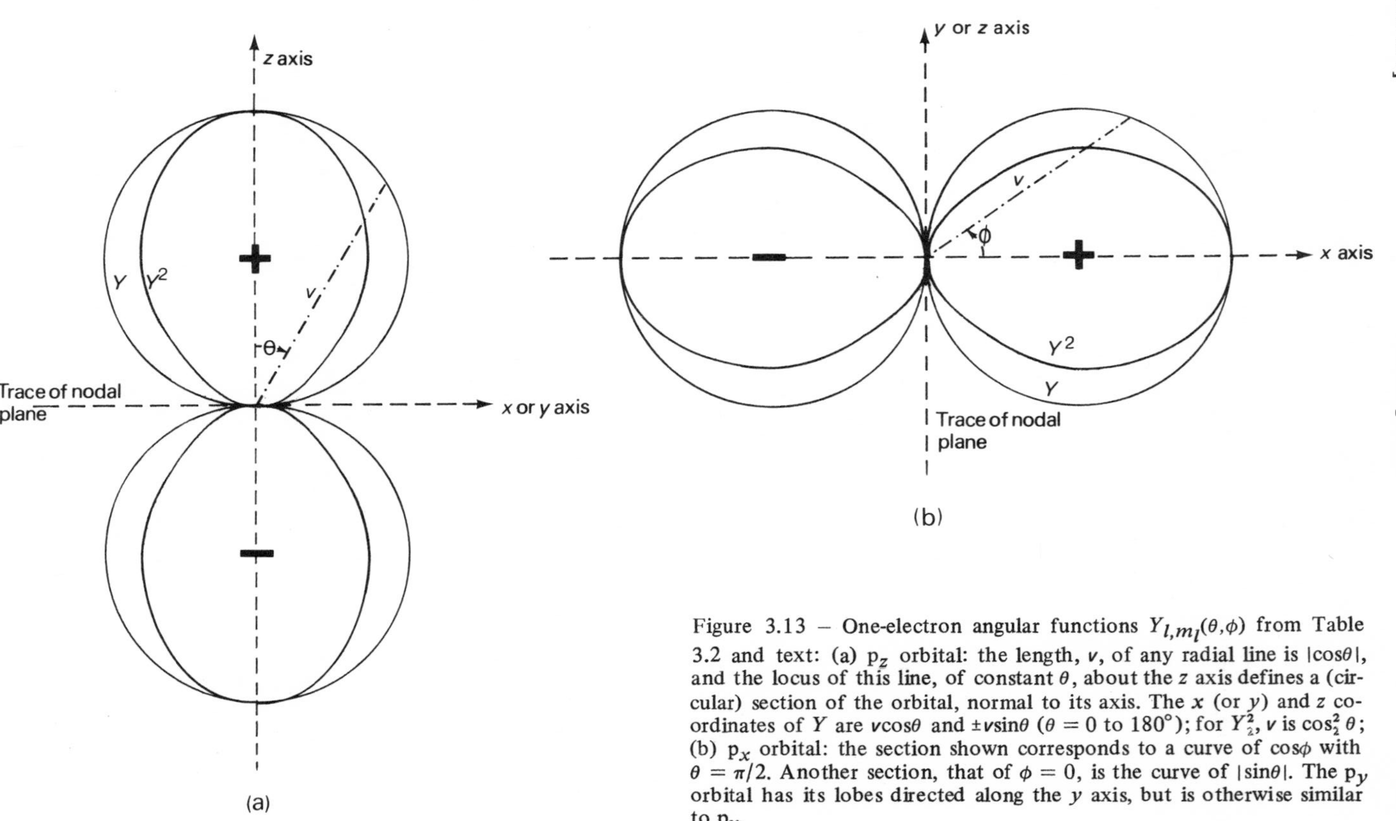


Figure 3.13 – One-electron angular functions $Y_{l,m_l}(\theta,\phi)$ from Table 3.2 and text: (a) $p_z$ orbital: the length, $v$, of any radial line is $|\cos\theta|$, and the locus of this line, of constant $\theta$, about the $z$ axis defines a (circular) section of the orbital, normal to its axis. The $x$ (or $y$) and $z$ co-ordinates of $Y$ are $v\cos\theta$ and $\pm v\sin\theta$ ($\theta = 0$ to $180°$); for $Y_2^2$, $v$ is $\cos_2^2\theta$; (b) $p_x$ orbital: the section shown corresponds to a curve of $\cos\phi$ with $\theta = \pi/2$. Another section, that of $\phi = 0$, is the curve of $|\sin\theta|$. The $p_y$ orbital has its lobes directed along the $y$ axis, but is otherwise similar to $p_x$.

density contour diagram in Figure 3.14, shown for a $2p_z$ orbital. We shall not often use this type of representation; we may note, however, that it emerges also from an X-ray crystallographic study of crystal and molecular structure (Figure 3.15). It should be noted that X-ray methods do not resolve electron density for separate orbitals. The X-ray electron density map for an atom represents an average density for the given atom in a structure: for this reason, the X-ray electron density contour maximum of an atom does not necessarily coinicide with the position of its nucleus, a significant factor in a discussion of the meaning of a bond length.

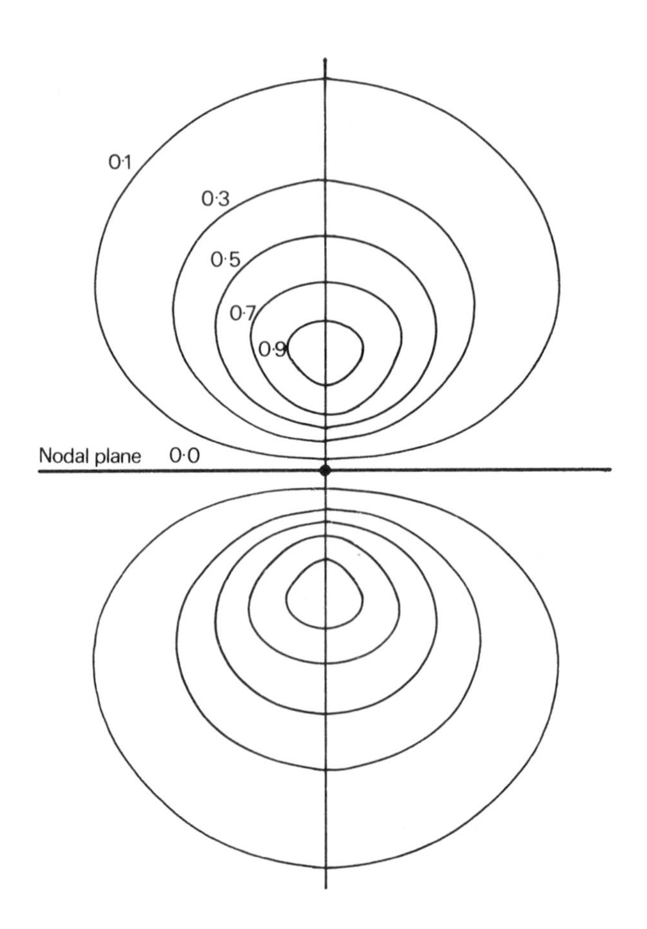


Figure 3.14 – Electron density contours for a $2p_z$ orbital of carbon as fractions of $\psi^2_{max}$. The 0.1 contour surface encloses about 66% of the $2p_z$ electron density; 90% is enclosed by the 0.03 contours.

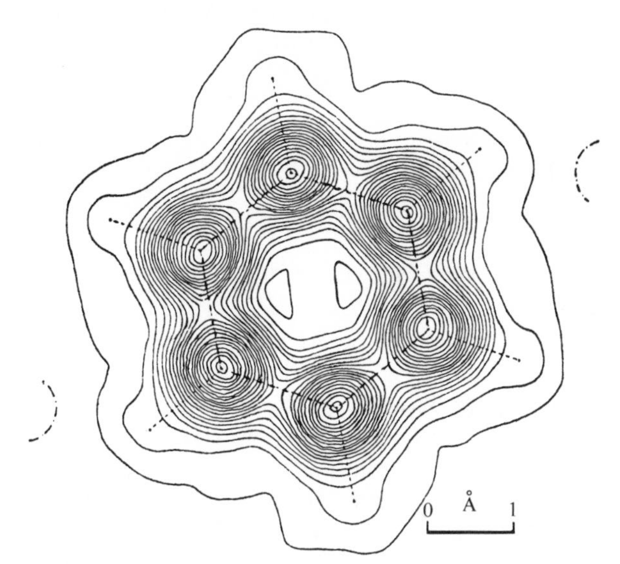


Figure 3.15 – Electron density map of benzene, in the plane of the molecule, obtained from X-ray diffraction studies on the solid. The contour intervals are 0.25 e $Å^{-3}$, and the zero contour is shown by –·–·– (after Cox, Cruikshank and Smith, and reproduced by permission of the Royal Society, London).

### 3.4.4 Orthogonality

An important feature of well designed atomic orbitals is that they are mutually orthogonal. This property is desirable if we are to be able to discuss them separately, for otherwise we would find that one contained part of another.

Mathematically, orthogonality is defined by

$$\int_{\substack{\text{all}\\ \text{space}}} \psi_i \psi_j \, \mathrm{d}\tau = \delta_{i,j} \ , \tag{3.90}$$

where $\delta_{i,j}$ is the Kronecker delta: $\delta_{i,j}$ is zero if $i \neq j$, and unity otherwise, for normalized orbitals. As an example, consider the 1s and $2p_x$ orbitals from Table 3.4.

$$\int_{\substack{\text{all}\\ \text{space}}} \psi_{1s} \psi_{2px} \mathrm{d}\tau \propto \int_0^\infty R_{1,0,0} R_{2,1,1} \, r^2 \mathrm{d}r \int_0^\pi \sin^2\theta \, \mathrm{d}\theta \int_0^{2\pi} \cos\phi \, \mathrm{d}\phi \ . \tag{3.91}$$

Without further elaboration we can see that the complete integral in (3.91) is zero, because $\int_0^{2\pi} \cos\phi \, \mathrm{d}\phi$ is zero. On the other hand, for the $2p_x$ orbital, we have, since† $\rho = (Z'/a_0)r$,

†Introducing $Z'$ for more generality.

$$\int_{\substack{\text{all}\\\text{space}}} \psi_{2p_x}^2 \, d\tau = [1/(32\pi)](Z'/a_o)^5 \int_0^\infty r^4 \exp(-Z'r/a_o) dr \int_0^\pi \sin^3\theta \, d\theta \int_0^{2\pi} \cos^2\phi \, d\phi \; . \tag{3.92}$$

The integrals on the right-hand side of (3.92) are, in turn, $\Gamma(5)/(Z'/a_o)^5$, 4/3 and $\pi$, so that the total product is unity.

### 3.4.5 Summary so far

The wave mechanics in this chapter has been treated in detail so far for a variety of reasons. Firstly, we obtain an idea of the complexity involved in an exact solution for the apparently simple hydrogen atom. It can be solved exactly only for one-electron atomic species such as H or $He^+$, but even for $H_2$ an exact solution is not attainable. We shall see later how good a solution we can obtain by approximate methods and, indeed, what we mean by a good solution.

The solutions for the electron in a box and the hydrogen atom have shown how the quantum concepts of energy quantization, quantum numbers, zero energy, degeneracy and atomic orbitals arise naturally from the theory. We shall not need to continue our study of the covalent bond in the same mathematical detail.

### 3.4.6 More about atomic orbitals

Each electron may be represented by a wave function, or atomic orbital $\psi$, such that $\psi^2$ is an approximate measure of the density distribution for this electron. Each atomic orbital is described by a set of quantum numbers: $n$, which determines the energy and size of the orbital; $l$, which determines its geometry; and $m_l$ determining which of the degenerate orbitals of the same $l$ value is being considered. Then, we must include electron spin. It is not possible to give a simple physical picture of electron spin: the spin quantum number $(m_s)^\dagger$ of ½ emerges from a relativisitic treatment of the Schrödinger equation, but we shall not go into this matter. It will be sufficient to accept that there are two components for the spin; a spin component of +1/2 is often called $\alpha$, and that of −1/2 is often called $\beta$. Thus, the 1s wave function of hydrogen in Table 3.4 may be written either as

$$\psi_{1,0,0,\frac{1}{2}} = (\pi a_0^3)^{-\frac{1}{2}} \exp(-\rho)\alpha \tag{3.93}$$

or

$$\psi_{1,0,0,-\frac{1}{2}} = (\pi a_0^3)^{-\frac{1}{2}} \exp(-\rho)\beta \; . \tag{3.94}$$

Two electrons with the same spin are said to have unpaired, or parallel, spins; the importance of spin will emerge more clearly when we consider the $H_2$ molecule.

† Conventionally, $m_s$ is a component resolved on the $z$ Cartesian axis.

The quantum numbers are governed by the Pauli Exclusion Principle, or antisymmetry rule, which states that no two electrons in the same atom can have the same four quantum numbers. Hence, an atomic orbital can accommodate a maximum of two electrons, provided that they have antiparallel, or paired, spins.

The atomic orbitals may be placed in order of their energies (in general, the energy depends on both $n$ and $l$):

$$1s < 2s < 2p < 3s < 3p \ldots$$

There are other orbitals which can be defined: thus with $n = 3, l = 2$ and $n = 4$, $l = 3$, for example, the corresponding orbitals are designated 3d and 4f, respectively.

In the hydrogen atom, orbitals of the same $n$ value have the same energy. The similar energies of other $n$s and $n$p orbitals, and especially the strong directional character of the p orbitals, plays an important part in determining stereochemistry, as we shall see when studying the ensuing sections.

## 3.5 AUFBAU PRINCIPLE

The description of atomic orbitals is linked closely with the periodic table of elements†. Starting with hydrogen, we can feed, one at a time, electrons into atomic orbitals, so completing the electronic configuration for each atom. This procedure constitutes the *aufbau,* or building-up, principle. Let us carry out an exercise for the first few elements in the periodic table. In Figure 3.16 each atomic orbital considered is represented by a rectangular cell, and electrons with antiparallel spins are indicated by ↑↓. The elements H, He, Li and Be present no difficulties; in boron, the 2p electron can occupy any one of the three degenerate orbitals. With carbon however there are two distinct possibilities; that shown, and another in which any one of the p orbitals contains two spin-paired electrons. The dilemma is resolved by Hund's rule, which states that degenerate or near-degenerate orbitals tend to be occupied preferentially singly by electrons with parallel spins, since electrons with parallel spins tend to avoid one another (see page 182).

We may note that the stability of the inert-gas configuration, the pillar of Lewis's electron-pair bond theory, is realized by a system of occupied orbitals, up to and including those governed by the principal quantum number $n$. The energy of an atomic orbital may be associated with the ionization energy (ionization potential) of an atom; the value of the first ionization energy of each element in Figure 3.16 is shown in parentheses against the element symbol.

†See the inside back cover.

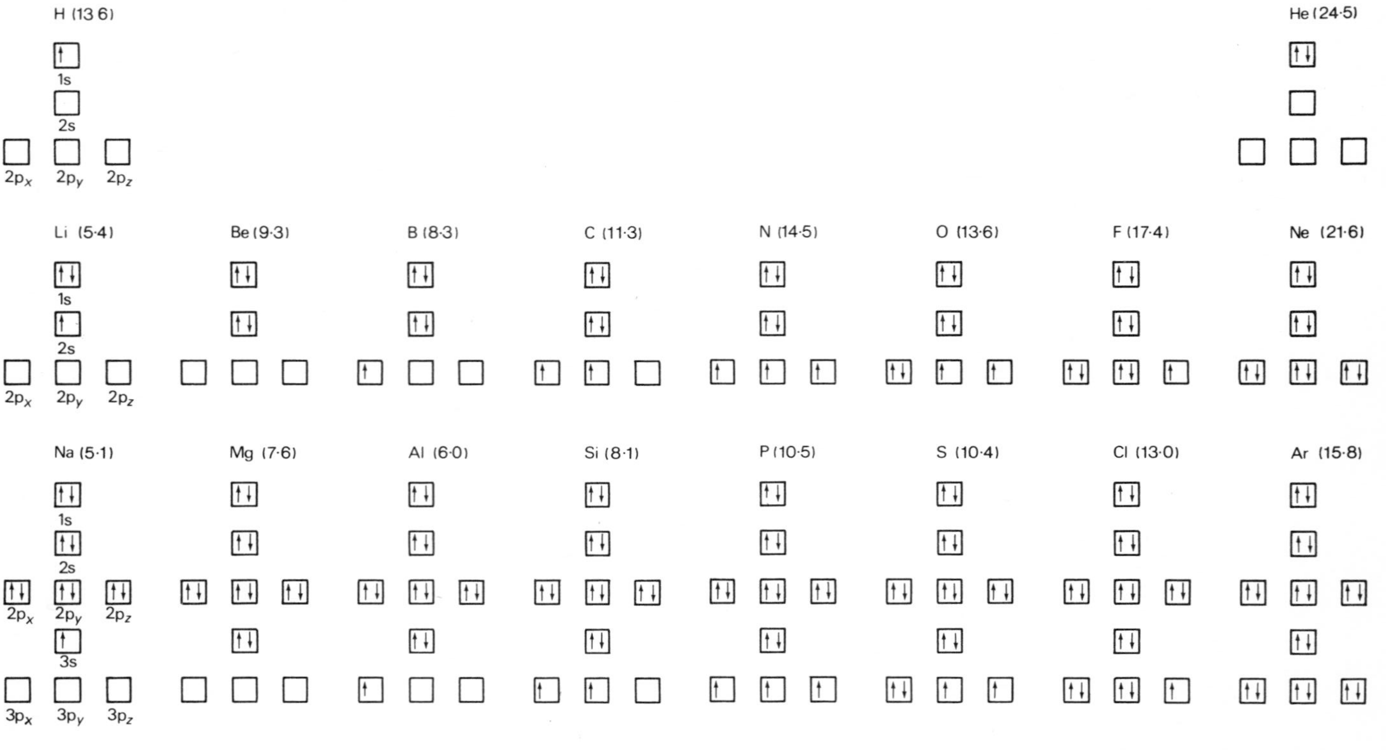


Figure 3.16 – Illustrating the Aufbau principle: ground state electron configurations of some elements with first ionization energies (eV) in parentheses.

With an obvious interpretation, we can write the electronic configurations of atoms in the manner of the following examples:

H $(1s)^1$
He $(1s)^2$
Li $(1s)^2(2s)^1$
—
—
C $(1s)^2(2s)^2(2p)^2$
—
—
—
Ne $(1s)^2(2s)^2(2p)^6$
Na $(1s)^2(2s)^2(2p)^6(3s)^1$ or $(Ne)(3s)^1$
—
—
Si $(Ne)(3s)^2(3p)^2$
—
—
—
Ar $(Ne)(3s)^2(3p)^6$
$Na^+(1s)^2(2s)^2(2p)^6$ or (Ne)

The periodic table is not completed exactly in the manner indicated in Figure 3.16. Between calcium, $(Ar)(4s)^2$, and zinc, $(Ar)(3d)^{10}(4s)^2$, there exists the first series of transition elements. Here, the 3d orbitals† are filled progressively from scandium to zinc. Even this process is not as simple as it seems: for example, atomic copper is written usually as $(Ar)(3d)^{10}(4s)^1$ and not $(Ar)(3d)^9(4s)^2$, as one might at first imagine. Detailed calculations show that the magnetic properties of copper metal are best explained by the distribution $(Ar)(3d)^{9.5}(4s)^{1.5}$.

Figure 3.16 refers to the ground state of atoms. By absorption of energy, electrons may be raised to higher energy levels than those occupied in the ground state. For example, an excited state of carbon, C*, may be represented as $(1s)^2(2s)^1(2p)^3$, in which a 2s electron has been promoted, transiently, to a 2p level; this situation is important in the bonding of carbon atoms, as we shall see later.

## 3.6 MORE THAN ONE ELECTRON

In a species containing two or more electrons, each electron comes under the influence of the potential fields of both the nucleus and the other electrons. The electrostatic interactions in the hydrogen molecule are indicated in Figure 3.17. Following (3.52), we write

$$\mathcal{H}\psi = E\psi \ , \tag{3.95}$$

†Suggested Reading: *Valence Theory* (Coulson or Murrel, Kettle & Tedder).

where $\mathcal{H}$ is given by

$$= -(h^2/8\pi^2 M)(\nabla_A^2 + \nabla_B^2) - (h^2/8\pi^2 m)(\nabla_1^2 + \nabla_2^2)$$

$$-[e^2/(4\pi\epsilon_0)]\ [(1/r_{A1}) + (1/r_{B2}) + (1/r_{A2}) + (1/r_{B1})$$

$$- (1/r_{12}) - (1/r_e)]\ . \quad (3.96)$$

The first term on the right-hand side of (3.96) describes the kinetic energy of the nuclei, each of mass $M$, the second term contains the kinetic energy of the electrons, each of mass $m$, and the third term lists the interparticle attractions. Now $M/m$ is approximately 1836, and we may neglect the term involving $1/M$ in comparison with the other terms. This treatment is known as the Born-Oppenheimer approximation, and may be assumed to apply herinafter. Effectively, it separates the kinetic energy of the nuclei from the electronic energy, and enables one to formulate the electronic energy as a function of $r_e$, the internuclear distance, for any given nuclear configuration. Figure 1.1 illustrates the type of curve so obtained.

It seems reasonable that the electronic energy is the property which will determine the size and shape of a molecule. Hence, we should like to be able to calculate curves such as Figure 1.1, so as to determine both $r_e$ and the dissociation energy, $D_e$, at least for diatomic molecules.

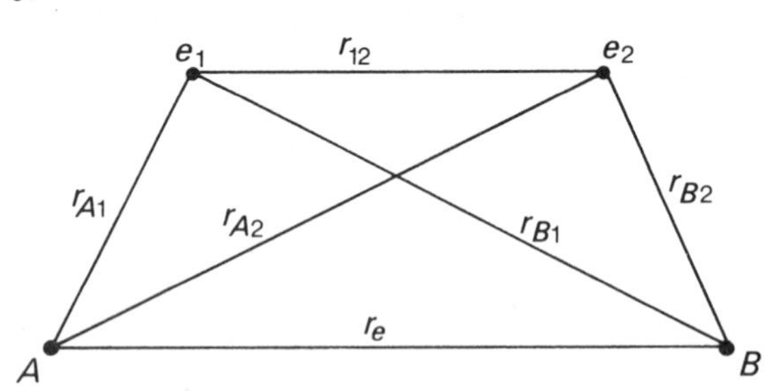


Figure 3.17 – Schematic arrangement of nuclei ($A$, $B$) and electrons ($e_1$, $e_2$) in the hydrogen molecule.

## 3.7 VARIATION METHOD

The wave equation (3.95) cannot be solved exactly for species with two or more electrons. Approximate methods must be employed in order to obtain the configuration of minimum energy for a system; we shall consider the variation method, an adaptation of Rayleigh's (1880) principle.

First we multiply both sides of (3.95) by $\psi^*$, or by $\psi$ if the wave function may be assumed to be real, and integrate over all coordinates, $d\tau$:

$$E = \int\psi\mathcal{H}\psi d\tau / \int\psi^2 d\tau\ . \quad (3.97)$$

We note in passing that we cannot modify (3.95) to read

$$E = \mathcal{H}\psi/\psi \; ; \tag{3.98}$$

$\mathcal{H}\psi/\psi$ is a constant, and not a function of the positional parameters implicit in $E$. A trial function ($\psi_t$) may be adopted for $\psi$, and from (3.97) we write the Rayleigh ratio, $\mathcal{E}$, as

$$\mathcal{E} = \int\psi_t\mathcal{H}\psi_t \, d\tau / \int\psi_t^2 \, d\tau \; . \tag{3.99}$$

In general, we expect that $\mathcal{E}$ will be greater (more positive) than $E$, and we may try other functions in the hope that $\mathcal{E}$ will tend to $E$. More conveniently, we can employ a technique known as the linear combination of atomic orbitals (LCAO).

### 3.7.1 LCAO method

Let a molecular wave function $\Phi$ be assumed to be compounded as

$$\Phi = c_1\psi_1 + c_2\psi_2 + \ldots + c_n\psi_n \; ; \tag{3.100}$$

$\psi_1$ to $\psi_n$ are called the basis set of atomic orbitals and $c_1$ to $c_n$ are variable parameters: $\mathcal{E}$ will be taken as a minimum when $\partial\mathcal{E}/\partial c_1 = 0, \partial\mathcal{E}/\partial c_2 = 0, \ldots, \partial\mathcal{E}/\partial c_n = 0$. The validity of the LCAO technique can be justified by the following argument. For a particular eigenfunction, $\psi_i$, we can write

$$\mathcal{H}\psi_i = \epsilon\psi_i \; , \tag{3.101}$$

where $\epsilon$ is the eigenvalue of $\psi_i$. Let a linear combination be formed:

$$\Phi = \sum_i c_i\psi_i \; . \tag{3.102}$$

Then

$$\mathcal{H}\Phi = \sum_i c_i\mathcal{H}\psi_i = \epsilon\sum_i c_i\psi_i = \epsilon\Phi \; . \tag{3.103}$$

Thus, $\epsilon$ is also the eigenvalue of the linear combination $\Phi$.

### 3.7.2 *H* and *S* integrals

If we substitute (3.100) in (3.99), restricting $n$ to 2, we obtain, since† $\int\psi_1\mathcal{H}\psi_2 d\tau = \int\psi_2\mathcal{H}\psi_1 d\tau$,

$$\mathcal{E} = \frac{c_1^2\int\psi_1\mathcal{H}\psi_1 d\tau + 2c_1c_2\int\psi_1\mathcal{H}\psi_2 d\tau + c_2^2\int\psi_2\mathcal{H}\psi_2 d\tau}{c_1^2\int\psi_1^2 d\tau + 2c_1c_2\int\psi_1\psi_2 d\tau + c_2^2\int\psi_1^2 d\tau} \; . \tag{3.104}$$

†By symmetry, at least for homonuclear diatomic molecules.

Let $H_{ij} = \int \psi_i \mathcal{H} \psi_j \mathrm{d}\tau$ and $S_{ij} = \int \psi_i \psi_j \mathrm{d}\tau$;

then,
$$\mathcal{E} = \frac{c_1^2 H_{11} + 2c_1 c_2 H_{12} + c_2^2 H_{22}}{c_1^2 S_{11} + 2c_1 c_2 S_{12} + c_2^2 S_{22}} \,. \tag{3.105}$$

If generally $\psi_i$ and $\psi_j$ are separately normalized, then $\int \psi_i \psi_j \mathrm{d}\tau \leqslant 1$, the equality sign holding for $i = j$; $S_{ij}$ is called the overlap integral and $H_{ij}$ the coulomb integral, for the functions $\psi_i$ and $\psi_j$. Figure 3.18 shows two 1s orbitals with different degrees of overlap; $S_{ij}$ has a significant value only in the overlap (shaded) region. We shall soon recognise that the bonding strength of two orbitals depends upon the degree of overlap, having regard to the signs of $\psi_i$ and $\psi_j$.

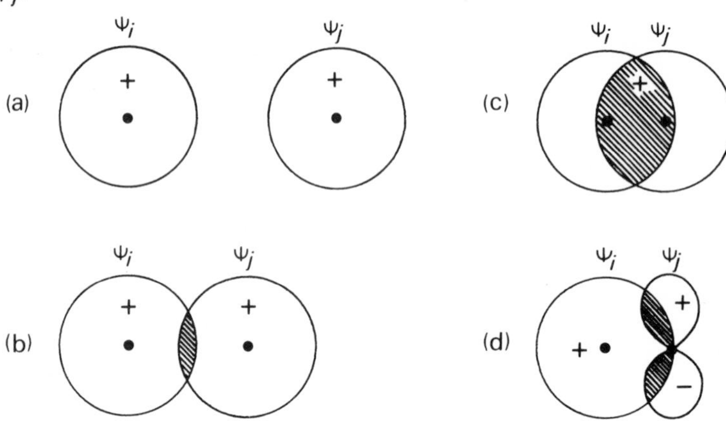


Figure 3.18 – Overlap of atomic orbitals: bonding arises through overlap of regions of the same sign, as in (b) and (c).

In applying the variation principle to (3.105), we need to equate $\partial\mathcal{E}/\partial c_1$ and $\partial\mathcal{E}/\partial c_2$ to zero. Thus, we obtain,

$$\partial\mathcal{E}/\partial c_1 = 0 = \frac{(c_1^2 S_{11} + 2c_1c_2S_{12} + c_2^2S_{22})(2c_1H_{11} + 2c_2H_{12}) - (c_1^2H_{11} + 2c_1c_2H_{12} + c_2^2H_{22})(2c_1S_{11} + 2c_2S_{12})}{(c_1^2S_{11} + 2c_1c_2S_{12} + c_2^2S_{22})^2} \,. \tag{3.106}$$

Rearranging (3.106):

$$(2c_1H_{11} + 2c_2H_{12}) = \frac{(c_1^2H_{11} + 2c_1c_2H_{12} + c_2^2)(2c_1S_{11} + 2c_2S_{12})}{(c_1^2S_{11} + 2c_1c_2S_{12} + c_2^2S_{22})} \,, \tag{3.107}$$

or
$$c_1H_{11} + c_2H_{12} = E(c_1S_{11} + c_2S_{12}) \,. \tag{3.108}$$

Hence
$$c_1(H_{11} - ES_{11}) + c_2(H_{12} - ES_{12}) = 0 \,, \tag{3.109}$$

and, by a similar argument,

$$c_1(H_{12} - ES_{12}) + c_2(H_{22} - ES_{22}) = 0 \ . \tag{3.110}$$

Equations (3.109) and (3.110) are known as the secular equations for the system (Appendix 17): $E$ has been used in place of $\mathcal{E}$, since the correct solution of these equations leads to the true energy.

### 3.7.3 Hydrogen atom again

Since we have already solved the Schrödinger equation for the hydrogen atom, we shall apply the variation principle to the same species. We postulate a trial wave function:

$$\psi = \exp(-\zeta r) \ . \tag{3.111}$$

Since this function is independent of $\theta$ and $\phi$, we can use the radial part of $\nabla^2$ from (3.58) in constructing (3.99). Thus,

$$\partial^2\psi/\partial r^2 + (2/r)\partial\psi/\partial r = (\zeta^2 - 2\zeta/r)\exp(-\zeta r) \ , \tag{3.112}$$

and

$$\mathcal{E} = \frac{\int_0^\infty \{\exp(-\zeta r)[(-h^2/8\pi^2\mu)(\zeta^2 - 2\zeta/r) - e^2/(4\pi\epsilon_0 r)]\exp(-\zeta r)\} r^2 \mathrm{d}r}{\int_0^\infty r^2\exp(-2\zeta r)\mathrm{d}r} \ . \tag{3.113}$$

Proceedings from left to right within the braces:

$$\mathcal{E} = \frac{\int_0^\infty \{\exp(-\zeta r)[(-h^2\zeta^2/8\pi^2\mu)\exp(-\zeta r) + (h^2/8\pi^2\mu)(2\zeta/r)\exp(-\zeta r) - e^2/(4\pi\epsilon_0 r)\exp(-\zeta r)]\, r^2\}\mathrm{d}r}{\int_0^\infty r^2\exp(-2\zeta r)\mathrm{d}r} \ . \tag{3.114}$$

Then,

$$\mathcal{E} = \frac{(-h^2\zeta^2/8\pi^2\mu)\int_0^\infty r^2\exp(-2\zeta r)\mathrm{d}r + (h^2\zeta/4\pi^2\mu)\int_0^\infty r\exp(-2\zeta r)\mathrm{d}r - e^2/(4\pi\epsilon_0 r)\exp(-2\zeta r)\mathrm{d}r}{\int_0^\infty r^2\exp(-2\zeta r)\mathrm{d}r} \ . \tag{3.115}$$

Making use of the results on $\Gamma$-functions (Appendix 16), we obtain

$$\mathcal{E} = \frac{\{(-h^2/64\pi^2\mu\zeta)\Gamma(3) + (h^2/16\pi^2\mu\zeta)\Gamma(2) - (e^2/16\pi\epsilon_0\zeta^2)\Gamma(2)\}}{\Gamma(3)/8\zeta^3} \ . \tag{3.116}$$

Thus,

$$\mathcal{E} = h^2\zeta^2/8\pi^2\mu - e^2\zeta/4\pi\epsilon_0 \ . \tag{3.117}$$

We now differentiate $\mathcal{E}$ with respect to the parameter $\zeta$ and set the derivative equal to zero. Hence,

$$\zeta = 4\pi^2\mu e^2/4\pi\epsilon_0 h^2 \ . \tag{3.118}$$

From (3.84), we see that $\zeta$ is $1/a_0$. Substituting (3.118) in (3.117), the energy minimum is given by

$$E = -\mu e^4/(8n^2h^2\epsilon_0^2) \ . \tag{3.119}$$

Comparison with (3.85) shows that we have obtained the energy for the ground state ($n = 1$) of the hydrogen atom. This result has been obtained because we made a particularly fortunate (or cunning) choice of trial wave function. A one-electron wave function of this type is known as a Slater-type orbital (STO); such orbitals are discussed briefly in Appendix 9.

## 3.8 VALENCE BOND MOLECULAR MODEL

We draw again (Figure 3.19) the potential energy curve for a diatomic molecule, and divide it into three regions. When two atoms are brought together from infinite separation to form a molecule, they pass through a region of separated atoms until $r$ decreases to a bond-forming magnitude in the true molecule region. In the separated-atom region, the two atoms retain their individuality. When they are brought together we can imagine that they retain much of their character, and that the molecular bond is formed through an overlap of atomic orbitals. Once the bond has been formed, electron (1), originating from atom $A$, may be found around atom $B$, and *vice versa*. Since $\psi_A(1)\psi_B(2)$ and $\psi_B(1)\psi_A(2)$ are two possible combinations†, we can write‡ the joint wave function as

$$\Phi = c_1\psi_A(1)\psi_B(2) + c_2\psi_B(1)\psi_A(2) \tag{3.120}$$

†If, as in this case, two wave functions $\psi_A$ and $\psi_B$ are uncorrelated, then the total wave function $\psi$ is given by the product $\psi_A\psi_B$, with an eigenvalue $E$ given by $E_A + E_B$, as the following proof shows. For the separate wavefunctions, $\mathcal{H}_A\psi_A = E_A\psi_A$ and $\mathcal{H}_B\psi_B = E_B\psi_B$: the complete Hamiltonian, $\mathcal{H}$, is, following section 3.6, $\mathcal{H}_A + \mathcal{H}_B$. Then the complete wave equation is $\mathcal{H}\psi = E\psi$, for

$$\mathcal{H}_A\psi_A\psi_B = \psi_B\mathcal{H}_A\psi_A = \psi_B E_A\psi_A = E_A\psi_A\psi_B, \text{ and}$$

$$\mathcal{H}_B\psi_A\psi_B = \psi_A\mathcal{H}_B\psi_B = \psi_A E_B\psi_B = E_B\psi_A\psi_B, \text{ whence}$$

$$\mathcal{H}\psi = (\mathcal{H}_A + \mathcal{H}_B)\psi_A\psi_B = (E_A + E_B)\psi_A\psi_B = E\psi \ .$$

($\mathcal{H}$ is an *operator*, and, in general, $\mathcal{H}\psi \neq \psi\mathcal{H}$; $E$ is a scalar, and $E\psi = \psi E$).

‡Ignoring electron spin; we shall use $\Phi$ and $\chi$ for a complete wave function in, respectively, the valence bond and molecular orbital techniques.

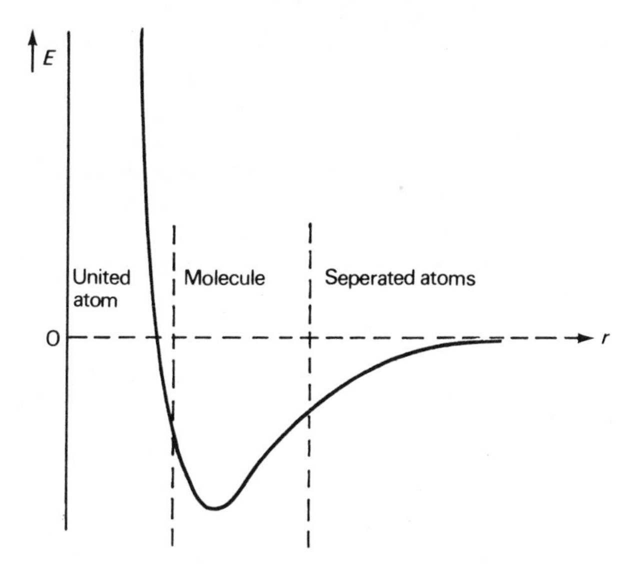


Figure 3.19 – Potential energy curve for a diatomic molecule, showing the regions of the united atom, true molecule and separated atoms.

Because the electrons are equivalent, and since probabilities are proportional to $\psi^2$, we can say that $c_1^2 = c_2^2$, whence

$$\Phi_{\pm} = \psi_A(1)\psi_B(2) \pm \psi_B(1)\psi_A(2) \ . \tag{3.121}$$

Heitler and London used this essentially atomic wave function in the molecule region, and by applying the variation principle to $\Phi_{\pm}$, obtained two energies $E_+$ and $E_-$ (Figure 3.20); $\Phi_+$ leads to bond formation, whereas $\Phi_-$ does not. The true curve lies below $E_+$, showing that $E_+$ is an approximation to the true energy. The calculated value of $r_e$ is 0.87 Å (experimental, 0.74 Å) and $D_e$ is −303 kJ mol$^{-1}$ (experimental, −458 kJ mol$^{-1}$); In this treatment, $\Phi$ is called a valence-bond (VB) wave function. Pictorially, electron (1) is paired with electron (2), the combination forming an electron-pair bond. We see the electron-sharing concept of Lewis in evidence but with a firmer and more quantitative foundation. Indeed, from (3.121) it is clear that electron (1) is sometimes around nucleus $A$ and other times around nucleus $B$, indicated by $\psi_A(1)$ and $\psi_B(1)$, respectively. Each of the two terms on the right-hand side of (3.121), considered alone, would not lead to a curve indicating a significant degree of bonding.

Do we consider that the agreement between $E_+$ and $E_{\text{true}}$ (Figure 3.20) is good? Our judgment must be related to the closeness of the calculated energy to the true value, having regard to the amount of effort put into the calculation. We could experiment further with our VB wave function. It is possible that both electrons could be around nucleus $A$ or nucleus $B$. With an obvious notation, we may write

$$\Phi_+ = \psi_A(1)\psi_B(2) + \psi_B(1)\psi_A(2) + \lambda[\psi_A(1)\psi_A(2) + \psi_B(1)\psi_B(2)]$$
$$= \psi_{\text{covalent}} + \lambda\psi_{\text{ionic}} \tag{3.122}$$

The compounding of wave functions of similar energy into a single wave function is often called resonance. Thus, (3.122) expresses a resonance† between the covalent and ionic wave functions. Detailed calculation shows that $\lambda \approx 1/6$ and that $D_e = 388$ kJ mol$^{-1}$.

By using a total wave function of fifty terms, a value for $D_e$ of $-457.99$ kJ mol$^{-1}$ has been obtained (best experimental, $-457.99 \pm 0.07$ kJ mol$^{-1}$). We may say that the approximation (3.121) is satisfactory for our purposes: it leads to a well defined minimum in the potential energy curve, and accounts for about two thirds of the bonding energy. We must not think that (3.121) represents a situation in which the electrons (1) and (2) have exchanged places; electrons are indistinguishable and cannot be enumerated. We shall take (3.121) as a first approximation to the correct VB wave function for the hydrogen molecule. The true function will be very like (3.121) but with more sophistication. We may note in passing that the term $H_{ij}$ in (3.105) is sometimes called the exchange integral, so we must be careful of too literal an interpretation of our simple wave function.

An equivalent approach to bond formation is through the molecular orbital method. It is of more general applicability to chemical problems, and forms a logical extension of the atomic orbital treatment for atoms.

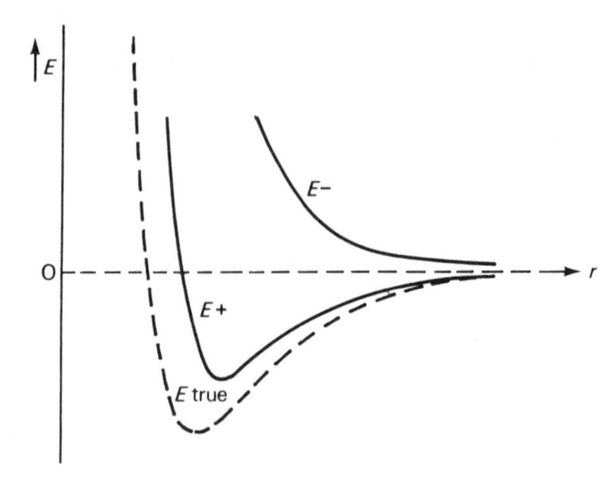


Figure 3.20 – Valence bond energies for $H_2$; $E_+$, $E_-$ and the ground state energy, $E_{\text{true}}$, are shown as a function of $r$.

†Resonance should not be interpreted in terms of an equilibrium mixture of two or more configurations, but rather as a quantum mechanical mixing process (see also section 3.9.8).

## 3.9 MOLECULAR ORBITAL MODEL

We shall introduce this molecular model with reference first to the hydrogen molecule ion $H_2^+$, which plays the same role for molecules as the hydrogen atom does for atoms. The wave equation for the single electron in $H_2^+$ can be solved within the Born-Oppenheimer approximation, and we shall study it through the LCAO technique.

The lowest energy configuration of a hydrogen atom is 1s, and we shall assume that the ground state MO of $H_2^+$ is somewhat similar. We let $\psi_A$ and $\psi_B$ represent the electron in the neighbourhoods of nuclei $A$ and $B$, respectively, and write the LCAO approximation, due originally to Hund and Mulliken, as

$$\chi = c_1 \psi_A + c_2 \psi_B \; . \tag{3.123}$$

The parameters $c_1$ and $c_2$ may be found formally by the variation method, but from symmetry, $\psi_A$ and $\psi_B$ must have equal weight, and, as in the VB treatment, $c_1^2 = c_2^2$. Hence, we obtain

$$\chi_\pm = \psi_A \pm \psi_B \; . \tag{3.124}$$

Following section 3.7.2 and Appendix 17 we have the secular equations,

$$(H_{11} - ES_{11}) + (H_{12} - ES_{12}) = 0 \tag{3.125}$$

and

$$(H_{12} - ES_{12}) + (H_{22} - ES_{22}) = 0 \; . \tag{3.126}$$

It is common to refer to $H_{ii}$ as a coulomb integral† $\alpha$, and to $H_{ij}$ as a resonance integral $\beta$. If the atomic orbitals are normalized, $S_{ii} = 1$, and calling $S_{12}$ $S$, we have

$$E = \frac{\alpha \pm \beta}{1 \pm S} \tag{3.127}$$

The result for $D_e$ is similar to that obtained by the VB method. If we consider $S$ to be small (not always a good approximation), then $E_\pm = \alpha \pm \beta$, and the difference between the two energy levels is $2\beta$ (Figure 3.21); $\chi_+$ is a bonding MO and $\chi_-$ is an antibonding MO. We can normalize $\chi$, following (3.34):

$$N_\pm^2 \int (\psi_A \pm \psi_B)^2 \mathrm{d}\tau = 1 \; . \tag{3.128}$$

Expanding:
$$1/N_\pm^2 = \int \psi_A^2 \mathrm{d}\tau + \int \psi_B^2 \mathrm{d}\tau \pm 2\int \psi_A \psi_B \mathrm{d}\tau \; , \tag{3.129}$$

whence
$$N_\pm = 1/[2(1 \pm S)]^{1/2} \; . \tag{3.130}$$

†This use of $\alpha$ and $\beta$ is Hückel's notation.

By comparison, the normalizing constant for (3.121) is $1/[2(1 \pm S^2)]^{1/2}$.

If we take next $H_2$, a similar scheme obtains. The energies of the MO's are again $\alpha \pm \beta$: now there are two electrons in the bonding MO, and so the total bonding energy would seem to be $2(\alpha + \beta)$. However, we have neglected inter-electronic repulsion terms, so that the calculated energy will be numerically too small if we use the same values for $\alpha$ and $\beta$ as in the hydrogen molecule ion, $[H_2]^+$. More detailed calculations are needed to take proper account of inter-electronic repulsion, but they lie outside the scope of this book‡.

We may note that the value of $2(\alpha \pm \beta)$ for $H_2$ is not, in fact, seriously in error (30 eV for $H_2^+$ and 53 eV for $H_2$), but this agreement must be regarded as fortuitous. We shall return to this discussion again in considering $\pi$-electron bonding.

The $H_2$ molecule can be approached from the united atom viewpoint. Consider again Figure 3.19; a united atom for a hydrogen molecule is helium. The two valence electrons are in a 1s orbital so that their spins are anti-parallel, by the Pauli principle. Imagine that the nucleus be divided into two halves: the two nuclei so formed will then repel each other, thus elongating the orbital in the direction of the molecular axis. The bond is determined by a pair of electrons, with opposite spins, encompassed by a molecular orbital (MO). Figure 3.22 illustrates the conceptual sequence of the events just described, and compares them with those for the VB method. Again, we are led to an LCAO description, like (3.123) for $H_2$.

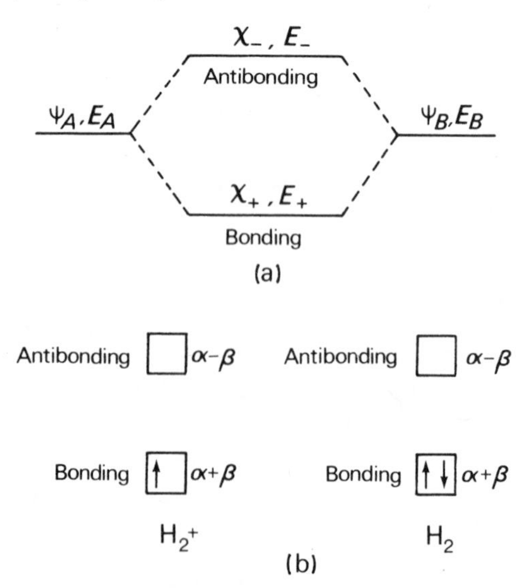


Figure 3.21 – $H_2^+$ and $H_2$; (a) Relative energy levels for bonding and antibonding MO's in $H_2^+$; (b) $H_2^+$ and $H_2$ compared.

‡Suggested Reading: *Quantum Mechanics* (Slater) or *Valence Theory* (Murrell, Kettle and Tedder).

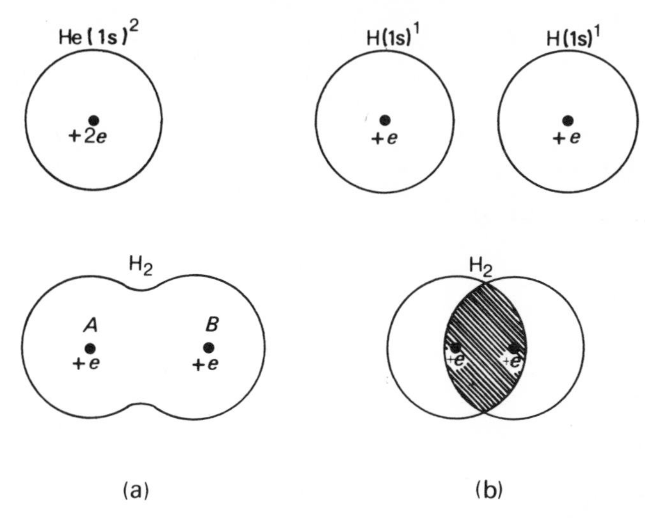


Figure 3.22 – Models for the $H_2$ molecule: (a) United atom He → $H_2$ (bi-centric MO); (b) Valence bond H + $H_2$ → $H_2$ (mono-centric AO's); the envelope of the two AO's would be very similar to that of the MO in (a).

### 3.9.1 Charge distribution in $H_2$

In the VB approximation, the electron density at any point is given, without proof here,

by
$$\rho_\pm = \psi_A^2 + \psi_B^2 \pm \frac{S}{1 \pm S^2}\Delta , \tag{3.131}$$

where $\Delta = 2\psi_A\psi_B - S(\psi_A^2 + \psi_B^2)$ and $S$ is the overlap integral. Taking $\psi = (\pi a_0^3)^{-1/2}\exp(-r/a_0)$ and multiplying by $e$, we obtain

$$\rho_\pm = \frac{e}{\pi a_0^3(1 \pm S^2)}\{\exp[-2r_A/a_0] + \exp[-2r_B/a_0] \pm 2S\exp[(-r_A + r_B)/a_0]\} . \tag{3.132}$$

The sum of the first two terms in braces is the superposition of two separate atoms. The third term is involved in the bond formation, and represents a distribution of density between the atoms in the overlap region. Figure 3.23 illustrates the several aspects of (3.132).

The MO approach is somewhat simpler. The normalized molecular orbital is, from (3.124) and (3.130),

$$\chi_\pm = (\psi_A \pm \psi_B)/[2(1 \pm S)]^{1/2} . \tag{3.133}$$

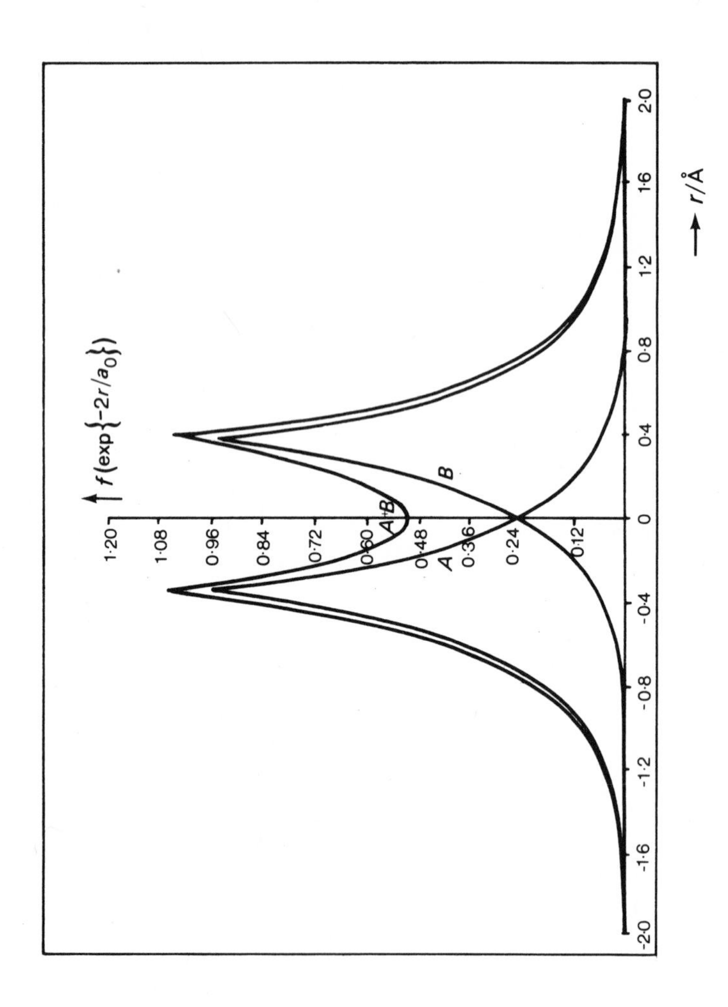
f(exp{−2r/a0})
1·20
1·08
0·96
0·84
0·72
0·60
0·48
0·36
0·24
0·12
A+B
A
B
-2·0
-1·6
-1·2
-0·8
-0·4
0
0·4
0·8
1·2
1·6
2·0
r/Å

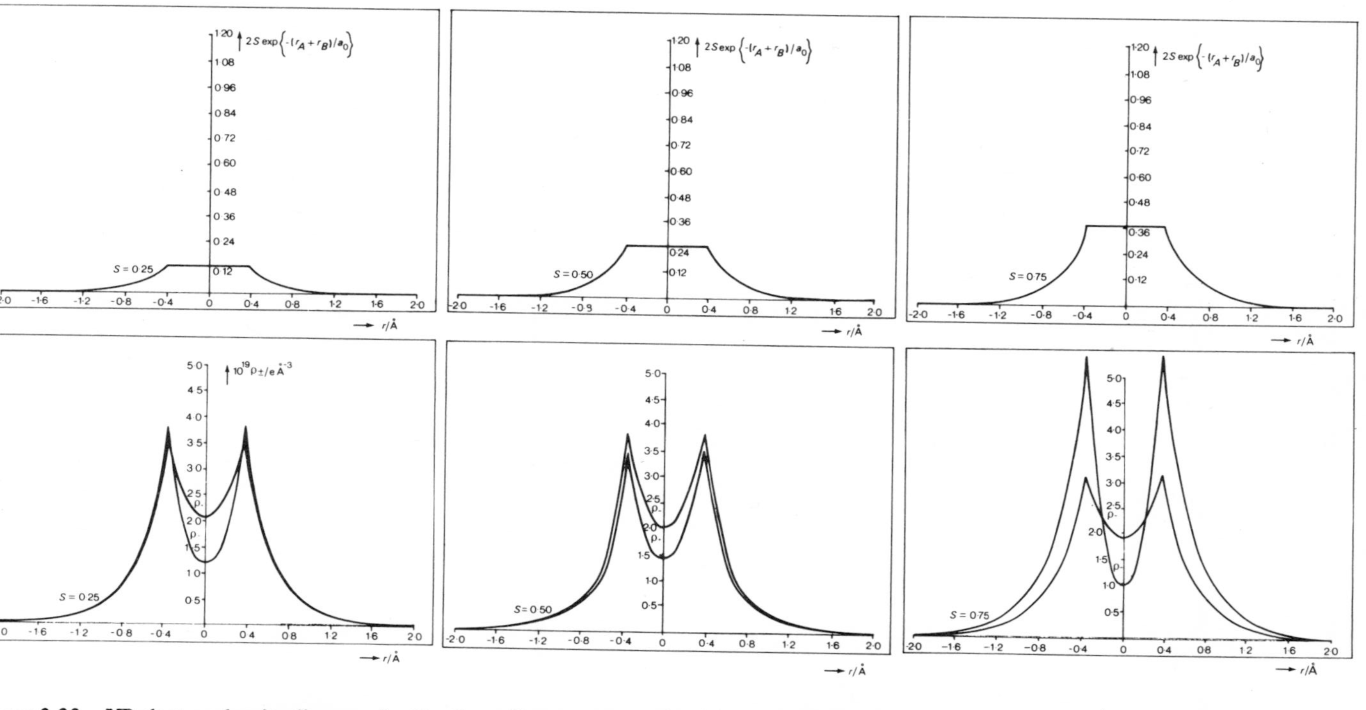


Figure 3.23 – VB electron density diagrams for $H_2$: the ordinates measure densities, and the abscissae measure distances ($r$) from the nuclei $A$ and $B$: (a) Separate atoms, $A$ and $B$, and their superposition $A + B$; no dependence on $S$; (b) Density from the overlap term, for $S = 0.25, 0.50$ and $0.75$; $r_A + r_B$ is constant between the two nuclei; (c) Complete electron density; superposition of (a) $(A+B)$ and (b).

The electron density, $\rho(A)$, around nucleus $A$ is therefore,

$$\rho_{\pm}(A) = (\psi_A^2 + \psi_B^2 \pm 2\psi_A\psi_B)/[2(1 \pm S)] \ . \qquad (3.134)$$

Since $\rho(A) = \rho(B)$, the total density $\rho_{\pm}$ is just $2\rho_{\pm}(A)$:

$$\rho_{\pm} = (\psi_A^2 + \psi_B^2 \pm 2\psi_A\psi_B)/(1 \pm S) \ . \qquad (3.135)$$

Writing this equation as

$$\rho_{+} = \psi_A^2 + \psi_B^2 + \frac{1}{1+S}\Delta \ , \qquad (3.136)$$

where $\Delta = 2\psi_A\psi_B - S(\psi_A^2 + \psi_B^2)$, shows that $\rho_+$ is given by the superposition of two separate atoms, $(\psi_A^2 + \psi_B^2)$, together with the density along the bond, sometimes called the difference density, since it is the density remaining after the atomic densities around the two nuclei have been subtracted from the molecular electron density. Figure 3.24 is a difference density diagram for $H_2$. Equation (3.136) differs from (3.131) only in the form of the overlap function; the plot of the important density function, $\rho_+$, is quite similar.

### 3.9.2 Bonding, antibonding and symmetry

Equation (3.124) was developed without reference to the choice of atomic orbital. Thus, we can combine an AO of an atom $A$ with another AO of an atom $B$ and obtain a bonding and an antibonding molecular orbital, as illustrated in Figure 3.22.

Homonuclear diatomic molecules exhibit symmetry, and this symmetry is seen also in their orbitals, thus permitting a useful classification of molecular orbitals:

(a) Orbitals which are symmetrical about the molecular axis are called $\sigma$-type MO's (Figure 3.25a–d);

(b) Orbitals which have a nodal plane containing the molecular axis are called $\pi$-type MO's (Figure 3.25e,f);

(c) Orbitals may be described as even, $g$, (Ger. *gerade*) or as odd, $u$, (Ger. *ungerade*) with respect to inversion across the centre of the molecule (Figure 3.25a–f).

It may be noted the the $\sigma_g$ and $\pi_u$ MO's are bonding whereas the $\sigma_u$ and $\pi_g$ MO's are antibonding.

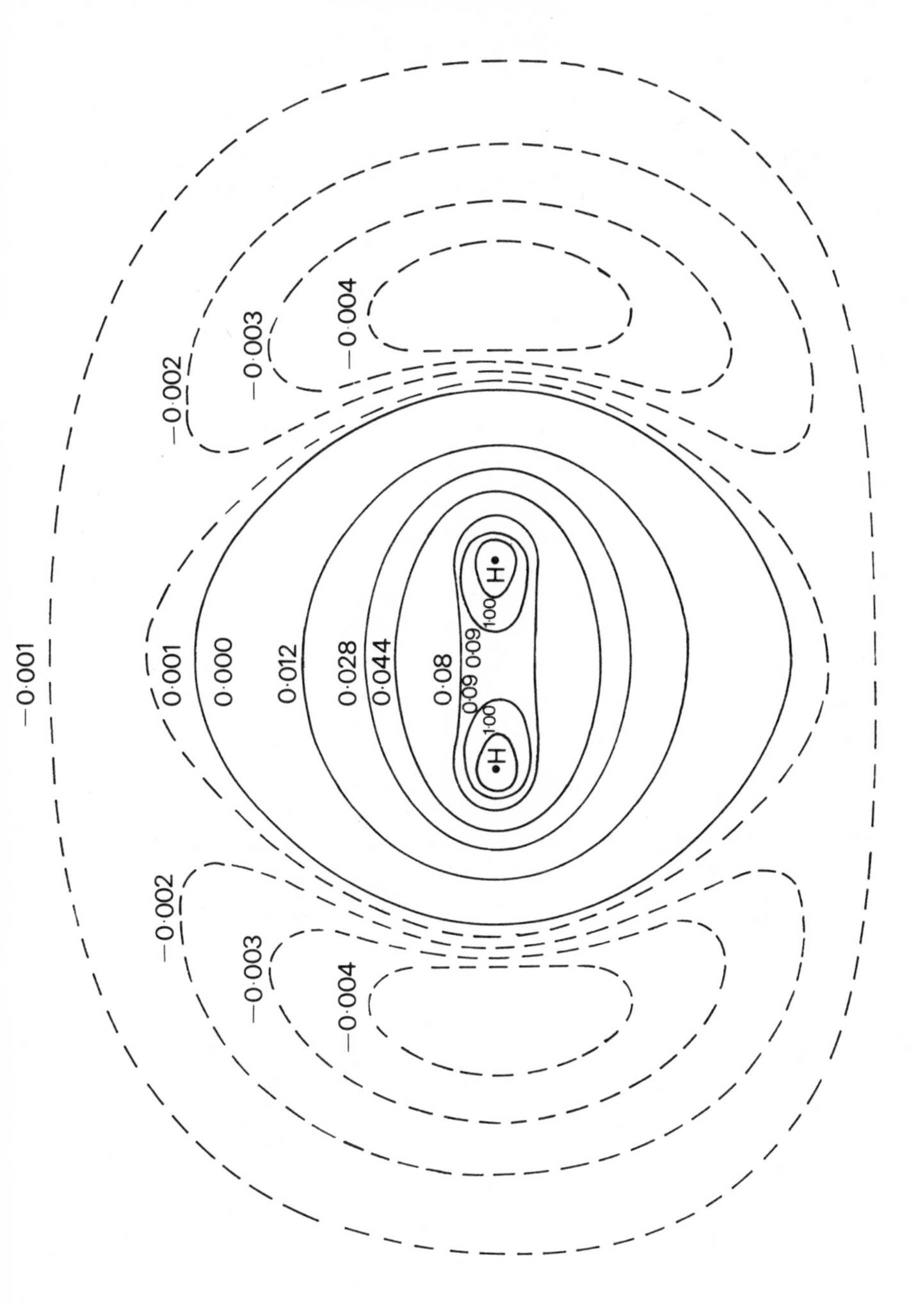


Figure 3.24 – Electron density difference map between the $H_2$ molecule and two H atoms; contours in electron units (after Bader and Henneker, and reproduced by permission of the American Chemical Society).

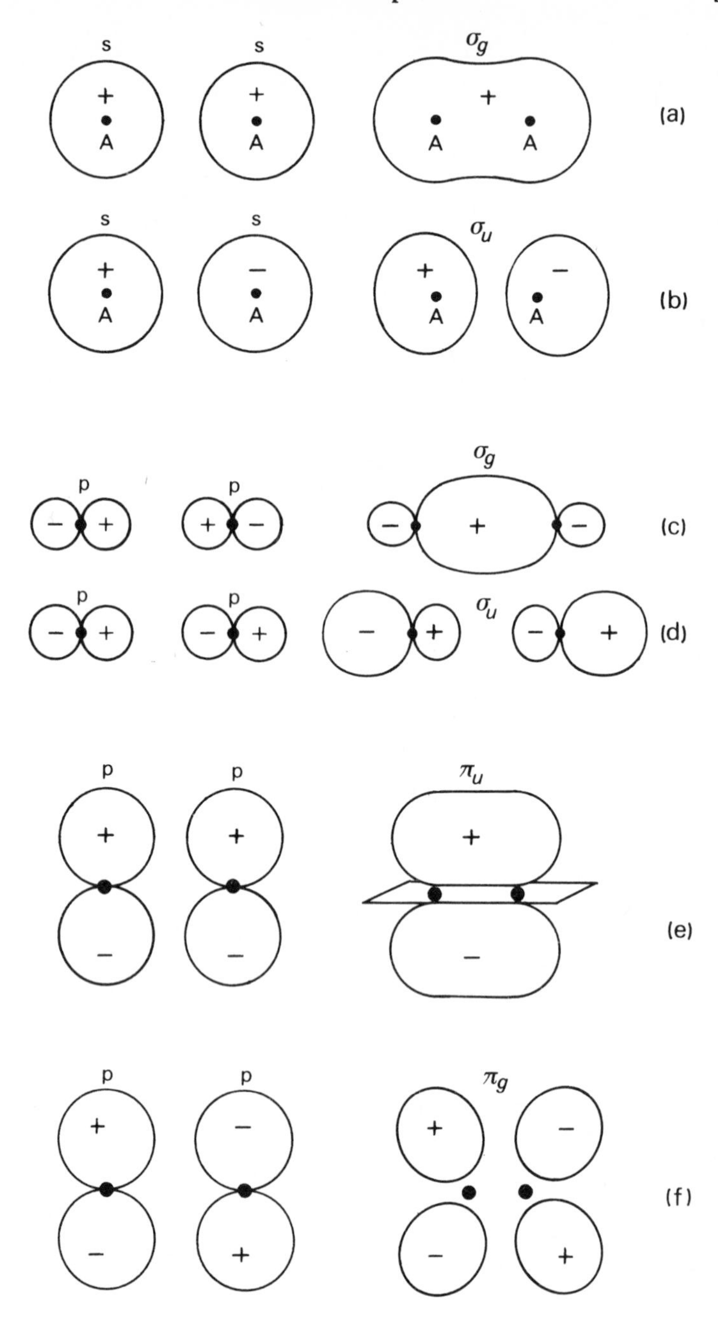


Figure 3.25 – Schematic diagrams of MO's ($\chi$-functions) of differing types and symmetries formed by linear combinations of AO's ($\psi$ functions).

### 3.9.3 The spin factor

Wave functions such as (3.121) and (3.124) treat only the spatial part of the complete wave function, which itself comprises the product of space and spin terms. The functions $\Phi_+$ and $\chi_+$ are symmetrical with respect to the interchange of electrons, and the Pauli principle tells us that the complete wave function is antisymmetric; hence the antisymmetry must reside in the spin term.

There are two possible spin modes, usually called $\alpha$ and $\beta$. We have seen how the wave function $\Phi_+$ in (3.121) leads to a bonding situation, in the lower energy state. Furthermore, the + sign leads to a single energy state (singlet), and the − sign leads to a triple energy state (triplet), as explained through Table 3.5. We can begin our analysis of spin by considering that electron (1) has spin $\alpha$, and that electron (2) has spin $\beta$; the spin wave function will be designated $\alpha(1)\beta(2)$. However, we must consider also the function $\alpha(2)\beta(1)$, and, since $\alpha(1)\beta(2)$ and $\alpha(2)\beta(1)$ are degenerate, the linear combinations $\alpha(1)\beta(2) \pm \beta(2)\alpha(1)$ are also permissible functions. The complete space and spin functions are summarized in Table 3.5. If we interchange the descriptors (1) and (2) throughout *all* the functions, we find that those involving spin terms (a) are multiplied by −1 (antisymmetric), whereas those involving spin terms (b) are unchanged (symmetric).

**Table 3.5 – Total valence bond wave function for** $H_2$

| Space terms | | Spin Term (a) | | Spin terms (b) |
|---|---|---|---|---|
| $\psi_A(1)\psi_B(2) + \psi_B(1)\psi_A(2)$ | X | $\alpha(1)\beta(2) - \beta(1)\alpha(2)$ | OR | $\alpha(1)\alpha(2)$<br>$\beta(1)\beta(2)$<br>$\alpha(1)\beta(2) + \beta(1)\alpha(2)$ |
| $\psi_A(1)\psi_B(2) - \psi_B(1)\psi_A(2)$ | X | $\alpha(1)\alpha(2)$<br>$\beta(1)\beta(2)$<br>$\alpha(1)\beta(2) + \beta(1)\alpha(2)$ | OR | $\alpha(1)\beta(2) - \beta(1)\alpha(2)$ |

Heisenberg (1926) showed that only antisymmetrical wave functions for electrons explained the spectra of the so-called *ortho*helium (triplet) and *para*-helium (singlet) species of this element. Thus, we reject the spin terms (b), and the two complete wave functions are, in order, singlet and triplet. The bond-forming wave function is therefore

$$\Phi_+ = [\psi_A(1)\psi_B(2) + \psi_B(1)\psi_A(2)]\ [\alpha(1)\beta(2) - \beta(1)\alpha(2)]\ . \quad (3.137)$$

The spin term corresponds to a total spin of zero: the spins are antiparallel, or paired, and again we note the relationship of our model to the Lewis electron-pair bond. Spin arises for MO's in a similar manner, but while it is necessary for us to know of spin and its effect on the wave function, we shall not develop it further.

### 3.9.4 Notation for MO's

The two MO's in (3.124) are labelled $1\sigma_g$ and $1\sigma_u$, and the ground state for $H_2$ is $(1\sigma_g)^2$. The most usual sequence of energies for homonuclear diatomic molecules is as follows; the molecular axis is taken in the $z$ direction.

$$1\sigma_g < 1\sigma_u < 2\sigma_g < 2\sigma_u < 1\pi_{xu} = 1\pi_{yu} < 3\sigma_g < 1\pi_{xg} = 1\pi_{yg} < 3\sigma_u < \ldots$$

Sometimes, however, the $3\sigma_g$ MO lies below the $1\pi_u$ degenerate pair with respect to energy. Figure 3.26 is a schematic representation of the molecular orbitals for a molecule $A_2$.

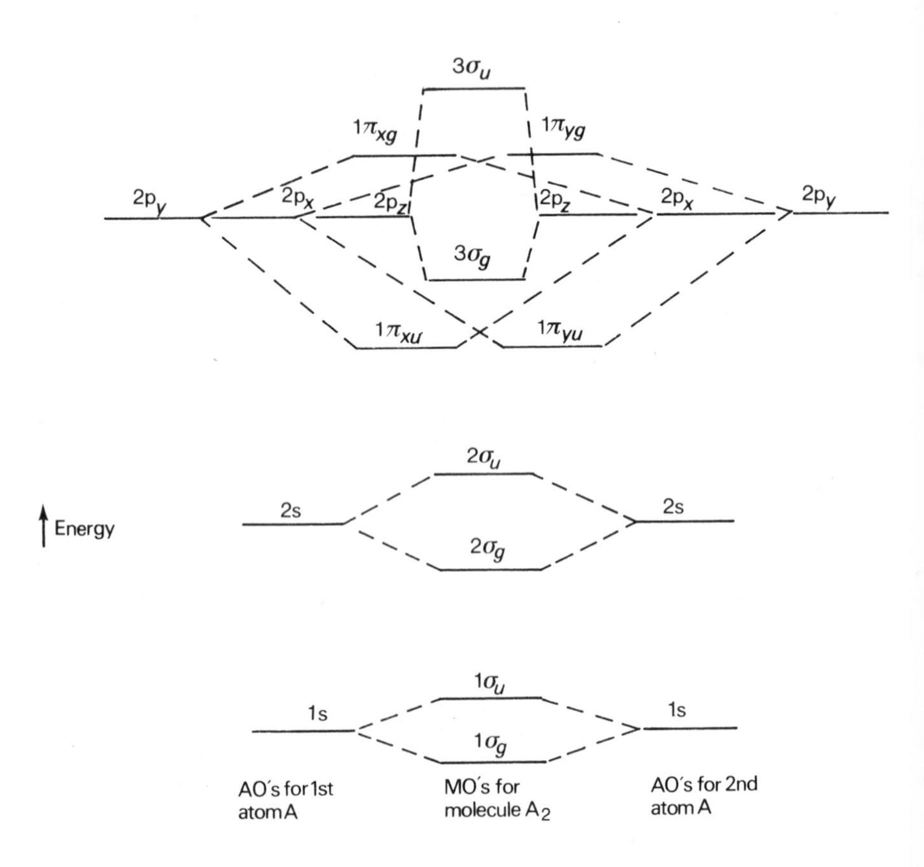


Figure 3.26 – MO's for a homonuclear diatomic molecule.

### 3.9.5 Some homonuclear diatomics: VB and MO models compared

Now that we have defined the order of energies for MO's, we can apply the *aufbau* principle to the problem of writing down the electronic configurations of simple diatomic molecules: we shall use this exercise to compare the VB and MO models.

Di-hydrogen, $H_2$ [H $(1s)^1$]

VB A quantum mechanical combination, or resonance hybrid, of the canonical forms

$$\text{H–H} \quad \text{H}^+\ \text{H}^- \quad \text{H}^-\ \text{H}^+$$

is postulated, with equal weights for the ionic forms so as to preserve a non-polar nature for the molecule but relatively much greater weight for the covalent form. Canonical forms are structures of similar energy, and quantum mechanical 'mixing', or resonance, leads to a structure more stable than any one of them.

MO Two electrons with paired spins occupy a $1\sigma_g$ molecular orbital.

Di-helium, $He_2$ [He $(1s)^2$]

VB No electrons from the two atoms are available for pairing, and so no canonical forms can be described. Hence, there is no resonance energy to overcome the repulsion between the complete (closed) shells of the helium atoms, and no bond is formed.

MO $He_2$ would be written $(1\sigma_g)^2(1\sigma_u)^2$. The combination of a bonding MO and the corresponding antibonding, MO leads to overall antibonding, (Figure 3.26) and so $He_2$ is not a stable entity.

Di-lithium, $Li_2$ [Li $(1s)^2(2s)^1$]

VB Li–Li, $Li^+Li^-$ and $Li^-Li^+$ are canonical forms. The 1s electrons are localized in atomic orbitals, but the 2s electrons form a bond pair.

MO $(1\sigma_g)^2(1\sigma_u)^2(2\sigma_g)^2$: the bonding arises from $(2\sigma_g)^2$.

Di-nitrogen, $N_2$ [N$(1s)^2(2s)^2(2p_x)^1(2p_y)^1(2p_z)^1$]

VB Three unpaired electrons reside in each nitrogen atom and, following Hund's rule, the spins are parallel. Three electron-pair bonds, one $\sigma$ and two $\pi$, are formed.

MO $(1\sigma_g)^2(1\sigma_u)^2(2\sigma_g)^2(2\sigma_u)^2(1\pi_{xu})^2(1\pi_{yu})^2(3\sigma_g)^2$. Again, we have the triple bond, one $\sigma$-type and two $\pi$-type.

Di-oxygen, $O_2$ [O $(1s)^2(2s)^2(2p_x)^2(2p_y)^1(2p_z)^1$]

VB The expectation of a simple double bond is precluded by the fact that

$O_2$ is paramagnetic, an effect which depends on the presence of unpaired electrons. The VB explanation is given in terms of one $\sigma$ bond and two 3-electron bonds:

$$O \overset{\cdot\cdot\cdot}{\underset{\cdot\cdot\cdot}{—}} O$$

MO Following Hund's rules, we assign one electron to each of the degenerate $1\pi$ pair of MO's, giving the configuration $(N_2)\ (1\pi_{xg})^1\,(1\pi_{yg})^1$ and explaining neatly the paramagnetic property of $O_2$.

If we let unpaired electron 1, at a point $\mathbf{r}_1$, occupy an orbital $\psi_a$, and electron 2 at $\mathbf{r}_2$ occupy an orbital $\psi_b$, then we would write the wave function as $\psi_a(\mathbf{r}_1)\psi_b(\mathbf{r}_2) - \psi_b(\mathbf{r}_1)\psi_a(\mathbf{r}_2)$. As the electrons approach each other, $\mathbf{r}_1$ tends to $\mathbf{r}_2$, and when $\mathbf{r}_1 = \mathbf{r}_2$ the wave function would vanish. The property of electrons with parallel spins of avoiding each other (Hund's rule) is called spin correlation.

In oxygen, the magnetic moment which corresponds to two unpaired electrons has the value of 2.8 $\mu_B$, where $\mu_B$ is the Bohr magneton.

Di-fluorine, $F_2$ [F $(1s)^2(2s)^2(2p_x)^2(2p_y)^2(2p_z)^1$]

VB A single electron-pair bond is formed, F—F. The ionic canonical forms are not significant because of the high ionization energy of fluorine, $F \longrightarrow F^+$ 1681 kJ mol$^{-1}$ (cp. Li, 520 and H, 1312 kJ mol$^{-1}$).

MO $(N_2)(1\pi_{xg})^2(1\pi_{yg})^2$, a stable molecule.

Di-neon, $Ne_2$ [Ne $(1s)^2(2s)^2(2p_x)^2(2p_y)^2(2p_z)^2$]

VB The second electron shell is complete, and there are no electrons available for bonding.

MO $(N_2)\ (1\pi_{xg})^2(1\pi_{yg})^2(3\sigma_u)^2$: reference to Figure 3.26 shows that the dominant antibonding orbital $(3\sigma_u)^2$ ensures that $Ne_2$ is not formed.

In this short section we have shown that the VB and MO models both give simple accounts of homonuclear diatomic molecules. Sometimes, as in the MO model of $O_2$, one picture is, perhaps, more easily accepted than the other.

### 3.9.6 Heteronuclear diatomic molecules

In this section, we shall consider the hydrides of the halogens and of some other elements. Taking hydrogen fluoride first, the VB picture is given in terms of the overlap of the 1s orbital of hydrogen with the $2p_z$ orbital of fluorine (molecular axis along $z$), as indicated by Figure 3.27. The canonical structures can be written as

$$H - F \qquad H^+\ F^- \qquad H^-\ F^+\ ,$$

and the bond-forming VB wave function is

$$\Phi_+ = \psi_{\text{covalent}} + \lambda_1 \psi_{\text{ionic}} + \lambda_2 \psi_{\text{ionic}} \ . \tag{3.138}$$

Calculation shows that $\lambda_2$ is negligible, as we have already intimated (see also section 3.9.7.); hence,

$$\Phi_+ = \psi_{\text{covalent}} + \lambda \psi_{\text{ionic}} \ , \tag{3.139}$$

and $\lambda$ is chosen so as to minimise the Rayleigh ratio (3.99).

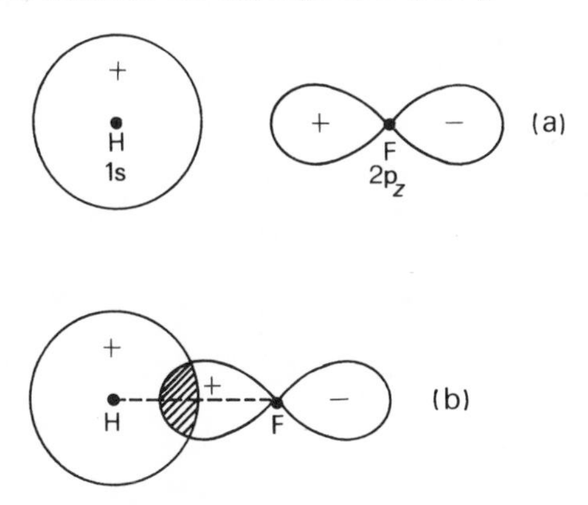


Figure 3.27 – Bond-forming AO's for HF: (a) separate AO's; (b) overlapped AO's.

We can make an approximate calculation of $\lambda$ by the following method. Let us assume that the dipole moment of the HF molecule arises through the two bonding electrons, that is, it is associated with $\psi_{\text{ionic}}$. The dipole moment, $p$, is given by (see Figure 3.28)

$$p = qed \ , \tag{3.140}$$

where $q$ is the excess fractional electronic charge considered to reside on each of the two atoms separated by a distance $d$. From experiments, $p$ is $6.07 \times 10^{-30}$ C m; since $d = 0.927$ Å, $q$ is 0.41. From (1.139), remembering that electron density is approximately proportional to the square of the wave function, we may take the covalent and ionic contributions to bonding in the ratio of $1{:}\lambda^2$. Thus, the fractional ionic character is given by $q$, and

$$q = \lambda^2/(1 + \lambda^2) \tag{3.141}$$

Hence in HF, $\lambda = 0.83$; Table 3.6 lists data for the halogen hydrides.

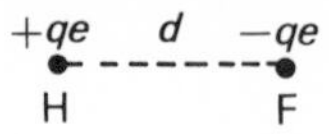

+qe d −qe
H F

Figure 3.28 – Dipole diagram for HF; $p = qed$.

**Table 3.6 – Ionic character of the halogen hydrides**

| | HF | HCl | HBr | HI |
|---|---|---|---|---|
| $10^{30}p$/C m | 6.07 | 3.44 | 2.77 | 1.50 |
| $10^{10}d$/m | 0.927 | 1.27 | 1.41 | 1.61 |
| $q$ | 0.41 | 0.17 | 0.12 | 0.058 |
| $\lambda$ | 0.83 | 0.45 | 0.37 | 0.25 |

From the table, we see that as the periodic group is descended from F to I, the ionic contribution becomes progressively smaller. We may say that the electronegativities (see 3.9.7) of the halogens decrease in the order F > Cl > Br > I.

The molecular orbital bonding wave function for HF may be written as

$$\chi_+ = c_1\psi_A + c_2\psi_B \ . \qquad (3.142)$$

The ratio $c_2^2/c_1^2$ determines the extent to which the electron density is transferred from atom $A$ to atom $B$. If $A$ is hydrogen and $B$ fluorine, the transfer is very large. Figure 3.29 shows the electron density contours for a number of heteronuclear diatomic species. We may note that in LiH the dipole is $Li^{q^+}H^{q^-}$, whereas in HF it is $H^{q^+}F^{q^-}$. It is clear that the MO for each species becomes fatter around the non-hydrogen species along the sequence LiH to HF. The electronegativities of the atoms increase along the periodic row Li to F. Alternatively we can say, in VB terms, that the magnitude of $\lambda$ increases in the same order.

The molecular orbitals of HF are shown schematically in Figure 3.30. The 1s AO of hydrogen is similar in energy to the 2p AO's of fluorine. If the molecular axis is taken along the $z$ direction, the bond arises from the MO wave function

$$\chi_+ = c_1\psi[\mathrm{H(1s)^1}] + c_2\psi[\mathrm{F(2p_z)^1}] \ . \qquad (3.143)$$

The bonding MO is $3\sigma$, and a normal single bond is obtained. The $2p_x$ and $2p_y$ AO's have $\pi$-symmetry; they remain unchanged, and contribute non-bonding, or lone-pair, electrons to the molecule. The ground state of HF can be written as

$$(1\sigma)^2(2\sigma)^2(3\sigma)^2(1\pi_x)^2(1\pi_y)^2 \ . \qquad (3.144)$$

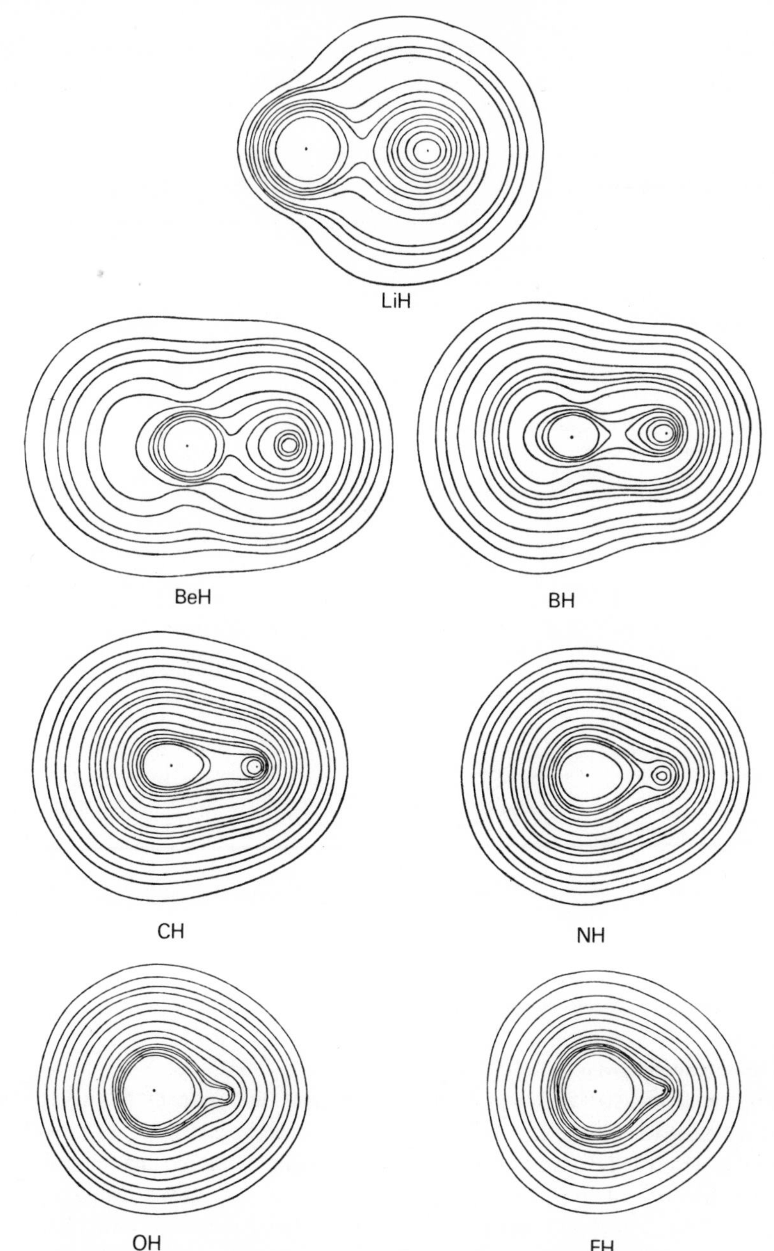


Figure 3.29 – Electron density contours for first-row diatomic hydrides ($A$H). All maps are drawn to the same linear scale. The nucleus $A$ is on the left in each map, and its innermost contours have been omitted for the sake of clarity (after Bader, Keaveney and Cade, and reproduced by permission of the American Institute of Physics.

We should note that, since the two atoms are dissimilar, we cannot use the *u,g* notation of Figure 3.26, because the molecule has no centre of symmetry. The value of $c_2^2/c_1^2$ is approximately 3.2.

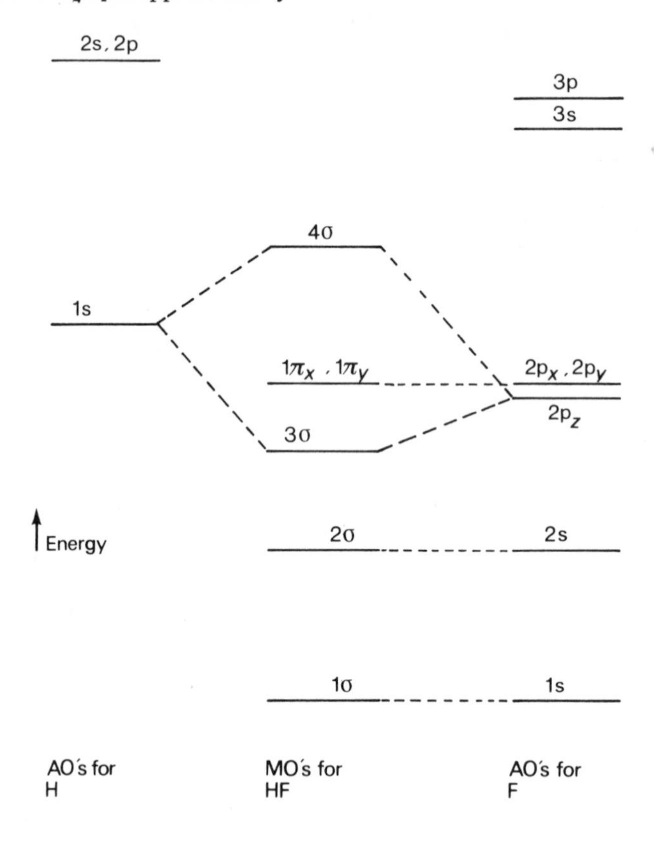


Figure 3.30 – MO's for HF: in the ground state, all except 4$\sigma$ are occupied by two spin-paired electrons.

### 3.9.7 Electronegativity

The electronegativity ($x$) of a species measures the tendency of that species to attract electrons. Thus the difference in the electronegativities of two species forming an essentially covalent bond should measure the ionic character of that bond. In the bond $A$—$B$, the difference in the electronegatives, $\Delta x_{AB}$, is given by $|x_A - x_B|$. Pauling's scale of relative electronegativities was drawn up empirically in 1932 by relating $\Delta x_{AB}$ to the geometric mean of the dissociation energies $D(A_2)$ and $D(B_2)$. Mulliken (1934) showed that a more straightforward measure of $x_A$ is given by $(I_A - E_A)$, the difference of the ionization energy and the electron affinity of $A$. It transpires that the Pauling and Mulliken scales are related by a numerical factor. Table 3.7 lists some of Pauling's results.

**Table 3.7 – Relative electronegativities (after Pauling)**

| | | | | | | |
|---|---|---|---|---|---|---|
| H | | | | | | |
| 2.1 | | | | | | |
| Li | Be | B | C | N | O | F |
| 1.0 | 1.5 | 2.0 | 2.5 | 3.0 | 3.5 | 4.0 |
| Na | Mg | Al | Si | P | S | Cl |
| 0.9 | 1.2 | 1.5 | 1.8 | 2.1 | 2.5 | 3.0 |
| K | Ca | Sc | Ge | As | Se | Br |
| 0.8 | 1.0 | 1.3 | 1.7 | 2.0 | 2.4 | 2.8 |
| Rb | Sr | Y | Sn | Sb | Te | I |
| 0.8 | 1.0 | 1.3 | 1.7 | 1.8 | 2.1 | 2.4 |
| Cs | Ba | | | | | |
| 0.7 | 0.9 | | | | | |

The relationship between $\Delta x$ and ionic character is not easily quantified; the best equation is that of Hannay and Smith (1946):

$$q = 0.16|\Delta x| + 0.035\Delta x^2 \quad , \tag{3.145}$$

and it is most reliable for $q < 0.5$. Figure 3.31 is a plot of (3.145), showing the data on the halogen hydrides from Table 3.6.

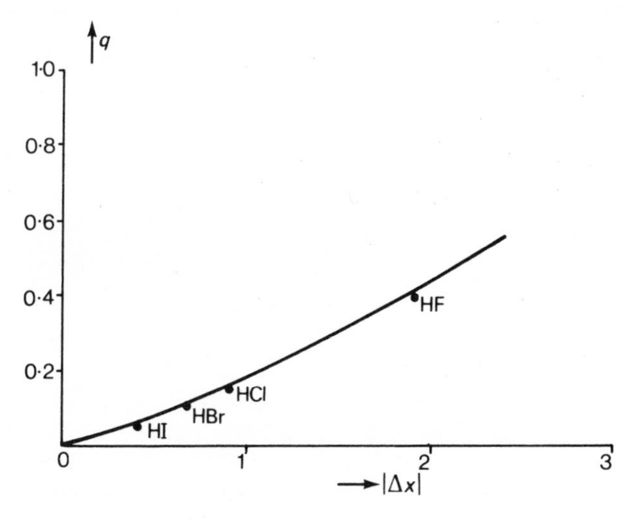


Figure 3.31 – Fractional ionic character ($q$) as a function of electronegativity difference $|\Delta x|$.

### 3.9.8 Hybridization

So far we have regarded atomic orbitals as largely unmodified in their combinations, in either the VB or the MO scheme of bond formation. In the water molecule, $H_2O$, we know, from experimental studies, that each O—H bond is 0.96 Å in length, and that the H—O—H angle is 104.3°. Now, oxygen is $(1s)^2(2s)^2(2p_z)^2(2p_x)^1(2p_z)^1$: we would expect divalency, with bonding between O(2p) and H(1s), and, from the orthogonality of the p orbitals, a bond angle of 90°. We could say that the two hydrogen atoms repel each other, thus opening out the angle between the p orbitals used in bonding, but let us approach this problem more slowly.

Consider two similar orbitals, say $p_u$ and $p_v$, their axes making an angle $\theta_{uv}$ with each other (Figure 3.32), and let another similar orbital, $p_w$, be directed orthogonally to $p_u$. Then $p_v$ can be resolved, like a vector, into components $p_v\cos\theta_{uv}$ along $p_u$ and $p_v\sin\theta_{uv}$ along $p_w$. The overlap integral is given by

$$S_{uv} = \int\psi(p_u)\psi(p_v)d\tau \ , \tag{3.146}$$

which can be written as

$$S_{uv} = \int\psi(p_u)\,[\psi(p_u)\cos\theta_{uv} + \psi(p_w)\sin\theta_{uv}]\,d\tau \ , \tag{3.147}$$

or

$$S_{uv} = \cos\theta_{uv}\int\psi^2(p_u)d\tau + \sin\theta_{uv}\int\psi(p_u)\psi(p_w)d\tau \ . \tag{3.148}$$

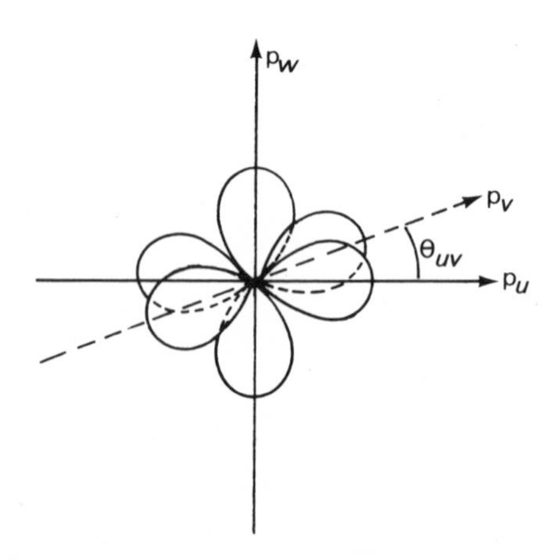


Figure 3.32 – Overlap integrals for two p orbitals: $p_v = p_v\cos^2\theta_{uv} + p_v\sin^2\theta_{uv} = p_u\cos\theta + p_w\sin\theta$.

If the orbitals are normalized (see p. 165), $\int\psi^2(p_u)d\tau = 1$, and $\int\psi(p_u)\psi(p_w)d\tau = 0$. Hence, $S_{uv} = \cos\theta_{uv}$, and is zero only where $p_u$ and $p_v$ are orthogonal, that is where $p_v$ coincides with $p_w$.

Suppose next that, in the water molecule, the oxygen atom, in forming a bond with hydrogen, makes use of its 2p and 2s AO's to produce a *hybrid* orbital. The orbital is different from either the 2s or the 2p: we shall formulate it as

$$\psi = \psi(\mathrm{p}) + \lambda\psi(\mathrm{s}) \ , \tag{3.149}$$

which implies that p and s contributions to the electron density occur in the ratio of $1:\lambda^2$ (see p. 193). We need two of these orbitals, and, if we wish them to be orthogonal, so that we may discuss them separately, then we require, with an obvious notation,

$$\int[\psi(\mathrm{p}_u) + \lambda\psi(\mathrm{s})]\,[\psi(\mathrm{p}_v) + \lambda\psi(\mathrm{s})]\,\mathrm{d}\tau = 0 \ . \tag{3.150}$$

Expanding:

$$\int\psi(\mathrm{p}_u)\psi(\mathrm{p}_v)\mathrm{d}\tau + \lambda\int\psi(\mathrm{p}_u)\psi(\mathrm{s})\mathrm{d}\tau + \lambda\int\psi(\mathrm{p}_v)\psi(\mathrm{s})\mathrm{d}\tau + \lambda^2\int\psi(\mathrm{s})\mathrm{d}\tau = 0 \ . \tag{3.151}$$

From previous results, the terms on the right-hand side are, successively, $\cos\theta_{uv}$, zero, zero and $\lambda^2$, whence

$$\cos\theta_{uv} = -\lambda^2 \ . \tag{3.152}$$

If $\theta_{uv}$ is not 90°, (3.152) shows that it must be greater than 90°. We know that, for $H_2O$, $\theta_{uv}$ is 104.3°; hence, $\lambda$ is approximately 0.5.

We have used part of the 2s closed shell in making each hybrid orbital. Energy has to be expended in opening the 2s shell, but it is more than compensated by the increased overlap of the hybrid orbitals. The remaining electrons from the 2s, $2p_x$ and $2p_y$ orbitals occupy another orthogonal† hybrid orbital; they constitute non-bonding, or lone-pair, electrons. Their off-centre distribution, together with that from those in the $2p_z$ orbital, contributes strongly to the polar character of the water molecule. This hypothesis is supported by the fact that the dipole moment of $H_2O$ is $6.00 \times 10^{-30}$ C m whereas that of $F_2O$ is only $0.67 \times 10^{-30}$ C m; the stronger polarity of the F—O bond of $F_2O$ reduces the lone-pair contributions. Figure 3.33 illustrates some of our discussion on the water molecule.

Consider next methane, $CH_4$. Carbon is $(1s)^2(2s)^2(2p)^2$: if it is to form four bonds, then four electrons from the $n$=2 level must be involved. Hybrid orbitals may be invoked, since the four electrons available are s and p types. The s : p

†Orthogonallity is defined by (3.90); it does not imply, necessarily, 90° angles between the axes of orbitals.

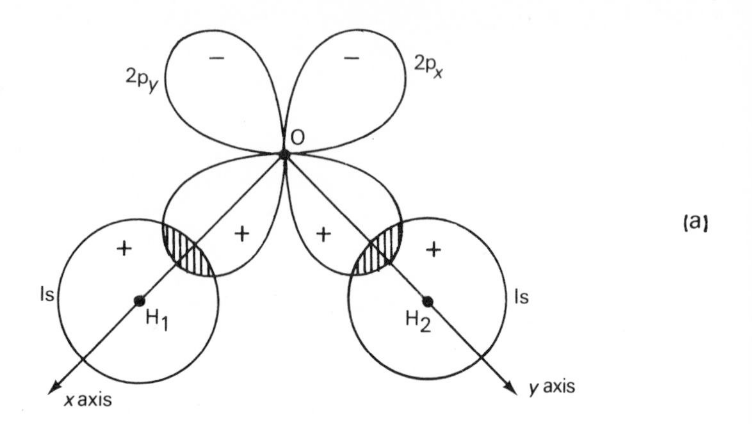


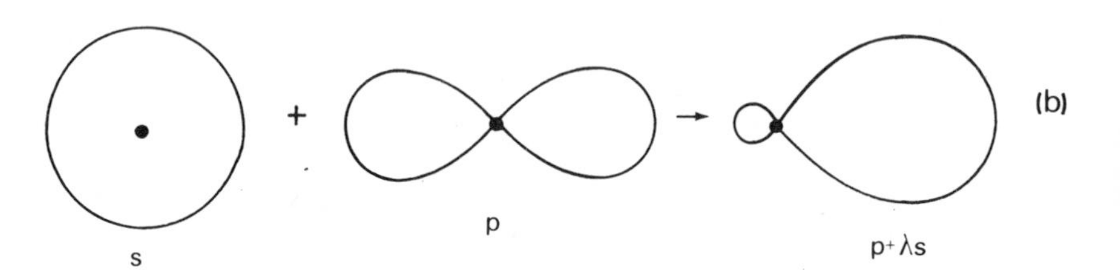


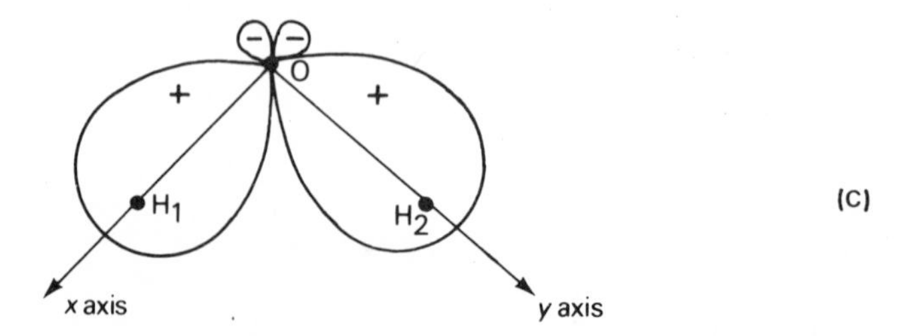


Figure 3.33 – Water molecule: (a) bonding through $O(2p_{x,y})$ and 2H(1s), leading to a bond angle of 90°; (b) sp hybrid orbital; (c) bonding with sp hybrids leads to a bond angle of 104.3° with $\lambda = 0.5$. The two hybrid orbitals in $H_2O$ are nearly independent of each other: if $H_2$ is replaced by, say, $CH_3$, the O—$H_1$ bond electrons are largely unaltered, thus providing a basis for characteristic bond lengths and energies. In $H_2O$, the third sp hybrid, containing the lone-pair electrons, is directed orthogonally to the other two hybrids.

character will be in the ratio of 1 : 3, and we shall call them $sp^3$ hybrid orbitals. From the previous discussion, we can label the hybrids

$$\psi(p) + \lambda\psi(s) , \tag{3.151}$$

where $\lambda$ must be $1/\sqrt{3}$. Since $\cos\alpha = -\lambda^2$, it follows that $\alpha$ is 109.47°, the tetrahedral angle. Hence, in methane, carbon can form four equivalent bonds with $sp^3$ hybrid orbitals whose axes make angles of 109.47° with one another (Figure 3.34). The energy needed to promote carbon to the valence state, $(1s)^2(2s)^1(2p_x)^1(2p_y)^1(2p_z)^1$, or C*, has been estimated to be approximately 430 kJ $mol^{-1}$; the average C—H bond energy is approximately 450 kJ $mol^{-1}$, so that bond formation through hybrid orbitals seems energetically feasible.

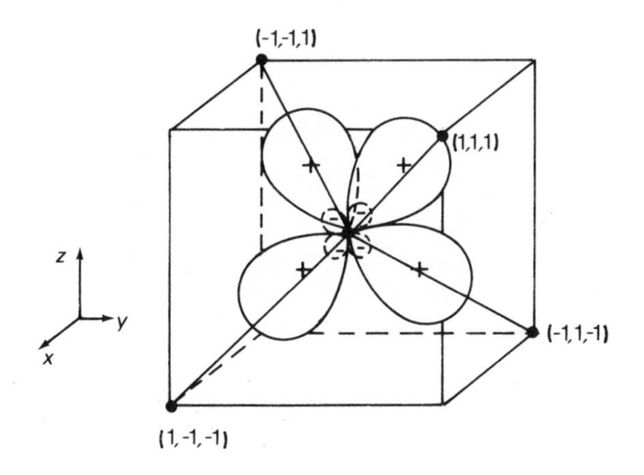


Figure 3.34 – Methane: $sp^3$ hybrid orbitals, showing their axes in relation to a cube. The carbon atom is at the centre of the cube, and the coordinates of the corners of the cube that lie on the orbital axes are given with respect to an origin on C.

By similar arguments, we can show that $sp^2$ bonds in, for example, ethene $C_2H_4$ have, ideally 120° valence angles. The C—C and C—H bonds are $\sigma$-type: the two electrons remaining on each carbon atom occupy a 2p orbital normal to the molecular plane, and the two such orbitals on the adjacent carbon atoms overlap to form a $\pi$-bond (Figure 3.35). In ethyne (acetylene) $C_2H_2$, the C—C and C—H bonds between two sp hybrids are again of the $\sigma$-type, and the remaining electrons form two $\pi$-bonds (Figure 3.36). The $\pi$-bonds are the source of reactivity in compounds with multiple bonds.

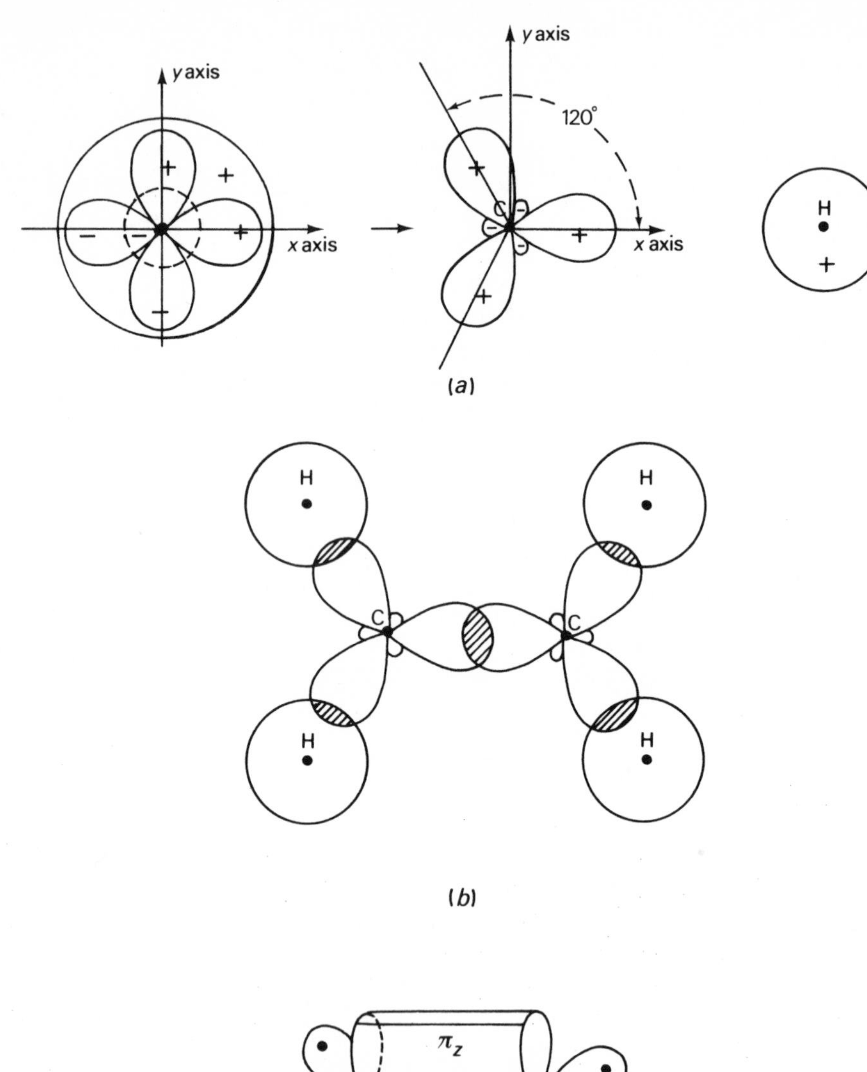


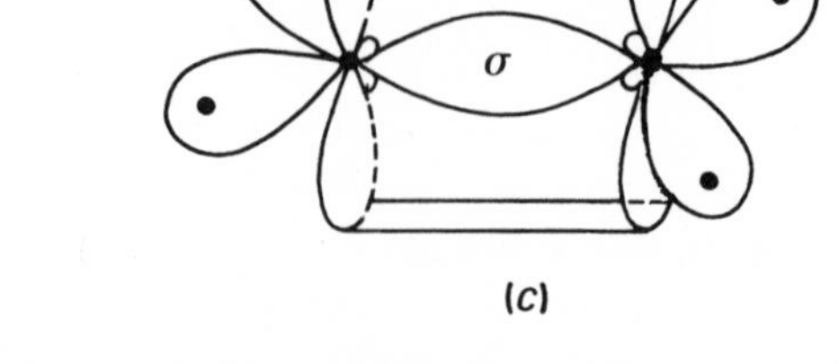


Figure 3.35 – Ethene: (a) Formation of an $sp^2$ carbon hybrid from 2s, $2p_x$ and $2p_y$ orbitals, and the 1s hydrogen AO, (b) overlap of orbitals, the $p_z$ orbital is directed normal to the plane of $C_2H_4$; (c) $\sigma$ and $\pi$-MO's in $C_2H_4$.

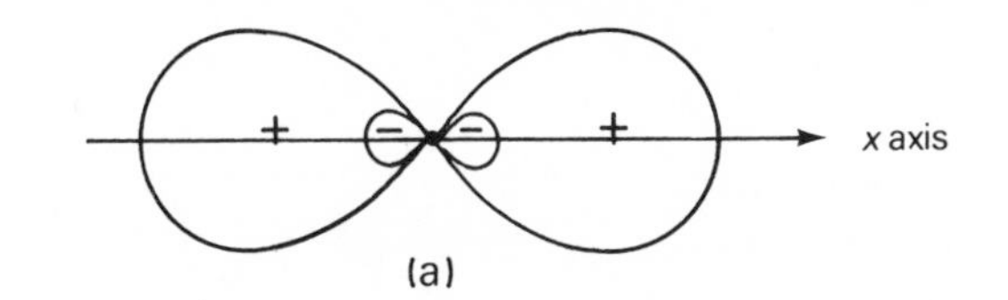


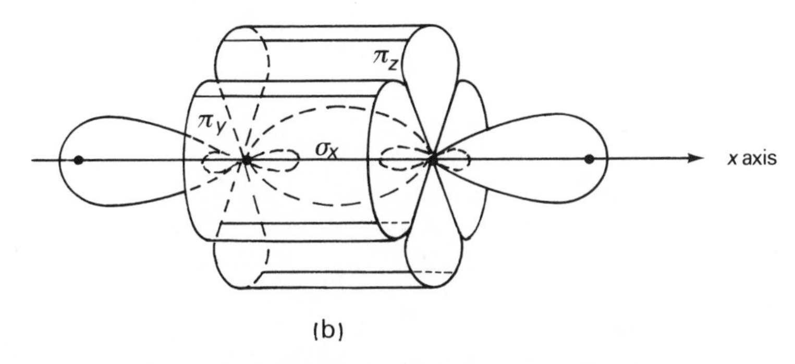


Figure 3.36 – Ethyne: (a) Linear sp carbon hybrid orbital, (b) the $2p_y$ and $2p_z$ AO's are mutually perpendicular with the $x$ axis, (c) $\sigma$ and two $\pi$ MO's in $C_2H_2$.

Hybridization is an essential feature of the VB model of bonding. It maintains an emphasis on the electron-pair bond in different situations. Through hybridization we can envisage greater overlap than with pure AO's without increasing repulsion forces, because hybrid orbitals have a non-spherical distribution of electron density. We must be careful not to raise hybridization to the level of reality. We do not know that it exists in the manner in which we have described it. Its value is that it allows us to modify our picture of the LCAO VB model so as to improve its explanation of experimental results.

In MO theory, hybridization is more a convenience than a necessity. It need not appear explicitly at all. Since our MO is formed by the LCAO method, the amounts of, say, s and p character in a bond are given by the coefficients $c_1, c_2, \ldots, c_n$. MO theory emphasises the tendency towards delocalization of electrons, that is, their distribution over the molecule as a whole.

### 3.9.9 Delocalized bonds

Buta-1,3-diene has the classical formula $CH_2{=}CH{-}CH{=}CH_2$, with conjugated double bonds (=—=— . . .); it appears to have one single bond and two double bonds. However, when we consider the experimentally-determined molecular geometry, we find the bond lengths $C\overset{1.35}{=}C\overset{1.48}{-}C\overset{1.35}{=}C$ (in Å). So the double bonds are a little longer than in ethene (1.33 Å), and the single bonds a little shorter than in ethane (1.53 Å). We describe this effect by saying that the

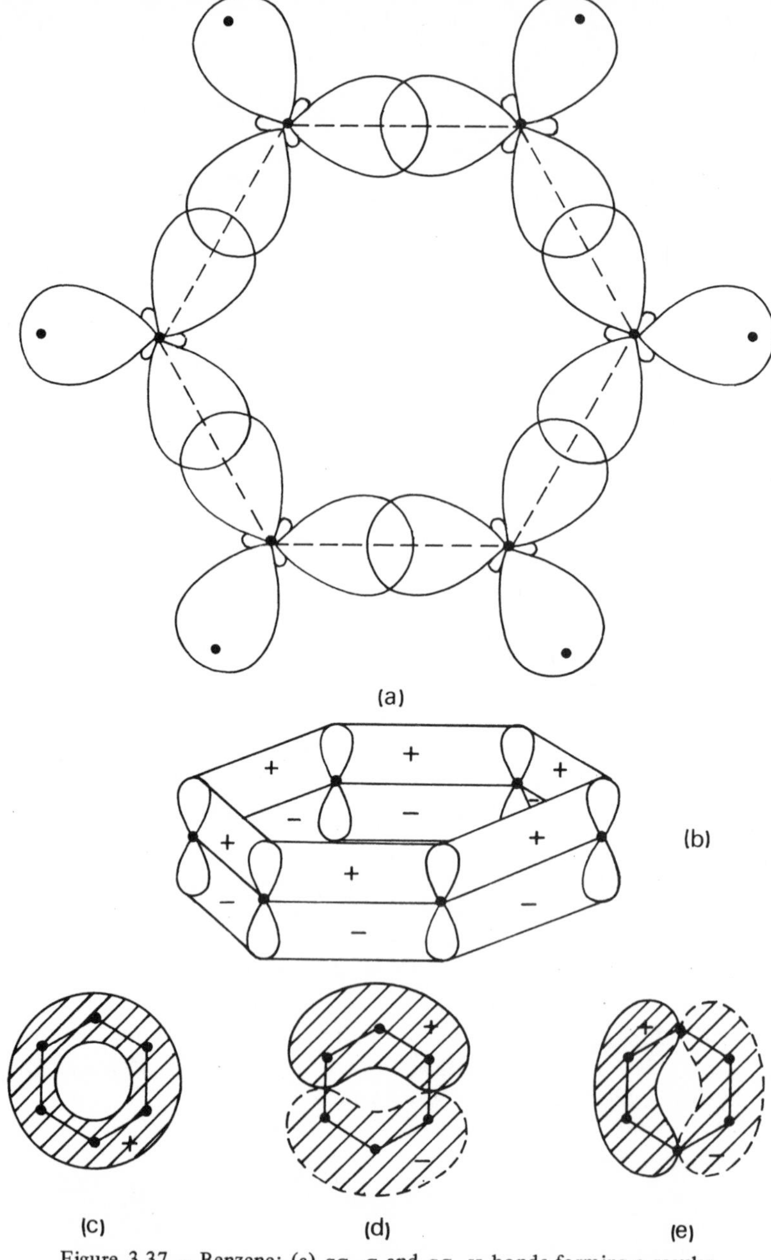


Figure 3.37 – Benzene: (a) $\sigma_{C-C}$ and $\sigma_{C-H}$ bonds forming a regular plane hexagon; the $2p_z$ orbitals are directed normal to the plane of $C_6H_6$; (b) double streamer $\pi$ orbital; (c), (d), (e) the same double streamer and two degenerate $\pi$ orbitals of next highest energy, respectively, seen in projection on to the plane of the ring.

$\pi$-electrons are partially delocalized over the molecule. If the electrons were fully delocalized, then we should expect that the C—C bond lengths would be equal. This situation is realised in benzene.

Benzene, $C_6H_6$, the molecule through which aromatic character is often defined, typifies a fully conjugated system of double bonds. The molecule is planar, with equal C—C (and C—H) bond lengths and 120° bond angles. This description implies $sp^2$ hybrid orbitals on the carbon atoms, forming $\sigma$-type C—C bonds, and $\sigma$-type C—H bonds with the 1s AO's of hydrogen. The six unpaired electrons occupy p orbitals which have their axes directed normally to the plane of the ring. These orbitals overlap with one another, forming, for the lowest energy $\pi$-type orbital, a complete double-streamer, above and below the ring plane (Figure 3.37). That the aromatic C—C bond is of a type intermediate between the single (1.53 Å) and double bonds (1.33 Å) is evidenced, among other properties, by its length, 1.40 Å. The $\pi$-electrons are completely delocalized over the whole molecule. It is this feature that gives benzene and other 'aromatic' compounds their unique place in organic chemistry.

## 3.10 COVALENT SOLIDS

Because we have considered a classification of solids based on the nature of the force mainly responsible for cohesion in the solid state, it follows that covalent solids will exhibit a disposition of covalent bonds in three dimensions. The number of such compounds is few.

The best example of a covalent solid is the diamond polymorph of carbon (Figure 1.14). Other elements in group IV of the periodic table, Si, Ge, Sn(grey), have the diamond structure type. Tin is dimorphous having, in addition to the grey form, a metallic structure type (white tin) (Figures 1.12 and 1.13). Lead, at the foot of group IVB, is a true metal and from carbon to lead we can trace a continuous change in bond type from almost wholly covalent to almost wholly metallic. This change is paralleled by a change in the electrical restivity ($\rho$) of the elements as shown below:

| | C(diamond) | Si | Ge | Sn(grey) | Sn(white) | Pb |
|---|---|---|---|---|---|---|
| $\rho$/ohm m | $5\times10^{12}$ | $2\times10^{3}$ | $5\times10^{-1}$ | $1\times10^{-5}$ | $1\times10^{-7}$ | $2\times10^{-7}$ |

Other covalent solids include compounds like silicon carbide (SiC)†, borazon (BN)† a polymeric form of boron nitride, quartz ($SiO_2$) (Figure 3.38) and titanium borides (TiB and $TiB_2$). We included blende ($\beta$-ZnS) and würtzite ($\alpha$-ZnS) under ionic compounds (Figures 2.13a and b). The bonds have a prominent covalent character, but whether they should be placed here instead of with ionic solids is a matter of choice. We have remarked that no classification of solids is without irregularities, a point that is borne out by the zinc sulphide structures.

† $\beta$-ZnS structure type.

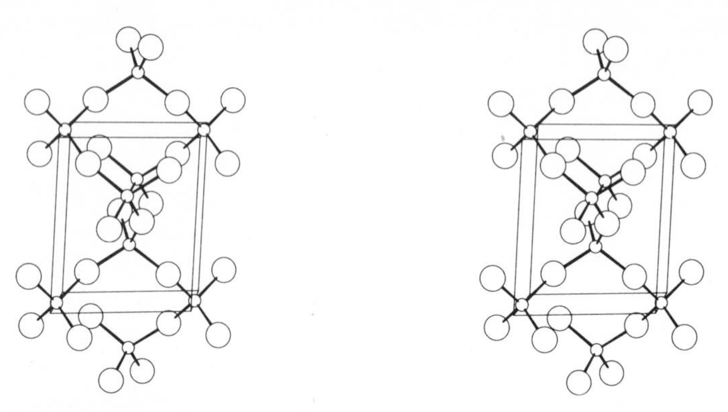

Figure 3.38 – Unit cell of the β-quartz ($SiO_2$) structure; circles in decreasing order of size are O and Si.

## 3.11 STRUCTURAL AND PHYSICAL CHARACTERISTICS OF THE COVALENT BOND

Covalent bonds exist between an atom and, generally, a small number of neighbours. Each bond is strongly directional (it is a vector), and we have discussed how this feature arises.

Covalent solids form strong, hard crystals, with low compressibility and expansivity, but with high melting point. In these properties they are very similar to ionic crystals. They are insulators, both in the solid and the molten state, a strong contrast to ionic solids. Covalent solids are, generally, chemically unreactive, and insoluble in all usual solvents. When discussing covalent solids, we shall omit organic compounds. They contain covalently bonded molecules, but the molecules are linked, in the solid state, by van der Waals' forces, which we shall consider in the next chapter.

## APPENDIX 14 SOLUTION OF A DIFFERENTIAL EQUATION

Consider the equation

$$d^2y/dx^2 + k^2y = 0 , \tag{A14.1}$$

where $k^2$ is a constant. Let $d^2/dx^2$ be represented by D. Then, we have

$$D^2y + k^2y = 0 . \tag{A14.2}$$

Consider next the equation

$$[(D - p_1)(D - p_2)]y = 0 . \tag{A14.3}$$

Expanding (A14.3):

$$D^2y - (p_1 + p_2)Dy + p_1p_2y = 0 \ . \qquad (A14.4)$$

Comparing (A14.2) and (A14.4), we see that they will be equivalent provided that $p_2 = -p_1$. Hence,

$$p_1^2 = -k^2 \ , \qquad (A14.5)$$

and the two roots are

$$p_1 = \pm ik \ . \qquad (A14.6)$$

Taking the terms in (A14.3) in turn,

$$(D - p_1)y = 0 \qquad (A14.7)$$

or

$$dy/y = p_1 dx \ . \qquad (A14.8)$$

On integrating (A14.8), we obtain

$$\ln y = p_1 x + \text{constant} \ . \qquad (A14.9)$$

Let the constant be $\ln A$, and using (A14.6), we have

$$y = A\exp(ikx) \ . \qquad (A14.10)$$

In a similar manner, we have also

$$y = B\exp(-ikx) \ , \qquad (A14.11)$$

and the complete solution may be written

$$y = A\exp(ikx) + B\exp(-ikx) \ . \qquad (A14.12)$$

## APPENDIX 15 SPHERICAL POLAR COORDINATES

### A15.1 Coordinates

The polar coordinates, $r$, $\theta$ and $\phi$, are defined through Figure 3.11 as

$$\begin{aligned} x &= r \sin\theta \cos\phi \ , \\ y &= r \sin\theta \sin\phi \ , \\ z &= r \cos\theta \end{aligned} \qquad (A15.1)$$

where $$r^2 = x^2 + y^2 + z^2\ . \tag{A15.2}$$

### A15.2 Volume element, $\mathrm{d}\tau$

In normalization problems, we may need to express a volume element $\mathrm{d}\tau$, $\mathrm{d}x\mathrm{d}y\mathrm{d}z$ in Cartesian coordinates, in polar coordinates. Consider the volume element $\mathrm{d}\tau$ shown in Figure A15.1; it corresponds to the quantity $\mathrm{d}x\mathrm{d}y\mathrm{d}z$. From the diagram, it is a straightforward matter to determine the magnitudes of the sides of the volume element, which may be taken to be parallelepipedal. Hence,

$$\mathrm{d}\tau = r^2\mathrm{d}r \sin\theta \ \mathrm{d}\theta\mathrm{d}\phi\ . \tag{A15.3}$$

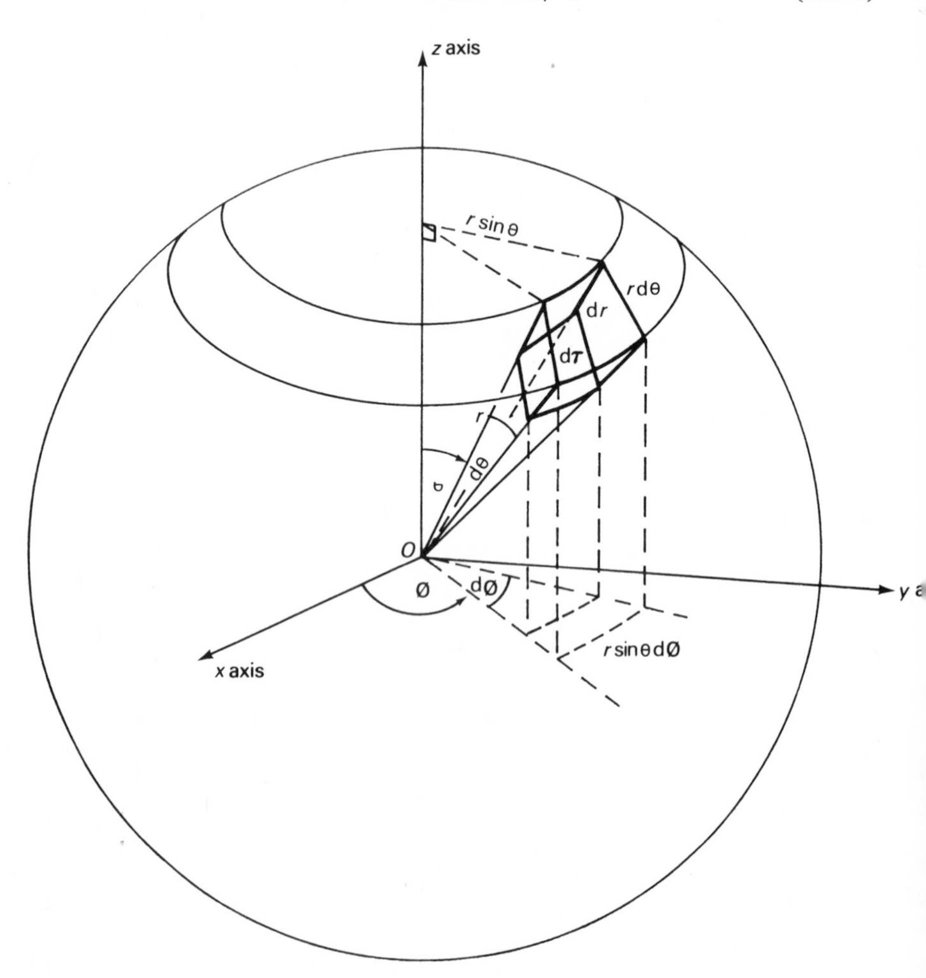


Figure A15.1 – Volume element $\mathrm{d}\tau$ in polar coordinates.

The limits of the variables, which correspond to $x$, $y$ and $z$ each between 0 and $\infty$, are

$$
\begin{aligned}
0 &\leqslant r \leqslant \infty \\
0 &\leqslant \theta \leqslant \pi \\
0 &\leqslant \phi \leqslant 2\pi \ .
\end{aligned}
\tag{A15.4}
$$

It is seen easily that these limits define a sphere of infinite radius.

### A15.3 Laplacian operator, $\nabla^2$

In solving the Schrödinger equation for the hydrogen atom, we used $\nabla^2$ in spherical polars rather than in Cartesian coordinates. The conversion of $\nabla^2$ from Cartesian to polar coordinates is lengthy but not difficult. From (3.77), we may write

$$\partial^2/\partial x^2 = [(\partial^2/\partial r^2)(\partial x/\partial r) - (\partial^2 x/\partial r^2)(\partial/\partial r)]/(\partial x/\partial r)^3 \ . \tag{A15.5}$$

From (A15.2), $\partial x/\partial r = r/x$, and $\partial^2 x/\partial r^2 = (x^2 - r^2)/x^3$. Hence,

$$\partial^2/\partial x^2 = (r/x)(\partial^2/\partial r^2) + (r^2 - x^2)(\partial/\partial r) \ . \tag{A15.6}$$

Equations for $\partial^2/\partial y^2$ and $\partial^2/\partial z^2$ are symmetrical. Hence, for the terms in $r$, we obtain

$$\partial^2/\partial x^2 + \partial^2/dy^2 + \partial^2/\partial z^2 = r^{-3}[r^3(\partial^2/\partial r^2) + 2r^2(\partial/\partial r)] \ , \tag{A15.7}$$

or

$$\nabla_r^2 = \partial^2/\partial r^2 + (2/r)\partial/\partial r \ . \tag{A15.8}$$

It is easy to show that the right-hand side of (A15.8) is equivalent to $(1/r^2)\dfrac{\partial}{\partial r}\left\{r^2\dfrac{\partial}{\partial r}\right\}$.

In a similar manner, using (A15.1) and remembering that $y/x = \tan\phi$, we can show that

$$
\begin{aligned}
\partial x/\partial\phi &= -r\sin\theta/\sin\phi \\
\partial^2 x/\partial\phi^2 &= r\sin\theta\cos\phi/\sin^2\phi \\
\partial y/\partial\phi &= r\sin\theta/\cos\phi \\
\partial^2 y/\partial\phi^2 &= r\sin\theta\sin\phi/\cos^2\phi \\
\partial z/\partial\phi &= \partial^2 z/\partial\phi^2 = 0 \ .
\end{aligned}
\tag{A15.9}
$$

Hence, using (3.77) again,

$$\nabla_\phi^2 = [1/(r^2\sin^2\theta)]\,\partial^2/\partial\phi^2 \ . \tag{A15.10}$$

It is left as an exercise to the reader to show that

$$\nabla_\theta^2 = (1/r^2)\partial^2/\partial\theta^2 + (\cos\theta/r^2)\partial/\partial\theta \ , \qquad (A15.11)$$

and is equivalent to $[1/(r^2\sin\theta)]\dfrac{\partial}{\partial\theta}\left\{\sin\theta\dfrac{\partial}{\partial\theta}\right\}$: it is useful to put $\tan\theta = \left\{\dfrac{x^2+y^2}{z}\right\}^{1/2}$; hence, $\nabla_r^2 + \nabla_\theta^2 + \nabla_\phi^2$, or $\nabla^2$, is given by (3.57).

## APPENDIX 16 GAMMA FUNCTION

The gamma function is useful in handling integrals of the type

$$\int_0^\infty x^n \exp(-ax^2)\,dx \ , \qquad (A16.1)$$

which occur in several areas of chemistry and chemical physics. The gamma function $\Gamma(n)$ may be represented by the integral equation

$$\Gamma(n) = \int_0^\infty t^{n-1}\exp(-t)\,dt \ . \qquad (A16.2)$$

The following particular results are important†

(a) For $n > 0$ and integral,

$$\Gamma(n) = (n-1)! \ ; \qquad (A16.3)$$

(b) For $n > 0$,

$$\Gamma(n+1) = n\Gamma(n) \ , \qquad (A16.4)$$

and if $n$ is also integral,

$$\Gamma(n+1) = n! \ ; \qquad (A16.5)$$

(c)

$$\Gamma(1/2) = \sqrt{\pi} \ . \qquad (A16.6)$$

As an example, we shall consider the solution of the integral

$$I = \int_0^\infty x^4\exp(-x^2/2)dx \qquad (A16.7)$$

Let $x^2/2 = t$, so that $x = (2t)^{1/2}$ and $dx = (2t)^{-1/2}\,dt$. Then,

$$I = 2\sqrt{2}\int_0^\infty t^{\frac{3}{2}}\exp(-t)dt \ . \qquad (A16.8)$$

Hence,

$$I = 2\sqrt{2}\,\Gamma(\tfrac{5}{2}), \text{ or } 3\sqrt{\pi/2} \ . \qquad (A16.9)$$

† Suggested Reading: *Mathematics* (Margenau and Murphy).

## APPENDIX 17 SOME ALGEBRA OF THE VARIATION METHOD

We re-state (3.109) and (3.110):

$$c_1(H_{11} - ES_{11}) + c_2(H_{12} - ES_{12}) = 0 \tag{A17.1}$$

and

$$c_1(H_{12} - ES_{12}) + c_2(H_{22} - ES_{22}) = 0 \ . \tag{A17.2}$$

Each of these equations can be solved for $c_2/c_1$, and, in order to obtain the same value for this ratio, we have

$$\frac{c_2}{c_1} = \frac{-(H_{11} - ES_{11})}{(H_{12} - ES_{12})} = \frac{-(H_{12} - ES_{12})}{(H_{22} - ES_{22})} \ . \tag{A17.3}$$

Re-arrangement of (A17.3) leads to the *secular equation*

$$(H_{11} - ES_{11})(H_{22} - ES_{22}) - (H_{12} - ES_{12})^2 = 0 \ , \tag{A17.4}$$

and it can be shown readily that (A17.4) can be represented by the determinantal form

$$\begin{vmatrix} H_{11}-ES_{11} & H_{12}-ES_{12} \\ H_{12}-ES_{12} & H_{22}-ES_{22} \end{vmatrix} = 0 \tag{A17.5}$$

The left-hand side of (A17.5) is called the *secular determinant*. The general requirement that a set of equations such as (A17.1) and (A17.2) have non-trivial solutions for $c_1$ and $c_2$ is that the corresponding secular determinant is zero.

### A17.1 Solution of determinants

Consider the determinant

$$D_1 = \begin{vmatrix} a_{11} & a_{12} \\ a_{21} & a_{22} \end{vmatrix} \tag{A17.6}$$

where, for our purposes, $a_{ij} = a_{ji}$. The expansion of this determinant is

$$D_1 = a_{11}a_{22} - a_{21}a_{12} \ . \tag{A17.7}$$

Consider next

$$D_2 = \begin{vmatrix} a_{11} & a_{12} & a_{13} \\ a_{21} & a_{22} & a_{23} \\ a_{31} & a_{32} & a_{33} \end{vmatrix} \ . \tag{A17.8}$$

$D_2$ may be factorized as

$$a_{11}\begin{vmatrix} a_{22} & a_{23} \\ a_{32} & a_{33} \end{vmatrix} - a_{12}\begin{vmatrix} a_{21} & a_{23} \\ a_{31} & a_{33} \end{vmatrix} + a_{13}\begin{vmatrix} a_{21} & a_{22} \\ a_{31} & a_{32} \end{vmatrix}, \qquad \text{(A17.9)}$$

or

$$a_{11}a_{22}a_{33} - a_{11}a_{32}a_{23} - a_{12}a_{21}a_{33} + a_{12}a_{31}a_{23} + a_{13}a_{21}a_{32} - a_{13}a_{31}a_{22} \ . \qquad \text{(A17.10)}$$

Higher rank determinants may be expanded in a similar manner.

## PROBLEMS TO CHAPTER 3

1. Draw dot diagrams for HF, $H_2O$ and $C_6H_6$.

*2. Millikan exposed a freshly-cut surface of sodium metal, in a vacuum, to monochromatic radiation from a quartz-mercury arc source. The photoelectrons emitted were collected in an oxidized copper Faraday cylinder. The current obtained for different values of an applied potential difference ($V$) was measured with an electrometer: in different experiments, differing wavelengths ($\lambda$) were used.

The deflexion ($\theta$) recorded by the electrometer is proportional to the photoelectric current, and $\theta$ increases as the p.d. becomes more negative at the metal. A field is set up in the space between the dissimilar Na and oxidised Cu materials, and a contact potential difference exists between them; in Millikan's experiments, it acts from Na to Cu, that is, Na is positive with respect to the Faraday cylinder.

| $\lambda = 5461$ Å | | $\lambda = 3650$ Å | | $\lambda = 3126$ Å | |
|---|---|---|---|---|---|
| V/volt | $\theta$ | V/volt | $\theta$ | V/volt | $\theta$ |
| −2.257 | 28 | −1.157 | 67.5 | −0.5812 | 52 |
| −2.205 | 14 | −1.105 | 36 | −0.5288 | 29 |
| −2.152 | 7 | −1.0525 | 19 | −0.4765 | 12 |
| −2.100 | 3 | −1.0002 | 11 | −0.4242 | 5.7 |
| | | −0.9478 | 4 | −0.3718 | 2.5 |

(i) For each wavelength, plot $\theta$ against the independent variable ($V$), and estimate the *minimum* applied voltage ($V_0$) which prevents the fastest-moving photoelectrons from reaching the Faraday cylinder.

(ii) An electron moving through a potential difference ($V$) aquires a kinetic energy of $Ve$ electronvolt. This energy is equivalent to $mv^2/2$, where $m$

and $v$ are the mass and speed of the electron. Show that the equation $V_0e = kv - \phi$ represents the variation of $V_0$ with frequency ($\nu$). Find the value of the constant $k$. What is $k$?

3. A proton and an electron, taken as point charges, are held at a distance not greater than $10^{-15}$ m by coulombic forces. Is this hypothesis feasible in the light of the uncertainty principle?
4. The Balmer series in the hydrogen spectrum was analysed originally in terms of wavelength ($\lambda$) through the equation

$$\lambda = K[n^2/(n^2 - 4)] \ ,$$

where $K$ is a constant, and $n$ is an integer greater than 2. Show that this equation is equivalent to (3.21), and find $K$ in terms of $R_H$. What is the energy of the spectral line, in the Balmer series, nearest to the red end of the visible spectrum?

*5. An electron is confined to a one-dimensional potential well of length $a$. Show that the eigenfunctions (3.41), with (3.44), are orthogonal.
6. What is the smallest value of kinetic energy for an electron in a cubical box of side $10^{-15}$ m?
7. Determine the average value of $(1/r)$ for an electron in the hydrogen 1s orbital (see Table 3.4). Calculate the average potential and kinetic energies for this electron.

*8. Calculate the probability of finding a hydrogen 1s electron in the volume bounded by $r = 1.10a_0$ to $1.11a_0$, $\theta = 0.20\pi$ to $0.21\pi$, $\phi = 0.60\pi$ to $0.61\pi$ ($\psi(1s)$ may be assumed to be constant over the small volume considered).
9. Write down the electronic configurations of N, Cl, $Cl^-$ and $K^+$.
10. Write down the ground state VB and MO descriptions for $Be_2$ and LiH.
11. The dipole moment of the water molecule is $6.00 \times 10^{-30}$ C m. Set up an equivalent dipole for the molecule, and calculate the total fractional ionic character in $H_2O$ (O-H = 0.96 Å, H-O-H = 104.3°). Compare the result with that obtained with (3.145).
12. Show that $sp^2$ hybridization of carbon leads to $\sigma$-bond angles of 120°.
13. Show that an $sp^x$ hybrid AO has a normalised wave function given by

$$\psi(sp^x) = (1 + x)^{-1/2}\{\psi(s) + x^{1/2}\psi(p)\}$$

14. Use the result in (13) to show that the $sp^3$ hybrid orbital has a normalised wave-function given by

$$\psi(sp^3) = \tfrac{1}{2}(s + \sqrt{3}\,p),$$

where we use s for $\psi(s)$ and p for $\psi(p)$. Refer to Figure 3.34. For the corner labelled 1,1,1, p may be resolved into the components $p_x, p_y, p_z$.

Show that $\psi(sp^3) = \frac{1}{2}(s + p_x + p_y + p_z)$

Formulate the other three $sp^3$ wave functions in a similar manner and show that they are mutually orthogonal.

*15. We shall assume that the important properties of ethene arise from its $\pi$ electrons. Then, we can treat ethene as a two-orbital problem, like $H_2$. We can write

$$\chi_\pi = c_1\psi_1 + c_2\psi_2 \, .$$

Show by the variation method that the $\pi$-bond energy ($E_\pi$) is $2\alpha + 2\beta$. Assuming that $H_{ij}$ for non-adjacent atoms = 0, and that $S_{ij} = 0$, show, by application of the variation method, that $E_\pi$ for buta-1,3-diene is $4\alpha + 4.472\beta$. By comparing this result with that for ethene, and taking $\beta$ as 64 kJ $mol^{-1}$, determine the delocalization energy for buta-1,3-diene.

16. What is the maximum degeneracy in the three-dimensional particle in a box eigenvalues if the maximum values of $n_x$, $n_y$ and $n_z$ are 2?

*17. The polymethene dye

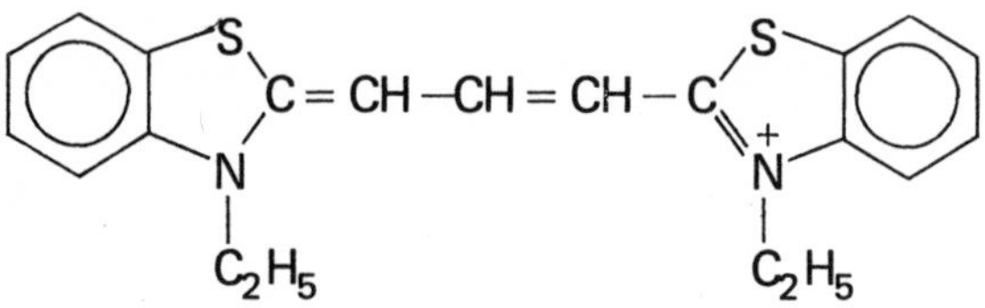

may be treated as the 'box'

$$-\ddot{N}-C=C-C=C-C=\overset{+}{N}-$$

where the mean bond length is $1.4 \times 10^{-10}$ m. The chain contains $2N+2$ $\pi$-electrons (two for each double bond and two for the neutral N atom), where $N$ is the number of double bonds, and they occupy the first $N+1$ MO's. The colour of the dye arises from the transition of an electron from the $N+1$ to the $N+2$ orbital. Show that the wavelength of this transition is given by

$$\lambda = \frac{3.297 \times 10^{12} \times L^2}{2N + 3} \text{ m} \, .$$

Calculate $\lambda$ and state the colour of the dye.

18. Show that the $2p_x$ and $2p_y$ atomic orbitals are orthogonal.

Chapter 4

# Van der Waals' Compounds

## 4.1 INTRODUCTION

In van der Waals' compounds, cohesion arises through forces which are weak compared with those present in ionic or covalent compounds. Generally, the energy of interaction is proportional to either $1/r^6$ or $1/r^3$ ($1/r$ in ionic and covalent compounds), where $r$ is the distance of closest approach to one another of the molecules or other discrete groups in a structure. The cohesive forces involved are often referred to as van der Waals' forces: in some compounds they are enhanced by hydrogen bonding or other attractive electrostatic processes. It is convenient to consider in this chapter solids which are formed from non-polar molecules, polar molecules and hydrogen-bonded molecules, as well as $\pi$-electron overlap compounds, clathrate structures and charge-transfer compounds. The differences between them, in this context, are in degree rather than in kind.

Van der Waals' attractive forces arise through dipolar mechanisms, even with non-polar molecules. We shall investigate the polarization of substances and the dipole moments of molecules and, for comparison, make reference to the gaseous and liquid states. Finally, we shall discuss the general structural and physical characteristics of molecular solids and consider possible sub-divisions of the very large number of compounds in this class.

## 4.2 ISOTROPIC DIELECTRICS

A simple theory of dielectrics will be introduced, applicable to gases, liquids and isotropic solids, in which the central feature is the polarization of atoms and molecules by an electrostatic field. We shall think of this form of polarization as a displacement of the electron density in a substance brought about by an electrostatic field. It will be necessary to draw upon results from the theory of electrostatics, and some important theorems are reviewed in Appendix 18.

Consider an isolated parallel-plate capacitor, initially in a vacuum, and carrying a charge density $\sigma$ (Figure 4.1). Its electrostatic field intensity, $E_V$, is given by

$$E_V = \sigma/\epsilon_0 \ , \tag{4.1}$$

where $\epsilon_0$ is the permittivity of a vacuum. The potential difference $V$ between the identical plates, $A$ and $B$, is $E_V l$, where $l$ is the perpendicular distance between the two plates.

Hence,
$$V = \sigma l/\epsilon_0 \ . \tag{4.2}$$

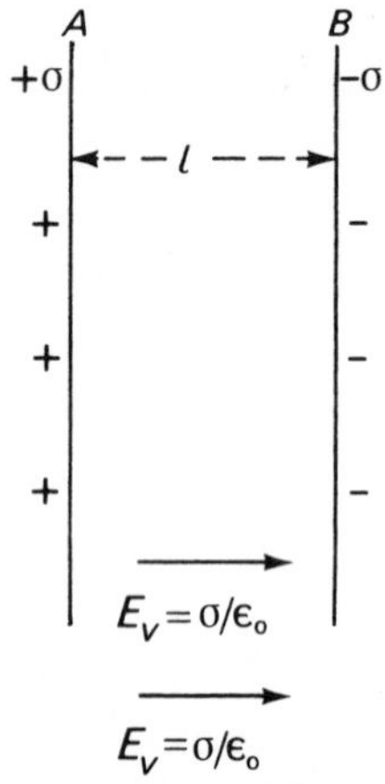

Figure 4.1 – Parallel-plate capacitor in a vacuum; the field has the same magnitude both inside and outside the capacitor.

The capacitance per unit area, $C$, is defined by

$$C = \sigma/V \ . \tag{4.3}$$

Hence, from (4.2) and (4.3)

$$C = \epsilon_0/l \ . \tag{4.4}$$

If the space between the plates of the capacitor is filled with a dielectric material, it is found that the electric field strength is given then, following (4.1), by

$$E = \sigma/\epsilon \ , \tag{4.5}$$

where $\epsilon$ is the permittivity of the dielectric. It is convenient to define a relative permittivity, or dielectric constant†, by

$$\epsilon_r = \epsilon/\epsilon_0 \ . \tag{4.6}$$

†Strictly $\epsilon_r$ depends upon the frequency of the electric field; its value is greater than unity.

Hence,
$$E = \sigma/(\epsilon_r\epsilon_0) , \tag{4.7}$$

and we see that, provided $\sigma$ remains constant, the field has been reduced by the factor $\epsilon_r$. Alternatively, if the capacitor is placed in a circuit of constant potential difference, then the charge density will be increased by the same factor.

The reduction in the electric field intensity arises from the polarization of the dielectric. Owing to the displacement of the electron density in the dielectric, dipoles will be induced in the medium and any permanent dipoles will tend to be aligned with the field. The net effect is the creation of an induced field $E_i$, opposing the applied field (Figure 4.2).

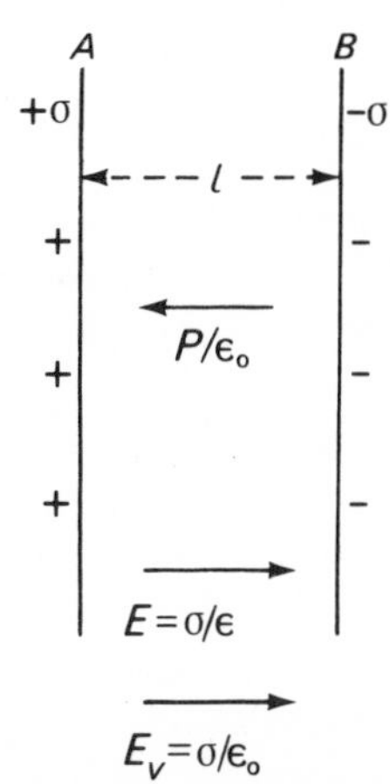


Figure 4.2 – Parallel plate capacitor filled with a dielectric; the resultant field, $E$, is less than the applied (vacuum) field.

We shall define the polarization $P$ of the dielectric as the total electrostatic dipole moment per unit volume of material. Thus $P$ has the nature of a charge density: for if the area of a plate of the capacitor is $a$, then the total moment arising from polarization is given by

$$p = Pal . \tag{4.8}$$

Since electrostatic moment is measured by the product of charge × distance, $P$ is evidently a surface charge density. As it opposes the quantity $\sigma$, the effective charge density on the capacitor plates is $(\sigma - P)$. Hence, from (4.1), the resultant field strength $E$ is given by

$$E = (\sigma - P)/\epsilon_0 . \tag{4.9}$$

Using (4.6), (4.6) and (4.9), we may write

$$E = (\epsilon E/\epsilon_0) - (P/\epsilon_0) , \tag{4.10}$$

or
$$E = \epsilon_r E - (P/\epsilon_0) \ . \tag{4.11}$$

Hence
$$P = \epsilon_0(\epsilon_r - 1)E \ . \tag{4.12}$$

Thus the presence of the dielectric in the capacitor produces a polarization which is $\epsilon_0(\epsilon_r - 1)$ times the real field. It should be noted that $E$ and $P$ are, strictly, vectors. When we need to use their vector properties the symbols will be given in bold type, as **E** and **P**.

We shall consider next the extent to which (4.12) may be applied, firstly to gases and then to the condensed states of matter.

## 4.3 POLARIZATION IN GASES

Except at high pressures, the molecules of a gas are sufficiently far apart for interactions between them to be negligible. Thus each molecule comes fully under the effect of the field in the dielectric. Polarization may take place through any of four mechanisms:

(a) The electron cloud around the nucleus of each atom is displaced, so that the centres of gravity of the positive and negative charges are altered (electron polarization).

(b) Bonded atoms or groups of atoms are displaced relative to one another; this effect is important for strongly polar molecules, since it involves alterations in the magnitudes of their bending and stretching modes of vibration (atom polarization).

(c) Permanent dipoles will tend to be aligned by the applied field, although this effect is opposed by the thermal motions of the molecules (oriental polarization).

(d) At high electric field strengths, species which can be polarized to differing extents in different molecular directions will tend to be aligned such that their direction of highest internal polarization tends to turn into the field. This effect will, for our purposes, be considered to be incorporated into the orientation polarization.

Thus, the above mechanisms (a) and (b) are concerned with translational effects and (c) and (d) with rotational effects on the electron density of atoms and molecules in the dielectric. We shall study first the electron polarization.

### 4.3.1 Electron polarization

Consider a molecule, initially non-polarized, placed in an electric field of strength $E$. As a result of electron displacement the molecule acquires a dipole moment $p$, which is proportional to the field strength:

$$p = \alpha E \ , \tag{4.13}$$

where $\alpha$ is the polarizability of the species. If there are $N$ identical molecules per unit volume then, since polarization has been defined as the total dipole moment per unit volume, we may write the electron polarization, $P_e$, as

$$P_e = Np = N\alpha_e E , \tag{4.14}$$

where $\alpha_e$ is the polarizability with respect to electron displacement alone. Since the dimensions of polarization are C m$^{-2}$ and those of field strength V m$^{-1}$, the dimensions of polarizability are F m$^2$.

Let us amplify the discussion by reference to a monatomic species, such as argon, which contains a central nucleus of charge $+q$ surrounded by an electron cloud of total charge $-q$. The electron density is assumed to be uniform within a spherical region of radius $r$ (Figure 4.3). When a field $E$ is applied to the atom, the electron density is displaced relative to the nucleus. The centres of gravity of the positive and negative charges no longer coincide, and a dipole is created within the atom (Figure 4.4).

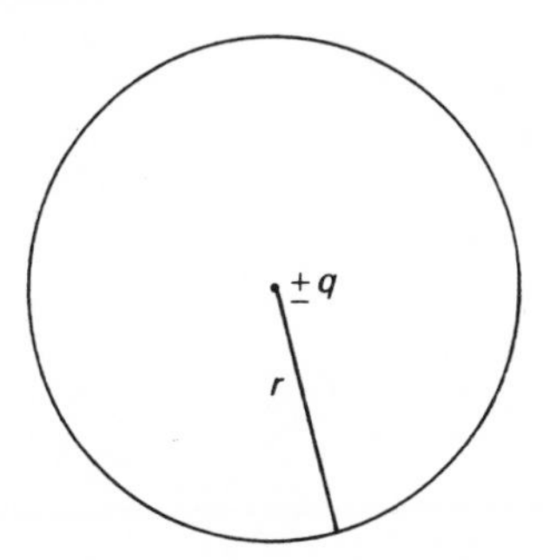


Figure 4.3 – Neutral atom, initially non-polarized, with the centres of gravity of the positive and negative charges coincident ($\pm q$).

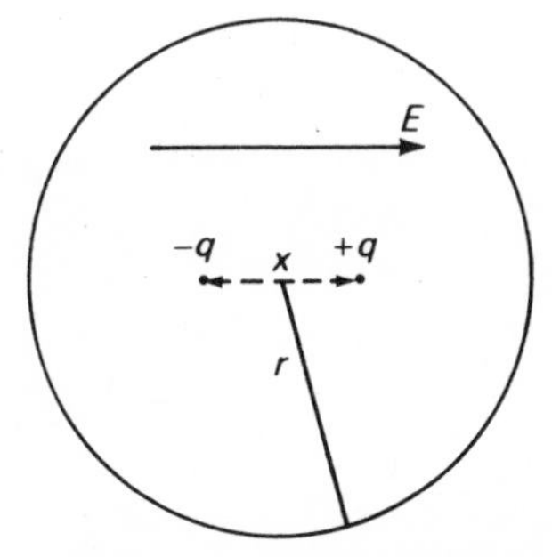


Figure 4.4 – In the presence of an applied field, E, the centres of gravity of positive and negative charge are separated to a distance $x$.

The electron density experiences a distorting force of magnitude $qE$ owing to the applied field, and an attractive force of $-qq/(4\pi\epsilon_0 x^2)$ between the nucleus and that part of the electron density which may be considered to be lying within a sphere of radius $x$, the distance of relative displacement of the centres of gravity of the opposite charges. Since the electron density has been assumed to be uniform, $q'$ is the volume ratio of the two spheres of radii $r$ and $x$ (Figures 4.4 and 4.5). Hence, at equilibrium, we write

$$qE = q^2(x/r)^3/(4\pi\epsilon_0 x^2) \ ,$$

or

$$E = qx/(4\pi\epsilon_0 r^3) \ . \qquad (4.16)$$

But $qx$ represents the moment, $p$, of the dipole created by the charge displacement within the atom. Hence,

$$p = 4\pi\epsilon_0 r^3 E \ . \qquad (4.17)$$

By comparing (4.17) with (4.13), it is evident that

$$\alpha_e = 4\pi\epsilon_0 r^3 \ . \qquad (4.18)$$

We may note that most literature reports values of $\alpha$ in cgs units: $\alpha = r^3$, with $r$ in cm. Such values may be converted to SI equivalents by the multiplying factor $4\pi\epsilon_0/10^6$ or approximately $1.113 \times 10^{-16}$. Values of $\alpha$ lie, generally, between $1 \times 10^{-40}$ and $10 \times 10^{-40}$ F m$^2$.

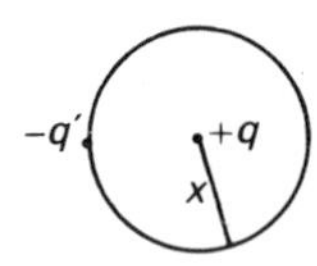


Figure 4.5 – Effective volume of the atom containing the dipole.

Using (4.14) and (4.18),

$$P_e = N4\pi\epsilon_0 r^3 E \ , \qquad (4.19)$$

and from (4.12), a general equation for polarization, we may write for the electron polarization

$$P_e = \epsilon_0(\epsilon_r - 1)E \ . \qquad (4.20)$$

Hence, and using (4.18),

$$(\epsilon_r - 1) = 4\pi N r^3 = N\alpha_e/\epsilon_0 \ . \qquad (4.21)$$

If $L$ is the Avogadro constant and $V_m$ the molar volume of the substance, then

$$N/L = 1/V_m = \rho/M \ , \qquad (4.22)$$

where $\rho$ is the density and $M$ the molar mass of the compound. Hence,

$$(\epsilon_r - 1)M/\rho = L\alpha_e/\epsilon_0 \, . \tag{4.23}$$

*Example: Gaseous argon*

$M = 0.0399 \text{ kg mol}^{-1}$, $\rho = 1.78 \text{ kg m}^{-3}$, $\epsilon_r = 1.000517$,
$L = 6.002 \times 10^{23} \text{ mol}^{-1}$, $\epsilon_0 = 8.854 \times 10^{-12} \text{ F m}^{-1}$.

$$\alpha_e = \frac{(1.000517 - 1) \times 0.0399 \times 8.854 \times 10^{-12}}{1.78 \times 6.022 \times 10^{23}}$$

$$\alpha_e = 1.7 \times 10^{-40} \text{ F m}^2.$$

### 4.3.2 Maxwell's theory

According to Maxwell's electromagnetic theory, the dielectric constant is related to the refractive index $n$, at comparable frequencies, by the equation

$$\epsilon_r = n^2 \, . \tag{4.24}$$

Since $\epsilon_r \approx 1$ for a gas, we may write

$$\epsilon_r^{1/2} = \{1 + (\epsilon_r - 1)\}^{1/2} = n \, . \tag{4.25}$$

Expanding to two terms and rearranging:

$$(\epsilon_r - 1) = 2(n - 1) \, . \tag{4.26}$$

### 4.3.3 Van der Waals' equation of state

The van der Waals' equation of state for a gas may be written

$$P = nRT/(V - b) - a/V^2 \, . \tag{4.27}$$

In this section, $P$ refers to the gas pressure. In the critical state of the gas, $(\partial P/\partial V)_T = (\partial^2 P/\partial V^2)_T = 0$. Hence, it may be shown† that the critical volume, $V_c$, is equal to $3b$, where $b$ represents four times the volume of the molecules (see Appendix 19). Thus,

$$V_c = 12v \, , \tag{4.28}$$

where $v$ is the volume of one molecule. In (4.21), the term $4\pi Nr^3$ represents three times the volume of spherical atoms of radius $r$ per unit volume of gas.

†Suggested Reading: *Physical Chemistry.*

Hence

$$(\epsilon_r - 1) = 3v \ , \tag{4.29}$$

or

$$(\epsilon_r - 1) = V_c/4 \ . \tag{4.30}$$

Since density is inversely proportional to volume, $V_c/1$ is equal to $\rho/\rho_c$, and

$$(\epsilon_r - 1) = \rho/(4\rho_c) \ , \tag{4.31}$$

where $\rho_c$ is the gas density in the critical state, a measurable property. In Table 4.1 we compare $(\epsilon_r - 1)$ with $2(n - 1)$ and $\rho/(4\rho_c)$.

**Table 4.1 – Properties of gases**

| | | $\rho$/kg m$^{-3}$ | $10^{-3}\rho_c$/kg m$^{-3}$ | $10^4(\rho/(4\rho_c))$ | $10^4(2(n-1))$ | $10^4(\epsilon_r - 1)$ |
|---|---|---|---|---|---|---|
| Monatomic, | Ar | 1.78 | 0.53 | 8.4 | 5.6 | 5.2 |
| non-polar | Xe | 5.90 | 1.15 | 12.8 | 14.0 | 14.7 |
| Polyatomic, | $O_2$ | 1.43 | 0.43 | 8.3 | 5.4 | 4.9 |
| non-polar | $CO_2$ | 1.98 | 0.46 | 10.8 | 9.0 | 9.2 |
| Polyatomic, | $NH_3$ | 0.77 | 0.23 | 8.4 | 7.5 | 72 |
| polar | $H_2O$ | 0.60 | 0.32 | 4.7 | 5.1 | 79 |

For the non-polar molecules there is a satisfactory agreement between all three of these measures of $\alpha_e$, *vide* (4.21), and we may conclude that each gives a satisfactory representation of the electron polarization in these substances. However for the polar molecules. $(\epsilon_r - 1)$ is of a different order of magnitude from either $\rho/(4\rho_c)$ or $2(n - 1)$. The dielectric constant is usually determined‡ for static, or low frequency, fields. In this circumstance the orientation of the permanent dipoles, an effect not yet included in this theory, makes a major contribution to the total polarization. The refractive index is a measure of the interaction of the high-frequency electric vector associated with light waves with the electron density of the compound. In other words it is concerned only with $P_e$. Not surprisingly, both $2(n - 1)$ and $\rho/(4\rho_c)$ show good agreement with $(\epsilon_r - 1)$, even for polar molecules: we must be careful to test our theory with the appropriate experimental data. It is evident that polarization is dependent upon the frequency of the applied field, but we shall return to this point later.

‡See Appendix 20.

### 4.3.4 Polarization in polar molecules

The total polarization $P$ of a molecule may be written as

$$P = P_e + P_a + P_o \ , \qquad (4.32)$$

where the subscripts refer respectively to the polarization mechanisms (a), (b) and (c+d) considered in Section 4.3. Atom polarization is small, often less than one tenth of $P_e$, and may be estimated by subtracting $P_e$ and $P_o$ from the total polarization; $P_o$ is large in polar molecules, often about one hundred times $P_e$, and must be determined explicitly.

In a homonuclear diatomic molecule $A_2$, there will be no permanent relative displacement of the effective centres of positive and negative charge, and the molecule is non-polar. In a heteronuclear diatomic molecule $AB$, the differing relative electronegativities cause electron density to be displaced towards the more electronegative centre, and a dipole is created. We note in passing that the literature reports values of dipole moments in D (Debye units): $1\text{D} = 10^{-18}$ cgs units. Such data are converted to SI values by multiplying by the factor $3.3356 \times 10^{-30}$.

Consider a molecule such as HF, for which the dipole moment is $6.3 \times 10^{-30}$ C m. If there are $N$ molecules per m$^3$, then $N$ is equal to $L/V_\text{m}$. Taking the molar volume, $V_\text{m}$, of a gas at 298.15 K as 0.02447 m$^3$ mol$^{-1}$, $N = 6.022 \times 10^{23}/0.02447 = 2.46 \times 10^{25}$ m$^{-3}$. Since the polarization is $Np$, we arrive at the result $P = 1.55 \times 10^{-4}$ C m$^{-2}$. From (4.9), we recall that the magnitude of the field capable of being produced through polarization is $P/\epsilon_0$ or $1.75 \times 10^7$ V m$^{-1}$ in this example. Hence if the molecules of HF in the given unit volume were aligned with their dipoles all parallel, they would develop a polarization field of about $1.8 \times 10^7$ V m$^{-1}$ (180 kV cm$^{-1}$), which is more than enough to lead to spontaneous dielectric breakdown by ionization of the gas. We know that this effect does not take place: hence not all dipoles can be simultaneously aligned with one another in the simple manner described.

The tendency for molecular orientation by an applied field is opposed by the thermal motions and collisions of the molecules at any given temperature above absolute zero. We shall next determine the effective fraction of polar molecules which may be considered to be aligned with the applied field at any given temperature.

### 4.3.5 Langevin's function

Consider an assembly of molecules of dipole moment $p$. We calculate first the potential energy of a dipole which lies with its axis at an angle $\theta$ to the applied field, $E$ (Figure 4.6). The potential energy $U$ of a dipole may be equated to the work done on the dipole in rotating it from a position of zero potential energy to any other position in the field. Let the dipole axis be initially perpendicular to $E$: its potential energy is zero because there is no component of $p$ in the direction of $E$. The dipole is then turned so that it makes an angle $\theta$ with $E$. The work done on the charge $+q$, resolved in the direction of $E$, is $-qE\sin(90 - \theta)$, or $-qEx$: the negative sign indicates a spontaneous movement of charge down the field gradient. Similarly, the work done on the charge $-q$ is also $-qEx$. Hence, $U$ is $-2qEx$, which, from Figure 4.6, is $-2qE(l/2)\cos\theta$. But, $ql$ is the dipole moment, $p$. Hence,

$$U = -pE\cos\theta \quad . \tag{4.33}$$

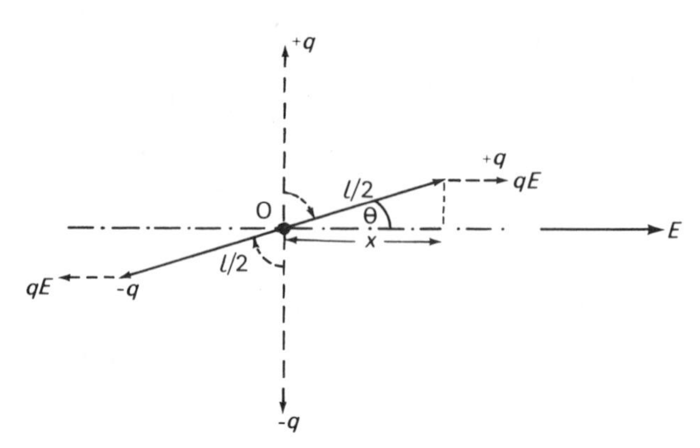


Figure 4.6 – Potential energy of a dipole.

The minimum (most negative) energy condition corresponds to $\theta = 0$. However, such alignment is broken up by thermal and mechanical movements of the molecules in the gas. The fractional number of dipoles lying at an angle $\theta$ to the field may be represented as the number lying within a solid angle $\delta\omega$. In other words it is the fraction of the dipoles the axes of which lie at angles to $E$ between $\theta$ and $\theta+\delta\theta$ (Figure 4.7). It is essential to realise that $\theta$ is not just an angle in the plane of the diagram; the important space lies between two coaxial cones of semi-vertical angles $\theta$ and $\theta+\delta\theta$.

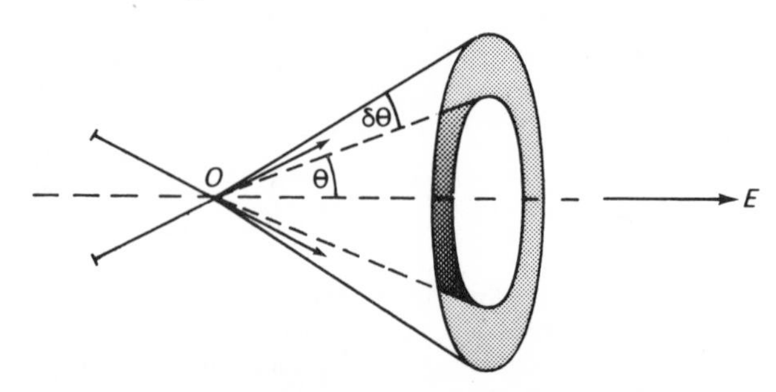


Figure 4.7 – Solid angle, $\delta\omega$, containing axes of dipoles.

A sphere of arbitrary radius $r$ is constructed with its centre $O$ at the mid-points of the dipoles (Figure 4.8). The generators of the two cones define an annular region of width $r\delta\theta$ on the sphere. The solid angle is defined as the ratio of the area of the annulus to the square of the distance from the annulus to the solid angle point. Thus, since $\delta\theta$ is small,

$$\delta\omega = 2\pi r(\sin\theta)r\delta\theta/r^2 , \tag{4.34}$$

or

$$\delta\omega = 2\pi(\sin\theta)\delta\theta . \tag{4.35}$$

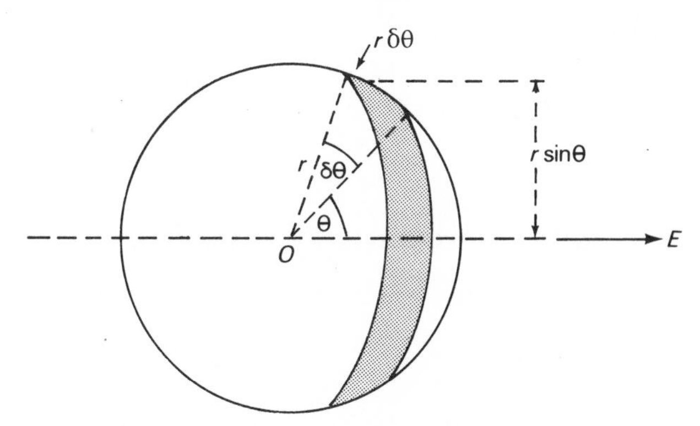


Figure 4.8 – Annulus of width $r\delta\theta$.

Clearly the solid angle is independent of $r$; the sphere is chosen simply because it is a convenient geometrical shape. For a zero field, the distribution of dipole axes is random, and the number of them per unit volume, $\delta N$, with their centres at $O$ in the solid angle $\delta\omega$, is proportional to the value $\delta\omega$; thus

$$\delta N = N_0 2\pi(\sin\theta)\delta\theta , \tag{4.36}$$

where $N_0$ is a constant dependent upon the total number of molecules under consideration. When $E$ is non-zero, the equation for $\delta N$ is weighted by a factor representing the energy needed to turn the dipoles into the angle $\theta$. Not all dipoles will require the same energy: in other words, different initial orientations have differing probabilities. We shall assume that a Boltzmann distribution (see Appendix 4) is applicable, with $U$ given by (4.33). Hence,

$$\delta N = N_0 2\pi(\sin\theta)\exp(pE\cos\theta/kT)\delta\theta . \tag{4.37}$$

If $\bar{p}$ be the average moment in the direction of the field, then $\bar{p} = p\overline{\cos\theta}$, since $p$ is really a constant and $\theta$ is the variable parameter, Thus,

$$P_o = N\bar{p} = Np\overline{\cos\theta} . \tag{4.38}$$

As a general result, the average value $\bar{X}$ of a distribution of values $X$, each having a probability function $\phi(X)$, is given by

$$\bar{X} = \int X\phi(X)\mathrm{d}X/\int\phi(X)\mathrm{d}X \ ; \tag{4.39}$$

it is like a weighted average. Applying this relationship to our problem:

$$\overline{\cos\theta} \ \frac{\int_0^\pi(\cos\theta)N_0 2\pi\sin\theta\exp(pE\cos\theta/kT)\mathrm{d}\theta}{\int_0^\pi N_0 2\pi\sin\theta\exp(pE\cos\theta/kT)\mathrm{d}\theta} \ . \tag{4.40}$$

Because of the construction used, the integration from 0 to $\pi$ gives the average value of $\cos\theta$ for all dipoles within the sphere.

To evaluate the integrals let $\cos\theta = y$, so that $\mathrm{d}\theta = -\mathrm{d}y/\sin\theta$, and let $pE/kT = a$. Then

$$\overline{\cos\theta} = \int_1^{-1} -y\exp(ay)\mathrm{d}y/\int_1^{-1} -\exp(ay)\mathrm{d}y \ , \tag{4.41}$$

or

$$\overline{\cos\theta} = \int_{-1}^{1} y\exp(ay)\mathrm{d}y/\int_{-1}^{1}\exp(ay)\mathrm{d}y \ . \tag{4.42}$$

Equation (4.42) may be written as

$$\overline{\cos\theta} = \mathrm{d}(\ln I)\mathrm{d}a \ , \tag{4.43}$$

where

$$I = \int_{-1}^{1}\exp(ay)\mathrm{d}y \ . \tag{4.44}$$

This result is verified easily by differentiation. Hence,

$$I = [\exp(a) - \exp(-a)]/a \ , \tag{4.45}$$

and

$$\ln I = \ln(1/a) + \ln(\sinh a) + \ln(2) \ . \tag{4.46}$$

Then

$$\mathrm{d}(\ln I)/\mathrm{d}a = -1/a + \cosh a/\sinh a \ , \tag{4.47}$$

and

$$\overline{\cos\theta} = \coth a - 1/a \ . \tag{4.48}$$

This equation is Langevin's function,

$$\mathcal{L}(a) = \coth a - 1/a \ . \tag{4.49}$$

Expanding $\coth a$:

$$\coth a = (1/a) + (a^3/3) - (a^3/45) + (2a^5/945) - + \ldots, \tag{4.50}$$

and neglecting terms in $a^3$ and higher powers,

$$\mathcal{L}(a) = a/3 \ . \tag{4.51}$$

Hence, by replacing $a$ by $pE/kT$,

$$\overline{\cos\theta} = pE/3kT \ , \tag{4.52}$$

and from (4.38)

$$P_o = Np^2E/(3kT) \ . \tag{4.53}$$

It may be noted that approximating $\exp(ay)$ in (4.42) to $(1+ay)$ leads to the same result, but without the knowledge of the magnitudes of the terms neglected.

Consider again the problem with which we began Section 4.3.4. If we say that dielectric breakdown will occur in a field of strength $10^7$ V m$^{-1}$, then since $pE/kT$ at 298 K is 0.015, the error in neglecting terms in coth $a$ of powers $\geqslant 3$ is only about $10^{-6}$: thus, the approximation may be deemed to be reasonable.

Experimental results show that if $P_o/pN$ is plotted as a function of $pE/kT$, or $a$, a curve of the type shown in Figure 4.9 is obtained. At small values of $a$, the slope is 1/3, in accordance with (4.53). The maximum value of $P_o/pN$ is 1, since the maximum value of $\overline{\cos\theta}$ is, from (4.40), unity. Considering again Table 4.1 and using (4.12) and (4.53), we have for $NH_3$ ($p = 5.0 \times 10^{-30}$ C m)

$$\epsilon_0(\epsilon_r - 1)E = Np^2E/3kT \ . \tag{4.54}$$

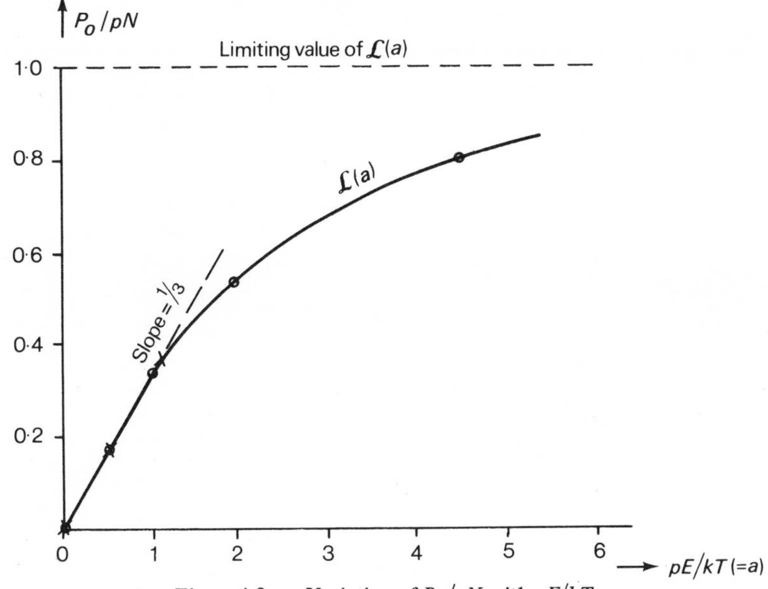


Figure 4.9 – Variation of $P_o/pN$ with $pE/kT$.

Hence,
$$(\epsilon_r - 1) = 62 \times 10^{-4} \ , \tag{4.55}$$

which is the major contribution to the total polarization in this gas.

### 4.3.6 Effect of frequency on polarization

In (4.32) we separated the total polarization into three main components, which may now be written

$$P = N\alpha_e E + P_a + Np^2E/(3kT) \ . \tag{4.56}$$

Figure 4.10 is a diagrammatic representation of these components.

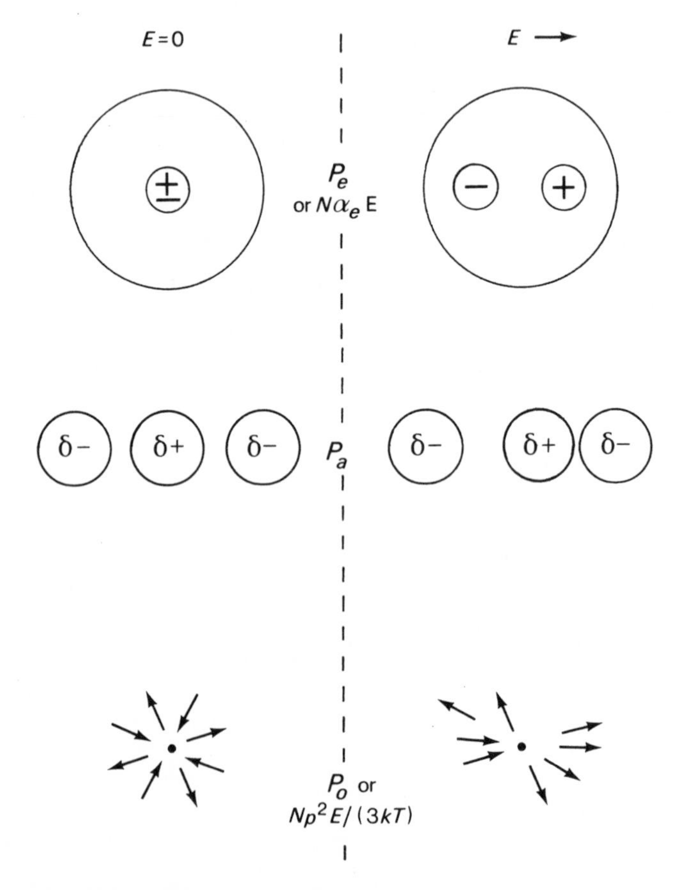


Figure 4.10 – Schematic representation of the components of the polarization process.

In a low frequency or a static applied field all of these processes contribute, in a polar molecule, to the total polarization. As the frequency increases the orientation and re-orientation of the whole molecule ceases to be able to follow the alternations of the field, and $P_o$ falls to zero. The contribution from the atom polarization effect continues until it too is unable to respond to the frequency of the field. Above a frequency of about $10^{15}$ Hz, which is the optical frequency range, only $P_e$ remains operative.

At a frequency where a polarization process ceases to follow the alternations of the field there is a discontinuity in the curve of polarization against frequency. In addition both $P_a$ and $P_o$ may show characteristic absorption effects, the so-called anomalous dispersion. Figure 4.11 shows schematically the frequency dependence of polarization for a polar molecule with single atomic and electronic characteristic absorption frequencies.

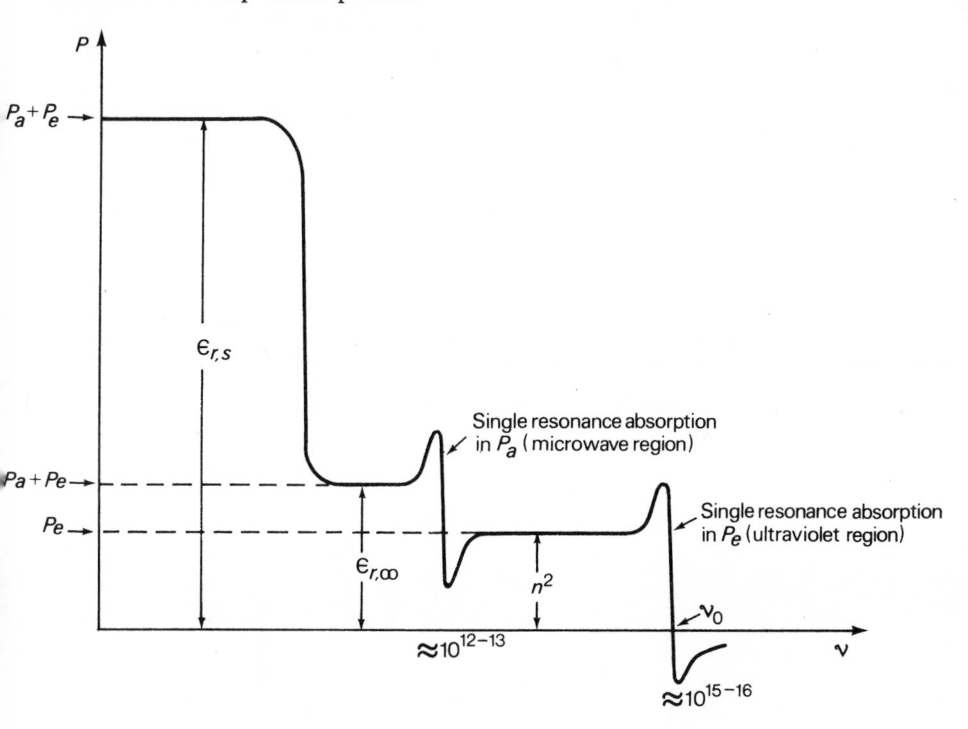


Figure 4.11 – Diagrammatic representation of the variation of polar ization $P$ with frequency $\nu$.

The total polarization and $P_e$ may be investigated experimentally through measurements of the static dielectric constant and the refractive index, respectively. $P_o$ may be obtained from dipole moment measurements, and then $P_a$ obtained by difference: we shall consider an example at a later stage. It may be noted in passing that the characteristic frequency $\nu_0$ on the dispersion curve is the same frequency that occurs in the detailed derivation of an expression for van der Waals' induced forces (see page 245). A simple treatment of these forces which are sometimes called dispersion forces, will be given later in this chapter.

## 4.4 POLARIZATION IN CONDENSED STATES

In our discussion of gases we assumed that the interactions between molecules were negligibly small. This situation will not normally be obtained in the condensed states (liquids and solids), and it is convenient to discuss them under two headings.

### 4.4.1 Non-polar molecules

In the condensed states of non-polar molecules, electrostatic interactions between neighbouring species are very slight, and we need consider only $P_e$. Figure 4.12 represents a body-centred cubic structure. The unit cell has a side of length $a$ and contains two of the species. If we assume close packing, then the volume $a^3$ is equal to $(4r/\sqrt{3})^3$ where $r$ is the effective radius of the species. The number of species, $N$, per unit volume is thus $1/(6.16r^3)$. From (4.21),

$$(\epsilon_r - 1) = 4\pi r^3/(6.16r^3) \ , \tag{4.57}$$

or

$$\epsilon_r \approx 3.0 \ . \tag{4.58}$$

Hence, from (4.24)

$$n \approx 1.73 \ . \tag{4.59}$$

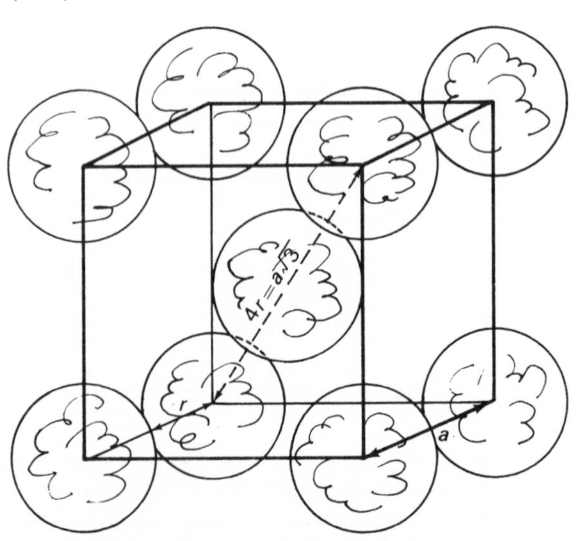


Figure 4.12 – Body-centred cubic unit cell of side $a$.

Many simple non-polar liquids and isotropic solids have values of refractive index close to 1.73: here are some examples: $Br_2$(l, 293 K), 1.76; S(l, 391 K), 1.88; $CS_2$(l, 163 K), 1.73; $P_4$(s, 298 K), 2.14; glass (s, 298 K), 1.50 – 1.95. In anisotropic (non-cubic) solids, the polarizability varies with direction in the solid. Hence $n$ also varies with direction, which gives rise to birefringence, or optical double refraction.

### 4.4.2 Polar molecules

When polar molecules are present in a condensed state, their dipoles interact strongly with one another. In this situation a polarization is induced in a molecule on account of the field due to the dipole of its neighbour. Let the induced dipole moment be $p'$. The field $E'$ produced in the dielectric by this dipole is given by

$$E' = p'/\alpha \; . \tag{4.60}$$

The field strength is thus commensurate with that which would produce a polarizability of $\alpha$, and cannot be ignored. In order to show how it increases the polarization effect we shall estimate next the resultant field on each polar molecule in a condensed state dielectric which is under the influence of an applied homogeneous field, $E_a$.

Within a parallel plate capacitor filled with a dielectric, consider a virtual spherical cavity around a molecule at $O$, the centre of the sphere. The radius $r$ of the sphere is sufficiently greater than the average molecular separation that we can assume that there are no molecular interactions inside the cavity (Figure 4.13). In isotropic materials, any atom or molecule is surrounded by other similar species forming a spherical region of nearest neighbours. We need to determine the resultant field in a spherical hole.

The applied field produces a polarization field in the dielectric in the usual manner. However, the resultant field, $E_r$, on a molecule at $O$ is made up from the following contributions:

(a) the applied field, $E_a$;
(b) the field due to the dielectric, excluding that in the virtual cavity, comprising
  (i) the polarization field, $-P/\epsilon_0$, and
  (ii) the field arising from the charge density on the surface of the cavity;
(c) the field due to the dielectric in the cavity, $E_i$. Thus,

$$E_r = E_a - P/\epsilon_0 + E_o + E_i = E + E_o + E_i \; . \tag{4.61}$$

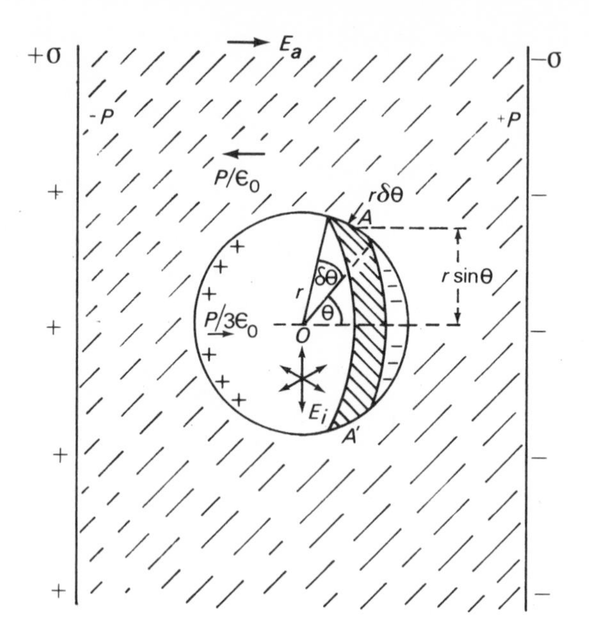


Figure 4.13 – Resultant field in a condensed polar dielectric material under the influence of an applied field, $E_a$.

*Calculation of $E_o$*

The dielectric outside the cavity is remote from the molecule at $O$, and may be regarded as a continuum. Hence, the field at $O$ due to the bulk dielectric is the same as that arising from the charge density on the walls of the cavity.

Consider an annular region of width $r\delta\theta$ on the sphere and subtending angles of $\theta$ and $\theta+\delta\theta$ at $O$. Since the element $r\delta\theta$ lies at an angle $\theta$ to the polarization field $P$, the apparent surface charge density at $A$ is $-P\cos\theta$.

The charge, $\delta q$, on the annulus is $-P(\cos\theta)2\pi r^2(\sin\theta)\delta\theta$, and its resulting coulombic field, $\delta E_o$, acting in the direction $OA$ is $\delta q/(4\pi\epsilon_0 r^2)$. This field may be resolved into components along and perpendicular to the direction of $E$ (Figure 4.14). The vertical components cancel but the horizontal components, $\delta q\cos\theta/(4\pi\epsilon_0 r^2)$, add to give the magnitude of the effect in the range $\theta$ to $(\theta+\delta\theta)$:

$$\delta E_o = (P\cos\theta)2\pi r^2(\sin\theta)\delta\theta(\cos\theta)/(4\pi\epsilon_0 r^2) \ . \tag{4.62}$$

For the whole sphere, (4.62) must be integrated between the limits $\theta = 0$ and $\theta = \pi$:

$$E_o = (P/2\epsilon_0)\int_0^\pi \cos^2\theta\sin\theta\,d\theta \ . \tag{4.63}$$

Hence
$$E_o = (P/2\epsilon_0)\,\frac{[-\cos^3\theta]_0^\pi}{3}\,, \tag{4.64}$$

or
$$E_o = P/(3\epsilon_0)\;. \tag{4.65}$$

*Calculation of $E_i$*

In Appendix 18 it is shown that the field due to a dipole of moment $p$ at a distance $r$ is given by

$$\mathbf{E} = \frac{(3\mathbf{p.r}\,\mathbf{r} - r^2\mathbf{p})}{4\pi\epsilon_0 r^5}\;. \tag{4.66}$$

All species in the virtual cavity have the same dipole moment, $p$. Let **p** and **r** be referred to orthogonal axes, $x$, $y$ and $z$: then

$$\mathbf{p} = \mathbf{i}p_x + \mathbf{j}p_y + \mathbf{k}p_z\;, \tag{4.67}$$

and
$$\mathbf{r} = \mathbf{i}x + \mathbf{j}y + \mathbf{k}z\;, \tag{4.68}$$

where **i**, **j**, and **k** are conventional unit vectors. Consider the $n$th dipole lying along the $z$ axis. Then, generally,

$$\mathbf{E}_n = \mathbf{i}E_{n,x} + \mathbf{j}E_{n,y} + \mathbf{k}E_{n,z}\;, \tag{4.69}$$

which, from (4.66) is given also by the expression

$$\begin{aligned}&[3(\mathbf{i}p_x + \mathbf{j}p_y + \mathbf{k}p_z)_n \cdot (\mathbf{i}x + \mathbf{j}y + \mathbf{k}z)_n(\mathbf{i}x + \mathbf{j}y + \mathbf{k}z)_n - \\ &\quad r^2(\mathbf{i}p_x + \mathbf{j}p_y + \mathbf{k}p_z)_n]/(4\pi\epsilon_0 r^5)\;.\end{aligned} \tag{4.70}$$

If the dipole lies along $z$, $p_x = p_y = 0$ and $p_z = p$. Hence,

$$E_{n,z} = [1/(4\pi\epsilon_0 r_n^5)](3z_n^2 - r^2 p_n)\;, \tag{4.71}$$

and for all particles in the sphere,

$$E_z = [1/(4\pi\epsilon_0 r^5)]p\sum_{n=1}^{N}(3z_n^2 - x_n^2 - y_n^2 - z_n^2)\;. \tag{4.72}$$

In an isotropic medium, all magnitudes of $x_n$, $y_n$ and $z_n$ are equally probable. Thus, $\overline{x_n} = \overline{y_n} = \overline{z_n} = 0$ and $\overline{x_n^2} = \overline{y_n^2} = \overline{z_n^2}$. From the second of these two conditions it follows, from (4.72), that $E_z$ is zero. Similar analyses may be used to show that $E_x$ and $E_y$ are also zero. The result means that because of the high symmetry, effectively cubic, of the isotropic dielectric, the internal field $E_i$ is zero. Thus from (4.61) and (4.65),

$$E_r = E + P/(3\epsilon_0) . \tag{4.73}$$

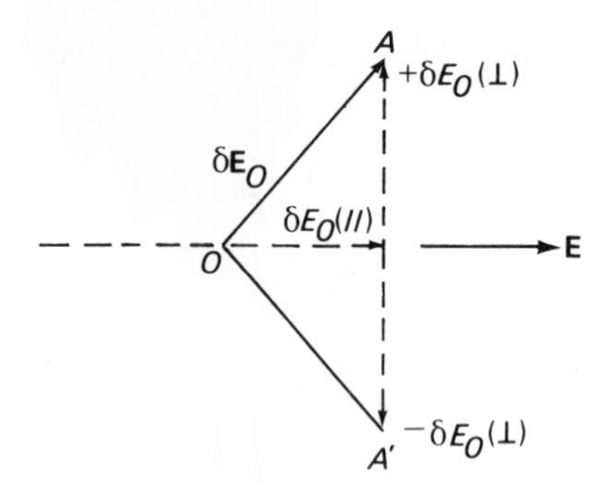


Figure 4.14 – Resolution of the field along $OA$ into components perpendicular and parallel to the applied field, **E**.

From (4.14), we may write

$$P = N\alpha E_r , \tag{4.74}$$

and substituting in (4.73),

$$P = N\alpha E + N\alpha P/(3\epsilon_0) , \tag{4.75}$$

or

$$P/E = N\alpha/\{1 + [N\alpha/(3\epsilon_0)]\} . \tag{4.76}$$

Using (4.12)

$$(\epsilon_r - 1) = N\alpha/\{\epsilon_0 - (N\alpha/3)\} . \tag{4.77}$$

Now

$$(\epsilon_r + 2) = (\epsilon_r - 1) + 3 , \tag{4.78}$$

or

$$(\epsilon_r + 2) = (N\alpha + 3\epsilon_0 - N\alpha)/\{\epsilon_0 - (N\alpha/3)\} . \tag{4.79}$$

Thus

$$\frac{(\epsilon_r - 1)}{(\epsilon_r + 2)} = N\alpha/(3\epsilon_0) , \tag{4.80}$$

and, using (4.22)

$$\frac{(\epsilon_r - 1)}{(\epsilon_r + 2)} \frac{M}{\rho} = L\alpha/(3\epsilon_0) , \tag{4.81}$$

where $\alpha$ may be regarded as the sum $(\alpha_e + \alpha_a + \alpha_o)$.

This relation is known as the Clausius-Mosotti equation. In appropriate experimental circumstances (see Section 4.3.2), $\epsilon_r$ may be replaced by $n^2$, the square of the refractive index, to give the Lorentz-Lorenz equation:

$$\frac{(n^2 - 1)}{(n^2 + 2)} \frac{M}{\rho} = L\alpha_e/(3\epsilon_0) \ . \tag{4.82}$$

These two equations are obeyed well for polar liquids, dilute solutions of polar substances in non-polar liquids, amorphous solids and cubic solids. We may consider (4.80) in the light of earlier results obtained in this chapter. Thus if the interaction between molecules is weak, we infer that $\epsilon_r \longrightarrow 1$. In this case, $(\epsilon_r + 2) \approx 3$, and (4.80) degenerates to (4.21). If $\epsilon_r \approx 4$, say, neglect of the strong interaction term gives a value for $(\epsilon_r - 1)$ which is twice too large. If $\epsilon_r$ is very large, say about 75, then the same approximation would make $(\epsilon_r - 1)$ twenty-five times too large.

### 4.4.3 Molar polarization

The left-hand side of (4.81) is termed the molar polarization†, $P_{\mathrm{m}}$. The molar polarization of a mixture is an approximately additive property of the components. Thus if we have a solution consisting of $n_A$ mol of component $A$ and $n_B$ mol of $B$, then

$$P_{\mathrm{m}}(AB) = x_A P_{\mathrm{m}}(A) + x_B P_{\mathrm{m}}(B) \ , \tag{4.83}$$

where the mole fraction $x_A$, for example, is given by

$$x_A = n_A/(n_A + n_B) \ . \tag{4.84}$$

In a similar manner, the left-hand side of (4.82) is called a molar refraction; it too, is an additive property.

*Example: Chlorobenzene*

Table 4.2 lists some results for dilute solutions of chlorobenzene (polar) in benzene (non-polar). Preliminary calculations to obtain $P_{\mathrm{m}}$ (chlorobenzene), extrapolated to infinite dilution, from the total molar polarization, (4.83), have been carried out already.

†Molar polarization is a chemical term; it is not a true polarization, and it has the units of a molar volume.

**Table 4.2** – Molar polarization of $C_6H_5Cl$ in $C_6H_6$

| $T$/K | $(10^3/T)$/K$^{-1}$ | $10^4P_m(C_6H_5Cl)$/m$^3$mol$^{-1}$ |
|---|---|---|
| 193 | 5.18 | 1.065 |
| 213 | 4.69 | 1.000 |
| 233 | 4.29 | 0.940 |
| 253 | 3.95 | 0.893 |
| 273 | 3.66 | 0.855 |
| 293 | 3.41 | 0.815 |
| 313 | 3.19 | 0.778 |
| 333 | 3.00 | 0.755 |

Now, from (4.81) $P_m$ represents the total molar polarization of the species and, by analogy with (4.32), we may write

$$P_m = P_{m,e} + P_{m,a} + P_{m,o} . \quad (4.85)$$

From (4.53), $\alpha_o = Np^2/3kT$, and using (4.85) with (4.81), we obtain

$$P_m = P_{m,e} + P_{m,a} + Lp^2/(9\epsilon_0 kT) . \quad (4.86)$$

Equation (4.86) is of the form

$$P_m = c + b/T . \quad (4.87)$$

Hence a graph of $P_m$ against $1/T$ should be a straight line of slope $Lp^2/(9\epsilon_0 k)$. The graph for the $C_6H_5Cl$ results is shown in Figure 4.15, and its slope is $1.43 \times 10^{-2}$ m$^3$ mol$^{-1}$ K. Hence,

$$p^2 = \frac{1.43 \times 10^{-2} \times 9 \times 8.854 \times 10^{-12} \times 1.38 \times 10^{-23}}{6.022 \times 10^{23}} \text{ C}^2 \text{ m}^2,$$

and the dipole moment, $p$, is $5.11 \times 10^{-30}$ C m.
From the graph at $10^3/T = 3.15$, $10^4 P_m = 0.775$. Hence the intercept $c$ is $3.25 \times 10^{-5}$ m$^3$ mol$^{-1}$; this quantity is the sum of $P_{m,a}$ and $P_{m,e}$. The refractive index for $C_6H_5Cl$ is 1.52. Using next the Lorentz-Lorenz equation (4.82):

$$P_{m,e} = (1.3104/4.3104)(112.56 \times 10^{-3}/1107) = 3.09 \times 10^{-5} \text{ m}^3 \text{ mol}^{-1}$$

Hence $P_{m,a} = 0.16$ m$^3$ mol$^{-1}$, a small fraction (1/20) of $P_{m,e}$.

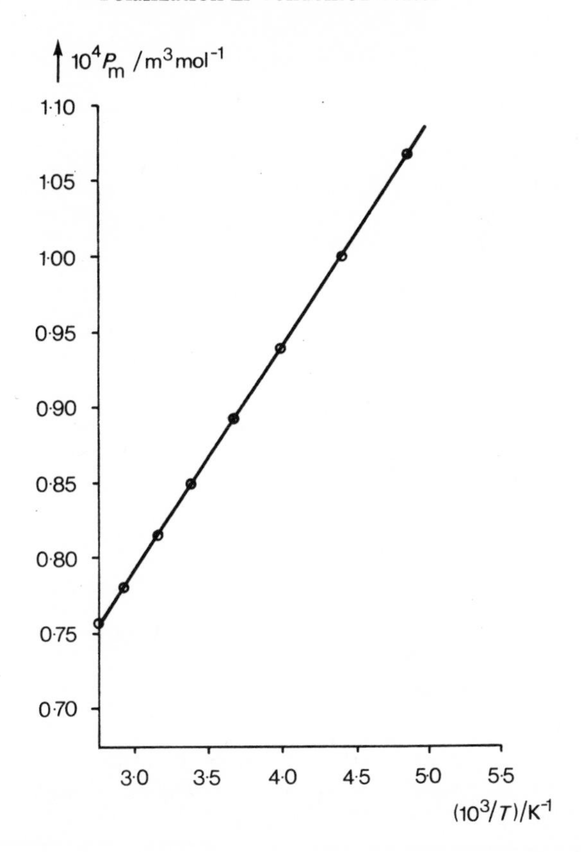


Figure 4.15 – Variation of the molar polarization ($P_m$) of chlorobenzene with temperature ($T$).

### 4.4.4 Effect of frequency in the condensed states

The dispersion of polarization in the condensed states follows the pattern found for gases. However owing to much higher viscosity, liquids show a much more rapid decrease of $P$ with increase in frequency than do gases. In the solid state, orientation effects may be absent altogether, since rotation of molecules by the applied field is hindered by strong intermolecular attractions and by steric effects. Water provides a good illustration.

The water molecule is strongly polar, $p = 6.0 \times 10^{-30}$ C m, and the attractive forces between the molecules, both in the liquid and the solid, are enhanced by hydrogen bonding. Figure 4.16 illustrates the dispersion of $\epsilon_r$ ($H_2O$) between 213 K and 293 K. It shows that the orientation effect falls off rapidly above $10^{10}$ Hz at 300 K, at about $10^3$ Hz at 250 K, and is hardly effective at all at 210 K.

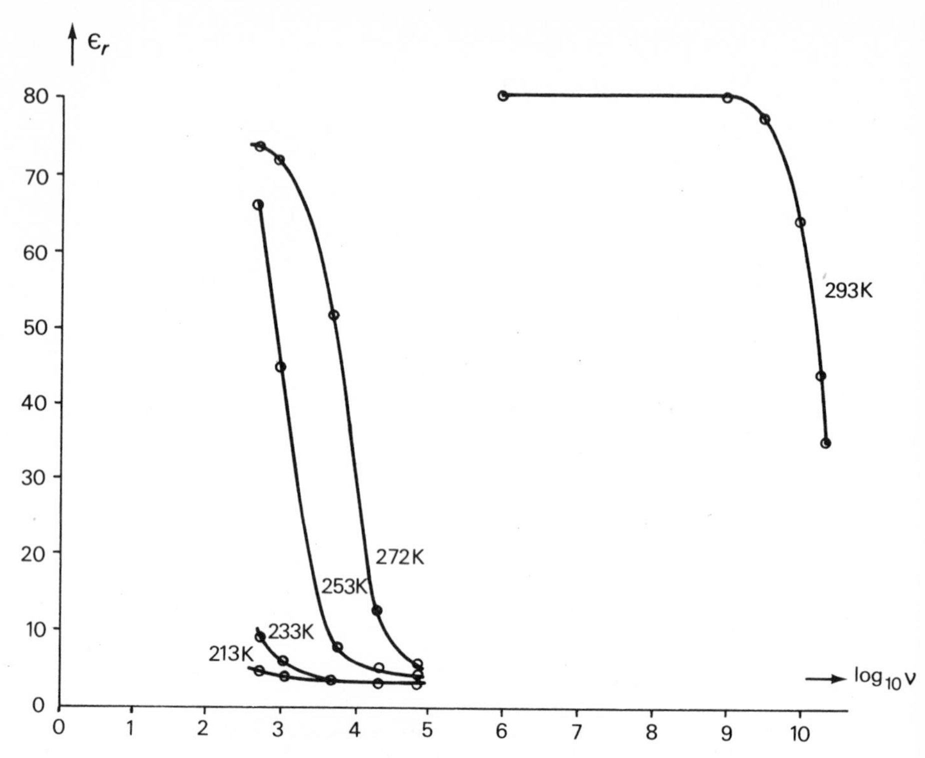


Figure 4.16 – Variation of relative permittivity ($\epsilon_r$) for ice and water with frequency ($\nu$) and temperature.

## 4.5 THE STATIC DIELECTRIC CONSTANT

### 4.5.1 Gases

From (4.56) and (4.81), but ignoring the strong interaction term $3/(\epsilon_r + 2)$ on the left-hand side of (4.81) we may write the total molar polarization of a gas as

$$P_m = (\epsilon_{r,s} - 1)(M/\rho) = (L/\epsilon_0)\{\alpha_e + \alpha_a + p^2/(3kT)\}\ , \qquad (4.88)$$

where $\epsilon_{r,s}$ is the static dielectric constant. The orientation contribution is strongly dependent upon temperature. This fact is well illustrated by methane and its chloro-derivatives: Figure 4.17 shows the variation of $\epsilon_{r,s}$ with $1/T$. Methane and tetrachloromethane have zero dipole moments, and hence their dielectric constants do not vary with temperature. The dipole moments of $CHCl_3$, $CH_2Cl_2$ and $CH_3Cl$ are $3.4 \times 10^{-30}$, $5.2 \times 10^{-30}$ and $6.2 \times 10^{-30}$ C m, respectively. Thus from the ratios of values of $p^2$, we predict the slopes of the graphs for these compounds (Figure 4.17) to be in the ratio 1.0:2.3:3.3, and this expectation is borne out by experiment.

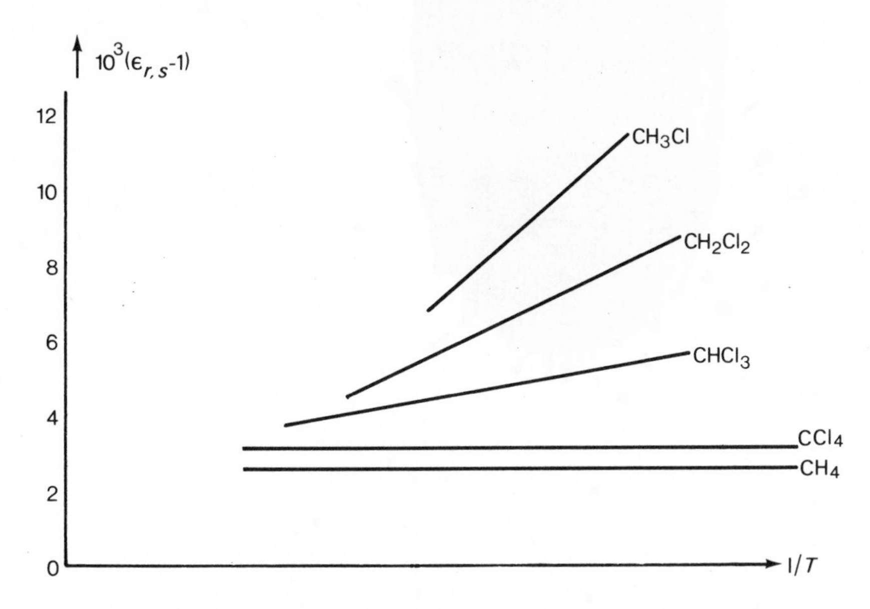


Figure 4.17 – Variation of static relative permittivity ($\epsilon_{r,s}$) of methane and the chloromethanes with temperature ($T$).

*Example: Hydrogen bromide*

As an example calculation on a polar gas we shall consider some experimental data on hydrogen bromide. Table 4.3 lists data for $\epsilon_{r,s}$ at three temperatures. From (4.88), we expect a straight line relationship between $P_m$ and $1/T$. The appropriate graph is shown in Figure 4.18; its slope is $Lp^2/(3k\epsilon_0)$, and its intercept is $(L/\epsilon_0)(\alpha_e + \alpha_a)$, or $(P_{m,e} + P_{m,a})$.

**Table 4.3 – $\epsilon_{r,s}$ data for HBr**

| $T$/K | $\epsilon_{r,s}$ | $(10^3/T)K^{-1}$ | $(10^6(\epsilon_{r,s}-1)M/\rho)/m^3\,mol^{-1}$ |
|---|---|---|---|
| 268 | 1.00323 | 3.73 | 70.88 |
| 294 | 1.00279 | 3.40 | 67.26 |
| 339 | 1.00223 | 2.95 | 62.21 |

From the graph the slope is $11.33 \times 10^{-3}\ m^3\ mol^{-1}\ K$ and thus $p$ is $2.63 \times 10^{-30}$ C m. The strong interaction term, $3/(\epsilon_{r,s} + 2)$, is about 0.9991: neglect of this term is of very small consequence on the final value of $p$, as the reader may care to verify. The refractive index of HBr is 1.00051. Following the previous example, but neglecting the term $3/(n^2 + 2)$, leads to the values $24.6 \times 10^{-6}$ and $4.1 \times 10^{-6}\ m^3\ mol^{-1}$ for $P_{m,e}$ and $P_{m,a}$, respectively.

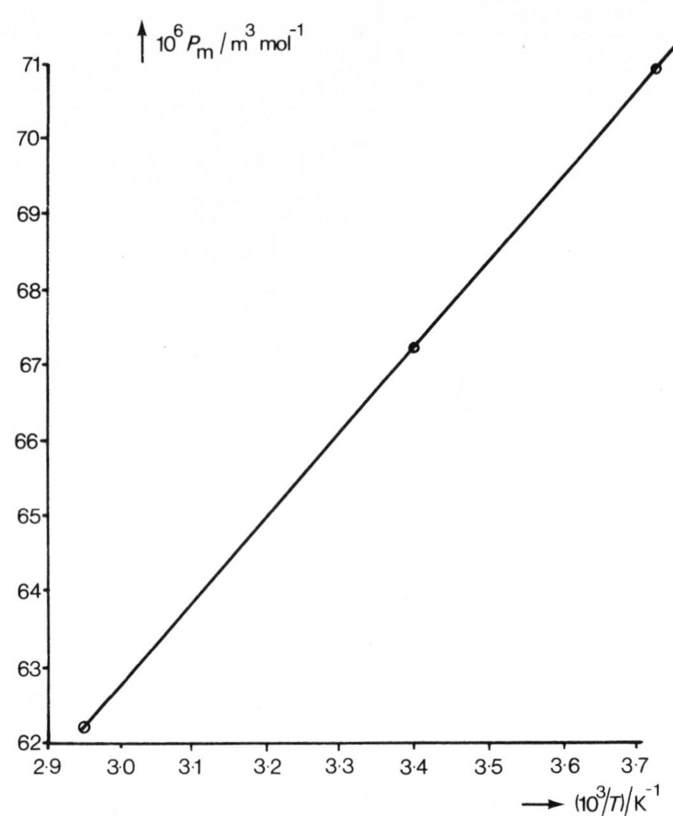


Figure 4.18 – Variation of the static relative permittivity ($\epsilon_{r,s}$) of gaseous hydrogen bromide with temperature ($T$).

### 4.5.2 Isotropic solids

Here we shall consider isotropic solids in three groups:

(a) those with $P_a = P_o = 0; P_e \neq 0$;
(b) those with $P_o = 0; P_e, P_a \neq 0$;
(c) those with $P_e, P_a, P_o \neq 0$.

In group (a), where static polarization arises from electron displacement only, we must be considering elements in their solid states. Hence, from (4.81),

$$\frac{(\epsilon_{r,s} - 1)M}{(\epsilon_{r,s} + 2)\rho} = L\alpha_e/(3\epsilon_0) \ . \tag{4.89}$$

This equation is difficult to verify for solid elements, because in order to vary $(M/\rho)$ the temperature must be altered, and the range of the experimental practicability of this procedure is small. However for this class of solid we may obtain a measure of $P_e$ from both $\epsilon_{r,s}$ and $n$. Thus for diamond, $\epsilon_{r,s}$ is 5.84 and $n^2$ is 5.843. Clearly (4.81) and (4.82) will give closely similar results for $P_e$.

Solids with more than one atomic species but possessing no permanent dipole moment may exhibit both $P_e$ and $P_a$ effects (group b). Ionic crystals are examples of substances with high values of $P_a$, as the following data on the potassium halides show:

| | KF | KCl | KBr | KI |
|---|---|---|---|---|
| $n^2$ | 1.90 | 2.22 | 2.43 | 2.81 |
| $\epsilon_{r,s}$ | 6.05 | 4.68 | 4.78 | 4.94 |

Perturbations of the vibrations of the positive and negative ions by an electrostatic field produce appreciable atom polarization effects. Since $P \propto (\epsilon - 1)$, the ratios of $P_a/P_e$ for these halides are 5.6, 3.0, 2.6 and 2.2, respectively, much in excess of the values found in non-ionic compounds. The large values of $P_a$ are associated with elastic ionic displacements: they decrease with a decrease in temperature because the amplitudes of the vibrations of the ions decrease.

### 4.5.3 Polarization in ionic compounds – a digression

In making a first-order correction to the cohesive energy of an ionic crystal, a term $-C/r^6$, representing an interaction between fluctuating dipoles, is added to the cohesive energy $U(r)$. The constant $C$ is given by

$$C = (S_{+-}c_{+-} + S_{++}c_{++}/2 + S_{--}c_{--}/2)/(4\pi\epsilon_0)^2 \tag{4.90}$$

where

$$S_{ij} = \sum_{\substack{i,j \\ i \neq j}} r_{ij}^{-6} \tag{4.91}$$

and

$$c_{ij} = (3eh\alpha_i\alpha_j/4\pi m_e^{1/2})\,[(\alpha_i/N_i)^{1/2} + (\alpha_j/N_j)^{1/2}]^{-1} \tag{4.92}$$

Equation (4.91) represents a 'lattice sum' interaction between like and unlike ions for all distances $r_{ij}$: for the NaCl structure type (Figure 1.2) $S_{+-} = 6.5952$ and $S_{++} = S_{--} = 1.8067$. In (4.92) $\alpha_i$ and $\alpha_j$ are the electron polarizabilities of the positive and negative ions respectively, and may be calculated with (4.82). The quantities $N_i$ and $N_j$ are the effective numbers of polarizable electrons in the ions. For $Li^+$, $Na^+$, $K^+$, $Rb^+$ and $Cs^+$, $N_i$ takes the values of 2, 10, 18, 23 and 31 respectively: for $F^-$, $Cl^-$, $Br^-$ and $I^-$, the $N_j$ values are 8, 17, 22 and 30 respectively.

*Example: Polarizability of* KCl

If we use the value of $n^2$ from above in (4.82), together with standard data in KCl, we have

$$\frac{1.22 \times 0.07456}{4.22 \times 1984} = \frac{6.022 \times 10^{23} \times \alpha(\text{KCl})}{3 \times 8.854 \times 10^{-12}} ,$$

and hence, $$\alpha(\text{KCl}) = 4.79 \times 10^{-40}\ \text{F m}^2 .$$

This value may be separated into components for the positive and negative ions by using the fact that ionic polarizabilities are approximately additive, together with an equation of Born and Heisenberg:

$$\alpha(\text{KCl}) = \alpha(\text{K}^+) + \alpha(\text{Cl}^-) , \tag{4.93}$$

and $$\alpha_i = A/(Z_i - S_i)^3 : \tag{4.94}$$

$A$ is a constant for isoelectronic ions (like $K^+$ and $Cl^-$), $Z_i$ is the atomic number of the $i$th ion and $S_i$ is the quantum mechanical screening constant (calculated from Slater's rules; see Appendix 9). Thus,

$$\alpha(\text{K}^+)/\alpha(\text{Cl}^-) = (6.10)^3/(7.75)^3 = 0.4876 ,$$

and $$\alpha(\text{K}^+) + \alpha(\text{Cl}^-) = 4.79 \times 10^{-40}\ \text{F m}^2 .$$

whence, $$\alpha(\text{K}^+) = 1.57 \times 10^{-40}\ \text{F m}^2 ,$$

and $$\alpha(\text{Cl}^-) = 3.22 \times 10^{-40}\ \text{F m}^2 .$$

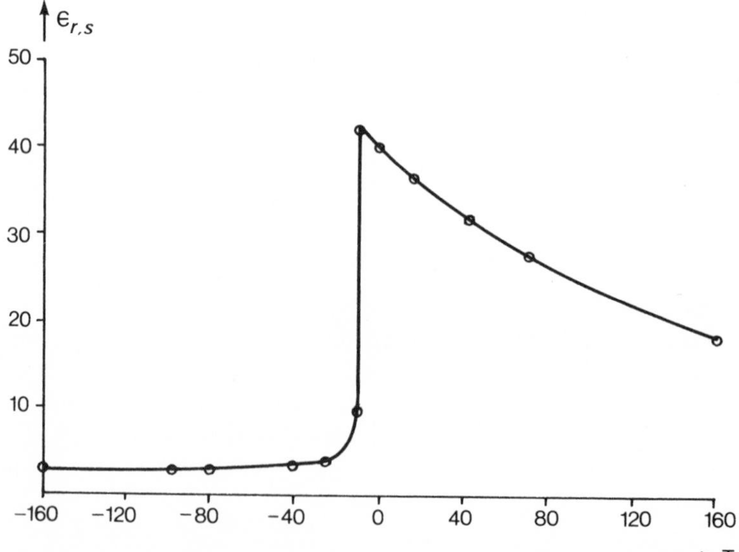


Figure 4.19 – Variation of the static relative permittivity ($\epsilon_{r,s}$) for nitrobenzene with temperature ($T$).

The effect of this type of calculation on the cohesive energy is illustrated in Chapter 6.

Solids in the third group (c), although possessing dipolar character, may not exhibit large values for $\epsilon_{r,s}$ because the molecular dipoles cannot easily, if at all, rotate into the field in the solid state. However as the crystal is brought towards its melting point temperature, alignment of dipoles is partially attained. Near the melting point, $\epsilon_{r,s}$ rises sharply and then falls off with further increase in temperature. The latter effect is a consequence of the thermal motion of the molecules in the liquid, as expressed by the term $p^2E/(3kT)$ in the Langevin theory. As an example, a diagrammatic representation of the behaviour of nitrobenzene is shown in Figure 4.19.

## 4.6 DIPOLE MOMENTS

Dipole moments provide infrormation on several interesting molecular features:

(a) the extent to which a molecule is permanently polarized;

(b) the geometry of the molecule, particularly bond angles; and

(c) approximate values for the charges on individual atomic species.

Additionally dipole moments are of considerable use as one criterion of the validity of theoretical calculations of equilibrium molecular geometries. The dipole moment of a molecule may be measured with good precision, and its calculated value is sensitive to both the magnitudes of the atomic charges and their relative spatial positions.

The dipole moment of a molecule is approximately the vector sum of the individual bond moments of the molecule, but it should be remembered that often an appreciable contribution to the dipole moment arises from non-bonding p orbitals of lone pairs, and not just from the displacement of charge along the $\sigma$-bond between two atoms. The water molecule is a good example of this effect (see page 199).

In an example, we determined the dipole moment for chlorobenzene from experimental data as $5.11 \times 10^{-30}$ C m. We can regard this value as a bond moment. Thus, in the sequence of substituted chlorobenzenes, I to VI, we can obtain other dipole moments, approximately, from the appropriate vector sums:

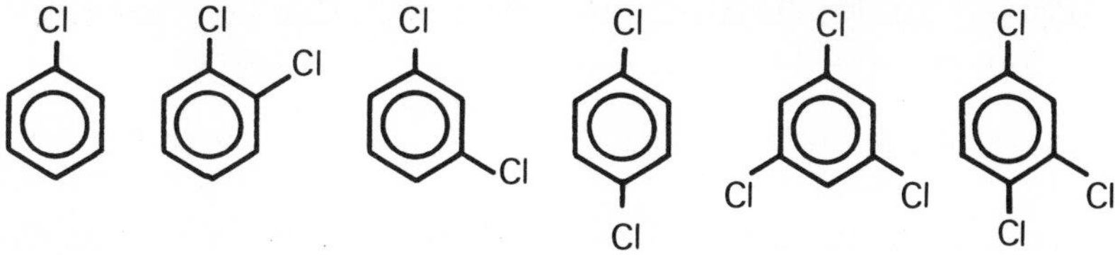

| | I | II | III | IV | V | VI |
|---|---|---|---|---|---|---|
| $10^{40}p$/C m | (5.1)† | 8.8(8.5)† | 5.1(5.4)† | 0 | 0 | 5.1(5.3)† |

†Experimental values.

The dipole moment of chloroethane is $6.8 \times 10^{-30}$ C m, larger than that for chlorobenzene, whereas one might have expected that the $\pi$-electrons would have intensified the effect of the negative charge on Cl in $C_6H_5Cl$. The 'opposite' result may be explained, in valence-bond terms, by the existence of resonance structures, VII to X, for chlorobenzene. Dipole moments are little used today in detailed structural studies, having been replaced by much more powerful diffraction and spectroscopic techniques.

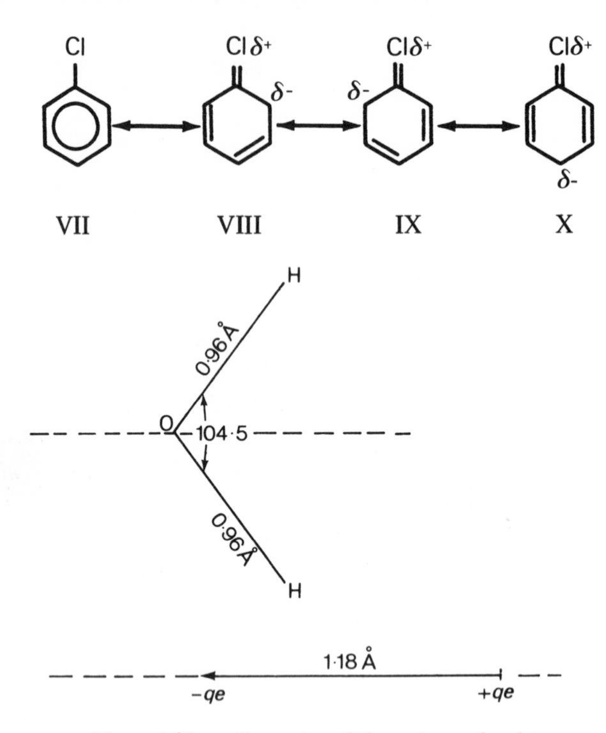


Figure 4.20 – Geometry of the water molecule.

The dipole moment may be used to obtain approximate values for the effective charges on the atoms in a molecule. Consider the water molecule (Figure 4.20). The dipole moment is $6.0 \times 10^{-30}$ C m: from the dipole diagram we may write

$$6.0 \times 10^{-30} = 1.602 \times 10^{-19} \times q \times 1.175 \times 10^{-10} ,$$

whence $q$, the charge on oxygen is 0.32 electrons. Hence the charge on each hydrogen atom is 0.16 electrons. It is interesting to note that the values obtained from a modern *ab initio* molecular orbital calculation are 0.36 and 0.18 electrons, respectively.

## 4.7 VAN DER WAALS' FORCES

In molecular compounds, the cohesive energies are often only about one tenth of those found for ionic and covalent solids. Van der Waals' attractive forces exist in all atomic and molecular assemblies, but they take on considerable importance in molecular solids where they provide the main mechanism for intermolecular attraction. In order to obtain an idea of the magnitude of these forces we shall look again at the example of argon.

Consider two argon atoms: each has a complete outer electron energy level, and the electronic distribution about its nucleus, averaged over a time which is large compared with the period of fluctuation of the electron density, has spherical symmetry. However, at any instant in time, a certain degree of asymmetry exists in the electron density and the atom behaves as a dipole of moment $p$. At a distance $r$ from this dipole, measured, for convenience, along its axis, a second argon atom experiences a field, $E$, given by

$$E = p/(4\pi\epsilon_0 r^3) \ . \tag{4.95}$$

The second atom is polarized by this field and acquires a dipole moment, $p'$, given by

$$p' = \alpha E \ . \tag{4.96}$$

From (4.33), the energy of the dipole of moment $p'$ is given by

$$U = -\alpha E^2 = -\alpha p^2/(16\pi^2\epsilon_0^2 r^6) \ . \tag{4.97}$$

Figure 4.21 illustrates this sequence of events. The importance of (4.97) is that it shows that $U$ is both attractive and depends upon $1/r^6$. A more detailed treatment shows that the second dipole will react with the first; for two particles, not necessarily of the same type, of polarizabilities $\alpha_1$ and $\alpha_2$, London (1930) deduced that the energy of attraction is given by

$$U_L = -0.75h\nu_0\alpha_1\alpha_2/(16\pi^2\epsilon_0^2 r_{12}^6) \ , \tag{4.98}$$

where $\nu_0$ is the (characteristic) frequency of fluctuation of the electron density and $r_{12}$ is the distance between the two particles under consideration. Whatever the direction of the instantaneous fluctuation of the electron density around a given atom, the resultant effect is of an attractive nature, thus effecting or enhancing cohesion.

Equation (4.98) may be regarded as satisfactory for distances, $r$, less than about $3 \times 10^{-8}$ m (300 Å). For distances greater than about $3 \times 10^{-7}$ m, the field generated at one atom takes approximately $3 \times 10^{-7}/(3 \times 10^8)$ s to reach its neighbours. This time, $10^{-15}$ s, is commensurate with the period of fluctuation of the electron density and consequently interference occurs. A detailed theoretical treatment shows that the energy of interaction is reduced, being then proportional to $1/r^7$.

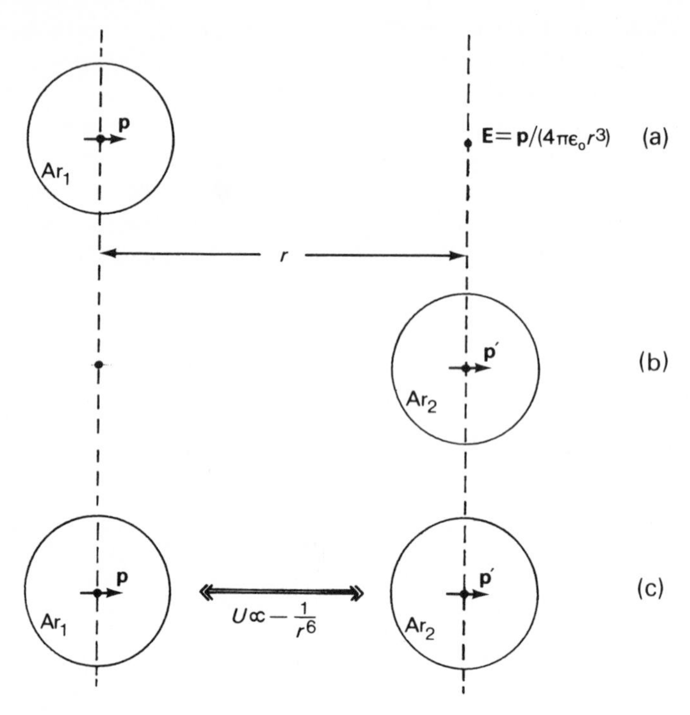


Figure 4.21 – Van der Waals' attraction: (a) Instantaneous dipole **p** in atom $Ar_1$ produces a field $\mathbf{E} = \mathbf{p}/(4\pi\epsilon_0 r^3)$ at a distance $r$ from it; (b) second argon atom at this point is polarized by this field, producing dipole **p′**, commensurate with **p**; (c) potential energy between the atoms, proportional to $-1/r^6$.

## 4.8 OTHER ATTRACTIVE FORCES

It was shown by Keesom that molecules with a permanent dipole moment $p$ interact further, giving rise to a cohesive energy which is proportional to $p^4/r^6$: actually $U_K = -(2/3)p^4/(16\pi^2\epsilon_0^2 kTr^6)$. This term cannot strictly represent a potential energy since it contains a dependence on temperature. The potential energy of two dipoles is proportional to $1/r^3$, and Keesom's result is obtained as an orientational average, assuming that the polarization of the molecules is isotropic. A further refinement by Debye took account of the interaction between the permanent and induced dipoles, leading to a term proportional to $\alpha p^2/r^6$; actually, $U_D = -2\alpha p^2/16\pi^2\epsilon_0^2 r^6$.

The three interactional energies, although small individually, may all be involved in cohesion. The energy needed to disperse the particles in a molecular solid into the gas phase may be represented approximately by the enthalpy of sublimation, $\Delta H_S$. It therefore provides an experimental quantity against which the energy model may be tested. The agreement between the results in columns 5 and 6 of Table 4.4 is good and lends support to the theory.

**Table 4.4** – Van der Waals' energies/kJ mol$^{-1}$

| | $U_K$ | $U_D$ | $U_L$ | Total | $-\Delta H_s$ |
|---|---|---|---|---|---|
| Ar | 0 | 0 | −8.5 | −8.5 | −8.4 |
| $H_2$ | 0 | 0 | −0.8 | −0.8 | −0.8 |
| CO | 0 | 0 | −8.4 | −8.4 | −7.9 |
| HCl | −3.3 | −1.0 | −16.8 | −21.1 | −20.0 |
| $H_2O$ | −36.4 | −1.9 | −9.0 | −47.3 | −47.3 |

## 4.9 STRUCTURAL AND PHYSICAL CHARACTERISTICS OF THE VAN DER WAALS' BOND

Van der Waals' forces can link an atom to an indefinite number of neighbours, and are undirected spatially. In the solid state of an inert gas, van der Waals' forces are the sole means of cohesion. In other molecular compounds we find relatively short covalent bonds between atoms, such as Cl-Cl (2.0 Å) in $Cl_2$ or C-C (1.40 Å) and C-H (1.08 Å) in benzene, but with characteristically longer (3.6 – 3.8 Å) contact distances between nearest non-bonded species.

The dependence of the van der Waals' energy on polarizability is shown by the trends in the melting point temperatures of the silicon tetrahalides, $SiX_4$:

| | $SiF_4$ | $SiCl_4$ | $SiBr_4$ | $SiI_4$ |
|---|---|---|---|---|
| $10^{40}\alpha_X$/F m$^2$ | 1.0 | 3.4 | 4.8 | 7.3 |
| Mp/K | 183 | 203 | 278 | 394 |

In these compounds the polarizability of the halogen increases more rapidly from F to I than does the intermolecular distance, with consequent enhancement of the cohesive energy. Over a corresponding range of molecular mass, say among the alkanes, the increase in melting point is only about 140 K.

Molecular compounds generally form soft, brittle crystals with low melting points and large thermal expansivities, both of these properties being usually very anisotropic. The electrical and optical properties of molecular solids may be said to be the aggregate of those of the component molecules, since the electron systems do not interact strongly in the solid state, and these properties are similar in the melt and in solution.

Molecular compounds are usually soluble in common solvents. The 'rule' like dissolves like is well known, but is deserving of some elaboration. Solubility is a chemical reaction between the solute and the solvent and, generally, large enthalpic effects are involved. These factors may be compounded into a free energy of dissolution, but while this parameter predicts solubility correctly, it is always more informative to consider the interplay of the enthalpy and entropy changes in a study of solubility.

In ionic solids, enthalpic effects are usually large because they involve strong interactions between the ions and the solvent. In molecular compounds however enthalpy changes are often very small, and the entropic component may become very important. We know that naphthalene dissolves in benzene but not in water, and that glucose dissolves in water but not in benzene to any significant extent.

Water is a hydrogen-bonded liquid, and a dissolution process in water as a solvent involves the breaking of some hydrogen bonds in order to accommodate the solute molecules. If the enthalpy change of solvation is very small numerically and the entropy gain relatively negligible, then the forces of attraction between the solvent molecules will not be overcome, and dissolution will not take place (napthalene and water). If, however a solvent for napthalene is chosen within which intermolecular attractions are very weak, the entropy gain of the solution over the (solvent + solute) system is dominant and dissolution occurs (napthalene and benzene). In the case of glucose and water, the hydrogen-bonded solvent is broken down because the polar —OH groups in the glucose molecules interact electrostatically with the water molecules. On the other hand the presence of benzene has insufficient effect on the intermolecular attractions in glucose and this compound is insoluble in benzene.

In all cases the balance of solubility is determined by the difference between the free energy of the solution and the free energy of the (solute + solvent) systems. In the few examples where sufficient thermodynamic data are available, the discussion given above can take on a quantitative significance.

## 4.10 CLASSIFICATION AND EXAMPLES OF VAN DER WAALS' SOLIDS

Van der Waals' solids are of divers types and extensive in number. Different authors recognise different numbers of types; here we shall enumerate four to form the basis of a classification:

(a) The 'inert' elements;
(b) elements in Periodic Groups VB – VIIB;
(c) small inorganic molecules; and
(d) organic compounds, including hydrogen-bonded compounds, clathrate compounds, charge-transfer complexes and $\pi$-electron overlap compounds.

Clathrate compounds exist only in the solid state, and may exhibit variable composition. There is often only slight interaction between the components, the host structure acting as merely a mechanical trap for the occluded molecule. A well known example of a host structure is that of 1,4-dihydroxybenzene (quinol), and this compound forms clathrates with small molecules such as $SO_2$, $CH_3OH$ and $CH_3CN$. The quinol/$CH_3CN$ compound has a small dielectric constant whereas that of the quinol/$CH_3OH$ clathrate is large. We may infer that the $CH_3CN$ molecules are in orientational or dynamic disorder, not strongly linked to the host structure. In the quinol/$CH_3OH$ clathrate however, the molecules of $CH_3OH$ are more or less locked in position, probably aided by hydrogen bonding. Ammonia, nickel cyanide and benzene form the clathrate compound illustrated in Figure 4.22. The $Ni^{2+}$ ions are coordinated octahedrally to two $NH_3$ molecules and four $[CN]^-$ ions, and also to 4 $[CN]^-$ ions in a square-planar array. The groups link to form a cage for the benzene molecule.

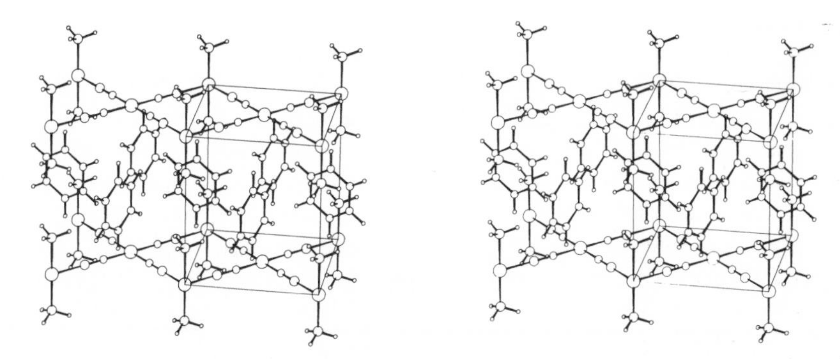

Figure 4.22 – Nickel cyanide-ammonia-benzene clathrate structure, with the unit cell outlined; circles in decreasing order of size are Ni, N, C and H. The shortest contact distances, C(benzene) – C(CN group) are 3.6 Å.

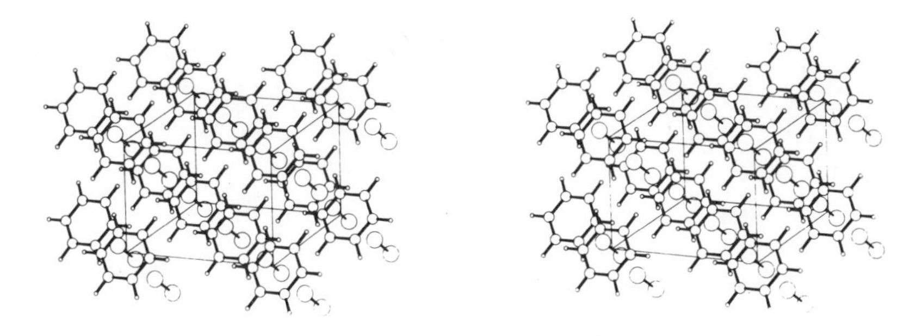

Figure 4.23 – Benzene-chlorine charge-transfer structure, with the unit cell outlined; circles in decreasing order of size are Cl, C and H. The shortest distances, Cl – benzene ring centre are 3.3 Å.

In contrast, benzene and chlorine form a compound $C_6H_6{\cdot}Cl_2$, a charge-transfer compound (Figure 4.23) in which stability arises from $\pi$-electron transfer from the benzene ring to the chlorine molecule. Bromine and benzene form a similar compound, but the benzene-iodine charge-transfer complex is known only in solution.

Electron overlap phenomena are important among aromatic hydrocarbons, such as anthracene (Figure 4.25f). The molecular planes are oriented such that the delocalized electrons in the $\pi$-orbitals of adjacent molecules can overlap, leading to enhanced stability. The following data show an interesting relationship between melting point and molecular mass:

| | Mp/K | Releative molecular mass (molecular weight), $M_r$ |
|---|---|---|
| Benzene, $C_6H_6$ | 279 | 78.1 |
| Cyclohexane, $C_6H_{12}$ | 280 | 84.2 |
| Naphthalene, $C_{10}H_8$ | 353 | 128.2 |
| Decahydronaphthalene, $C_{10}H_{18}$ | 230, 241† | 138.3 |
| Anthracene, $C_{14}H_{10}$ | 490 | 178.2 |
| Tetradecahydroanthracene, $C_{14}H_{24}$ | 335, 366† | 192.4 |

Other things being equal, increased molecular mass leads to increased melting point. However, the $\pi$-electron overlap in the aromatic compounds stabilizes them appreciably with respect to their alicyclic analogues. Biphenyl, $(C_6H_5)_2$ is another interesting compound: in the gas (free molecule) state, the planes of the two rings are twisted about the central C–C bond by an angle of 45° to each other, but in the solid they are coplanar and aligned approximately parallel from one molecule to another (Figure 4.24). Again, $\pi$-electrons are playing an important cohesive role in the solid.

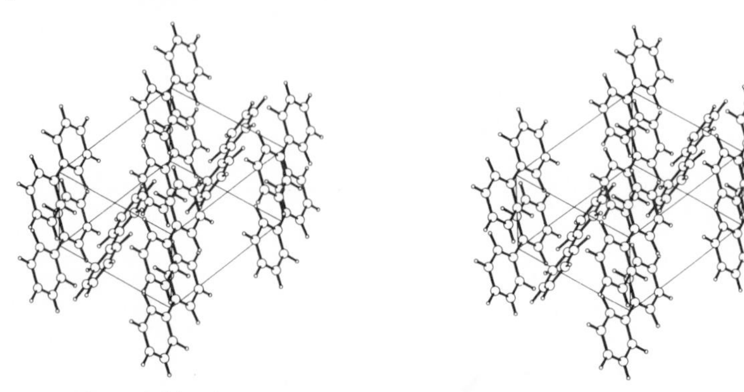

Figure 4.24 – Structure of biphenyl, $C_{12}H_{10}$, with the unit cell outlined; circles in decreasing order of size represent C and H. The shortest contact distances are 3.7 Å.

†Polymorphs.

### 4.10.1 Organic compounds

Organic compounds may be divided into three main shapes, equant, flat and long. Among these three types, we may consider both polar and apolar forms of bonding in the solid. In this way it is possible to devise a useful classification for the vast range of organic compounds that are known. Of course, as with ionic solids, the classification is not without some anomalies, but it is of considerable assistance in obtaining a general picture of this aspect of solid state chemistry. We shall treat this section only in outline, and illustrate the structural classes by diagrams. We consider first some of the structural units of organic compounds

In organic compounds, we can identify certain structural components that preserve their characteristic shapes, sizes and symmetries, within quite close limits, throughout a wide range of compounds: they are bond lengths, bond angles and, hence, complete molecular moieties.

Table 4.5 lists standard values for a range of bond lengths and bond angles. The value of a bond length or angle depends upon its environment, and the table shows this variation by means of the notation C4, C3, N2, O1, where the digit indicates the number of atoms directly bonded (the connectivity). Thus C4–C4 represents a C($sp^3$)–C($sp^3$) bond.

**Table 4.5 – Standard bond lengths and angles**

*Single bonds*

| Bond | Bond length/Å | Bond | Bond length/Å |
|---|---|---|---|
| H-H | 0.74 | C3-N2 | 1.40 |
| C4-H | 1.09 | C3-O2 | 1.36 |
| C3-H | 1.08 | C2-C2 | 1.38 |
| C2-H | 1.06 | C2-N3 | 1.33 |
| N3-H | 1.01 | C2-N2 | 1.33 |
| N2-H | 0.99 | C2-O2 | 1.36 |
| O2-H | 0.96 | N3-N3 | 1.45 |
| C4-C4 | 1.54 | N3-N2 | 1.45 |
| C4-C3 | 1.52 | N3-O2 | 1.36 |
| C4-C2 | 1.46 | N2-N2 | 1.45 |
| C4-N3 | 1.47 | N2-O2 | 1.41 |
| C4-N2 | 1.47 | O2-O2 | 1.48 |
| C4-O2 | 1.43 | | |
| C3-C3 | 1.46 | | |
| C3-C2 | 1.45 | | |
| C3-N3 | 1.40 | | |

**Table 4.5** *(C'td.)*

| Bond | Bond length/Å | Bond | Bond length/Å |
|---|---|---|---|
| *Single bonds* | | | |
| *Double bonds* | | | |
| C3–C3 | 1.34 | C2–O1 | 1.16 |
| C3–C2 | 1.31 | N3–O1 | 1.24 |
| C3–N2 | 1.32 | N2–N2 | 1.25 |
| C3–O1 | 1.22 | N2–O1 | 1.22 |
| C2–C2 | 1.28 | O1–O1 | 1.21 |
| C2–N2 | 1.32 | | |
| *Triple bonds* | | *Aromatic bonds* | |
| C2–C2 | 1.20 | C3–C3 | 1.40 |
| C2–N1 | 1.16 | C2–N2 | 1.34 |
| N1–N1 | 1.10 | N2–N2 | 1.35 |

*Bond angles*

| Apex atom | Geometry | Bond angle/deg | Example |
|---|---|---|---|
| C4 | Tetrahedral | 109.5 | $CH_4$ |
| C3 | Planar | 120 | $C_2H_4$ |
| C2 | Bent | 109.5 | –CHO |
| | Linear | 180 | HCN |
| N4 | Tetrahedral | 109.5 | $NH_4^+$ |
| N3 | Pyramidal | 107.5 | $NH_3$ |
| | Planar | 120 | $H_2N$–CHO |
| N2 | Bent | 109.5 | $H_2CHN$ |
| | Linear | 180 | HNC |
| O3 | Pyramidal | 109.5 | $H_3O^+$ |
| O2 | Bent | 104.5 | $H_2O$ |

As an example of a larger structural unit, consider the phenyl moiety: it is usually found to be planar, with C–C bonds of 1.40 Å, C–H bonds of 1.08 Å, and C–C–C angles and H–C–C angles all close to 120°. According to the nature and position of any substituents on the ring there may be small deviations from these ideal values.

Molecules in the solid state are linked by the van der Waals' forces which have been discussed in this chapter. It is found experimentally that intermolecular separations between similar atoms do not vary widely. Thus in the absence of hydrogen bonding, organic compounds containing C, H and O atoms exhibit non-bonded, or contact, distances of approximately 3.7 Å among a wide range of compounds. This feature leads to the development of van der Waals' radii for atoms, which represent the minimum distance of approach of atoms in neighbouring molecules in a crystal. This is an important parameter when determining the structure of a crystalline molecular solid. Table 4.6 lists these radii for a number of common elements. The van der Waals' radii can be correlated approximately with the size of the outer orbitals of atoms. Thus, a carbon $2p_z$ orbital containing 99% of the total $2p_z$ density extends from the nucleus to about 1.8 Å.

**Table 4.6** – Van der Waals' radii/Å

| Atom | Radius | Atom | Radius |
|---|---|---|---|
| H | 1.20 | N | 1.50 |
| O | 1.40 | F | 1.35 |
| P | 1.90 | S | 1.85 |
| Cl | 1.80 | As | 2.00 |
| Se | 2.00 | Br | 1.95 |
| Sb | 2.20 | Te | 2.20 |
| I | 2.15 | $-CH_3$ | 2.00 |
| Half-thickness of benzene ring | 1.85 | | |

Two atoms in neighbouring molecules may lie further away than the sum of their van der Waals' radii because of steric effects involving the molecule as a whole. In other compounds the van der Waals' radii sum may appear to be ignored: thus the distance between two oxygen atoms may be as small as 2.4 Å if they are hydrogen-bonded. We note that for those atoms that can exist as well defined ions, the van der Waals' radius is very similar to the ionic radius. This is in agreement with our discussion that the repulsive forces build up rapidly over a short range, because of the $1/r^6$ function. The effective size of bromine, for example, is about the same when the repulsion is balanced against the strong ionic attraction in KBr, as when balanced against the weak intermolecular attraction, in $C_6H_5Br$.

We conclude this chapter with a tabulated classification (Table 4.7) of organic compounds with illustrative examples. The reader should seek to extend his study of these classes of compounds through the standard reference works listed under Suggested Reading.

**Table 4.7 – Classification of molecular compounds**

| Molecular shape | Bonding class and characteristics | |
|---|---|---|
| | *Apolar* | *Polar* |
| EQUANT | Small, isometrically shaped molecules forming approximately close-packed structures. Examples are methane, ethane, hexachloroethane (Figure 4.25a), adamantane and cubane (Figure 4.25b). In some structures, such as methane, free rotation exists in the solids. | Small, isometrically shaped molecules, with dipolar or hydrogen-bonded interactions, rather than close-packing, dominating the structural configuration. Examples are urea, methanol, pentaerythritol (Figure 4.25c, methylamine and oxalic acid dihydrate (Figure 4.25d). |
| FLAT | Molecules lying with their planes nearly parallel. Staggered configurations may be adopted where $\pi$-electron overlap is possible. Examples are benzene (Figure 4.25e), hexamethylbenzene, anthracene (Figure 4.25f) and phthalocyanine. | Molecular packing dominated by dipolar or hydrogen-bonded interactions. Examples are 1,4-dinitrobenzene (Figure 4.25g), and 4-nitrophenol (Figure 4.25h). |
| LONG | Molecules lying in parallel or staggered configurations. Examples are octane (Figure 4.25i), hexane (Figure P4.1) and dicyanoethyne (Figure P4.2). With increasing temperatures, some paraffins, such as hexane, rotate in the solid state giving effective cylindrical symmetry to the molecule. | Long polar molecules, tending to associate in pairs through dipolar or hydrogen-bonded interactions. Alkan-ammonium salts are ionic, often with the carbon chains in free rotation. Examples are apidic acid, decanamide (Figure 4.25j), potassium caprate and propan-1-ammonium chloride. |

Figure 4.25 – Structures to illustrate the classification (Table 4.7) of organic compounds. In each example, the unit cell is outlined and circles in decreasing order of size represent atoms in the formula in decreasing atomic number. Hydrogen bonds are shown by double lines.

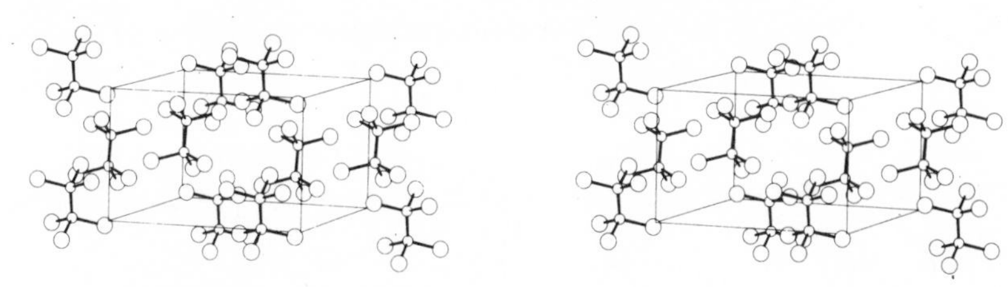

(a) Hexachloroethane, $C_2Cl_6$. Equant, apolar; shortest intermolecular contact distances, 3.7 Å.

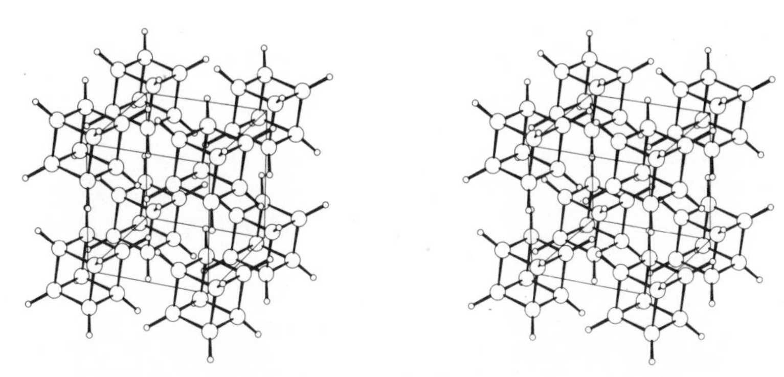

(b) Cubane, $C_8H_8$. Equant, apolar; shortest contact distance, 3.8 Å.

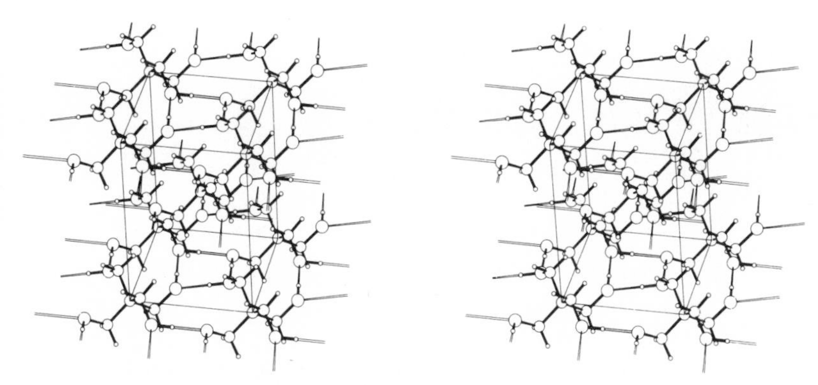

(c) Pentaerythritol, $C(CH_2OH)_4$. Equant, polar; O---H—O distances, 2.7 Å.

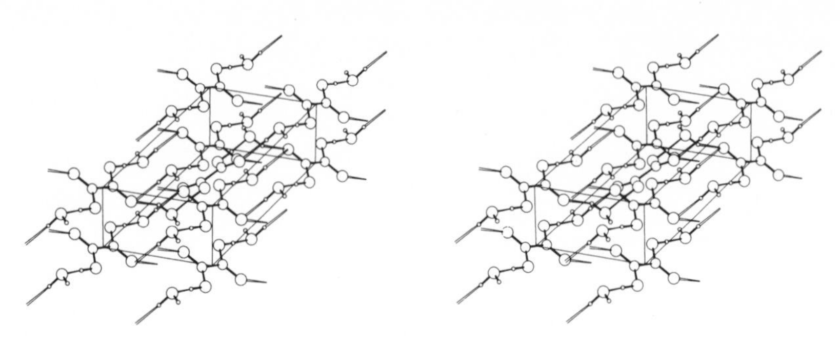

(d) Oxalic acid dihydrate. $(CO_2H)_2 \cdot 2H_2O$. Equant, polar; O---H---O distances, 2.5 Å.

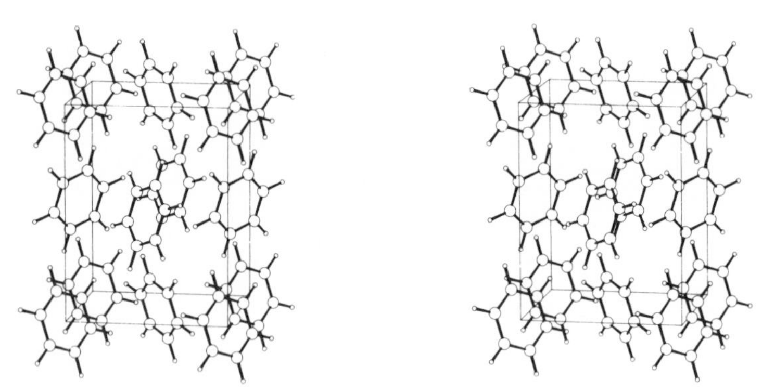

(e) Benzene, $C_6H_6$. Flat, apolar; shortest contact distances, 3.6 Å.

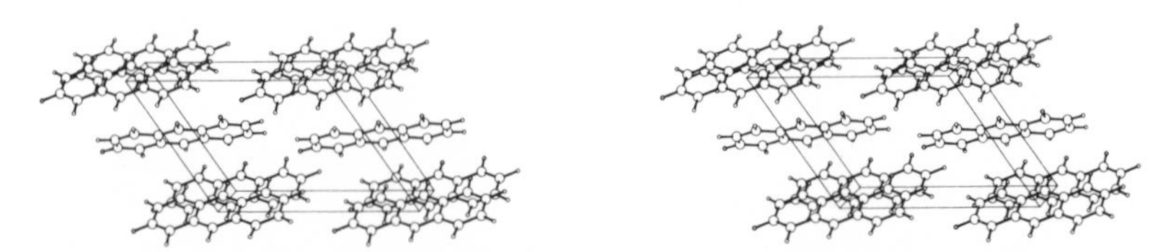

(f) Anthracene, $C_{14}H_{10}$. Flat, apolar; shortest contact distances, 3.6 Å.

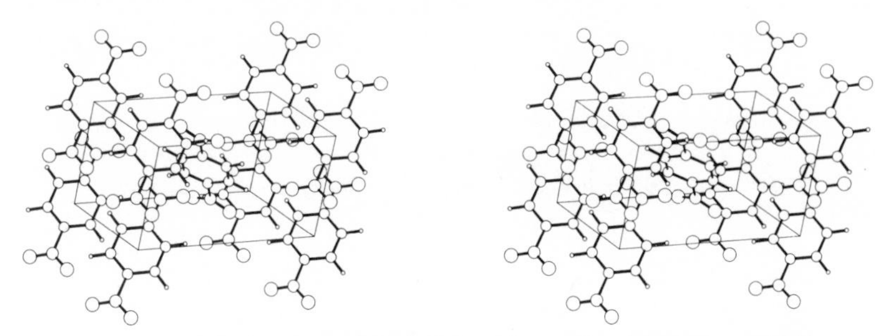

(g) 1,4-Dinitrobenzene, $C_6H_4(NO_2)_2$. Flat, polar; shortest contact distances, 3.2 Å.

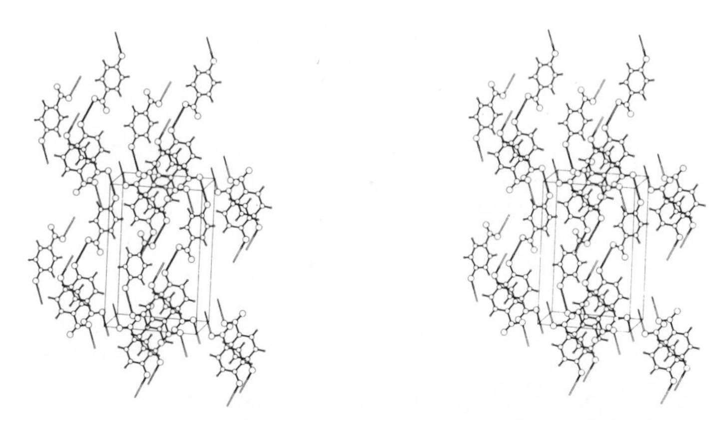

(h) 4-Nitrophenol, $C_6H_4NO_2OH$. Flat, polar; O--H--O intermolecular distances, 2.9 Å.

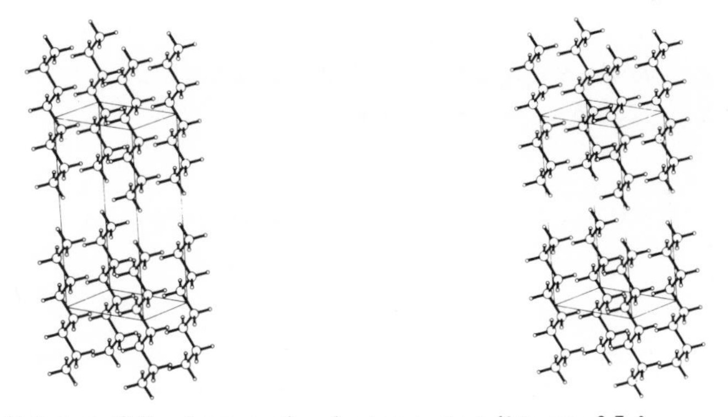

(i) Octane, $C_8H_{18}$. Long, apolar; shortest contact distances, 3.7 Å.

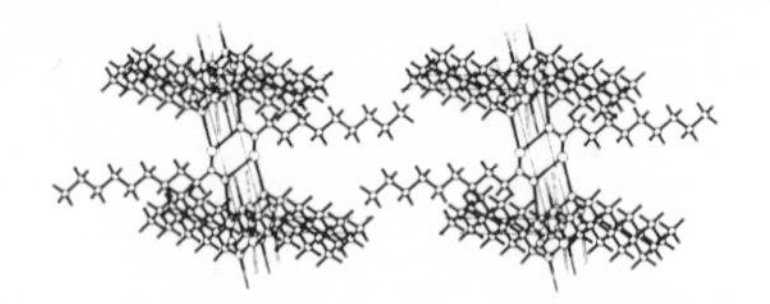
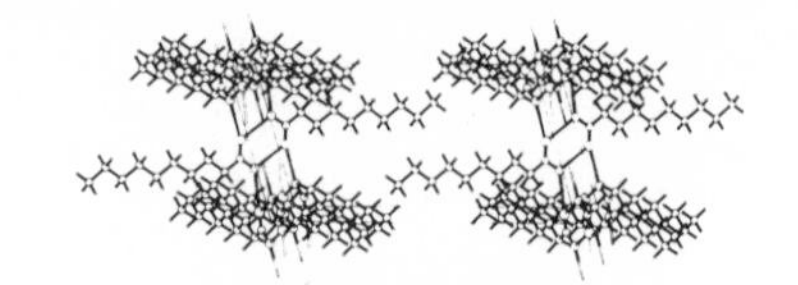

(j) Decanamide, $CH_3(CH_2)_8CONH_2$. Long, polar; O--H--N distances, 2.9 Å.

## APPENDIX 18 – SOME THEOREMS IN ELECTROSTATICS

### A18.1 Law of force

If two charges, $q1$ and $q2$, are at a distance $r$ apart, the force ($F$) between them is given by

$$F = q_1 q_2 / 4\pi\epsilon_0 r^2 \ , \qquad \text{(A18.1)}$$

where $\epsilon_0$ is the permittivity of a vacuum. If $q_1$ and $q_2$ include both magnitude and sign, then a negative $F$ indicates a force of attraction. It has been shown experimentally that the inverse power of $r$ is $2 \pm 10^{-9}$, but the inverse square law cannot otherwise be proved. If $q$ is in C and $r$ is in m, then $F$ is in J $m^{-1}$ or N. In order to illustrate the vector nature of $F$, we may write (A18.1) as

$$\mathbf{F} = q_1 q_2 \mathbf{r} / 4\pi\epsilon_0 r^3 \ . \qquad \text{(A18.2)}$$

### A18.2 Potential and field of a charge

If we wish to specify the force acting on a test charge without indicating the source of the charges responsible for the force, we say that a force **F** acting on a charge $q$ implies the existence of a field **E** at $q$, given by

$$\mathbf{E} = \mathbf{F}/q \ . \qquad \text{(A18.3)}$$

In order that $q$ should not itself perturb the field, we can write

$$\mathbf{E} = \lim_{q \to 0} \mathbf{F}/q \ , \qquad \text{(A18.4)}$$

but the definition of (A18.3) will often suffice for our purposes. From (A18.2),

$$\mathbf{E} = q\mathbf{r}/4\pi\epsilon_0 r^3 \ . \qquad \text{(A18.5)}$$

Let a unit test charge be moved from $A$ to $B$ (Figure A18.1), under the influence of a charge $q$ at $O$. In moving the distance ds from $P$ to $Q$, the field exerts a force **E** and does work **E**.d**s**, where

$$\mathbf{E}.\mathrm{d}\mathbf{s} = \frac{q \cos\theta \, \mathrm{d}s}{4\pi\epsilon_0 r^2} = \frac{q \, \mathrm{d}r}{4\pi\epsilon_0 r^2} \; . \tag{A18.6}$$

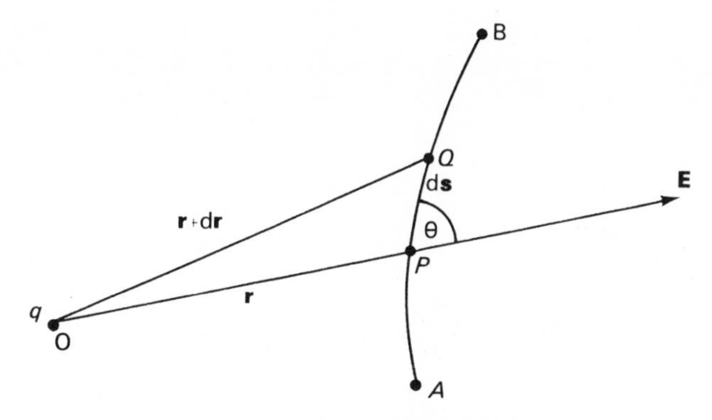


Figure A18.1 – Electrostatic field and potential.

The total work done ($V$) by the field in moving the test charge from $A$ to $B$ is thus,

$$V = \int_B^A \frac{q \, \mathrm{d}r}{4\pi\epsilon_0 r^2} = \frac{q}{4\pi\epsilon_0 r_A} - \frac{q}{4\pi\epsilon_0 r_B} \tag{A18.7}$$

The terms on the right-hand side of (A18.7) define the difference in potential between $A$ and $B$ due to the charge $q$. We can say that the work *we do* ($V'$) in moving the test charge is *minus* the work done by the field:

$$V' = \frac{q}{4\pi\epsilon_0 r_B} - \frac{q}{4\pi\epsilon_0 r_A} \; . \tag{A18.8}$$

If $A$ is at infinity, the potential at a point is the work we do in bringing a unit charge from infinity to the point. Thus, in general, dropping the suffixes, the potential at a point distant $r$ from a charge $q$ is given by

$$V = q/4\pi\epsilon_0 r \; . \tag{A18.9}$$

The difference in potential between $P$ and $Q$ (Figure A18.1) is given by

$$\mathrm{d}V = V_Q - V_P = - \text{ work done by field } = -\mathbf{E}.\mathrm{d}\mathbf{s} \; , \tag{A18.10}$$

and

$$V_B - V_A \; -\!\textstyle\int_A^B \mathbf{E}.\mathrm{d}\mathbf{s} \tag{A18.11}$$

We can resolve **E**.ds into Cartesian coordinates so that

$$\mathrm{d}V = -(E_x\mathrm{d}x + E_y\mathrm{d}y + E_z\mathrm{d}z) \,, \tag{A18.12}$$

whence

$$E_x = -\frac{\partial V}{\partial x}\,, \quad E_y = -\frac{\partial V}{\partial y}\,, \quad E_z = -\frac{\partial V}{\partial z}\,. \tag{A18.13}$$

We may write

$$\mathbf{E} = -\mathrm{grad}(V) \,, \tag{A18.14}$$

from which it follows that the operator grad is given by

$$\mathrm{grad} = \mathbf{i}\frac{\partial}{\partial x} + \mathbf{j}\frac{\partial}{\partial y} + \mathbf{k}\frac{\partial}{\partial z}\,, \tag{A18.15}$$

where **i**, **j** and **k** are unit vectors along the $x$, $y$ and $z$ Cartesian reference axes.

From (A18.9), we have

$$\mathrm{grad}(V) = \frac{q}{4\pi\epsilon_0}\,\mathrm{grad}\left(\frac{1}{r}\right) = \frac{q}{4\pi\epsilon_0}\left(-\frac{1}{r^2}\right)\mathrm{grad}\,(r)\,. \tag{A18.16}$$

But

$$\mathrm{grad}(r) = \mathbf{i}\frac{\partial r}{\partial x} + \mathbf{j}\frac{\partial r}{\partial y} + \mathbf{k}\frac{\partial r}{\partial z}\,, \tag{A18.17}$$

and

$$r^2 = x^2 + y^2 + z^2\,. \tag{A18.18}$$

Hence,

$$\mathrm{grad}(r) = \frac{1}{r}(\mathbf{i}x + \mathbf{j}y + \mathbf{k}z) = \frac{\mathbf{r}}{r}\,, \tag{A18.19}$$

whence

$$\mathrm{grad}(V) = -\frac{q\mathbf{r}}{4\pi\epsilon_0 r^3} = -\mathbf{E}\,. \tag{A18.20}$$

### A18.3 Potential of an ideal dipole

Consider a dipole (Figure A18.2) consisting of charges $\pm q$ separated by a distance $a$, such that the dipole vector is positive in the direction − to + (in chemistry, the opposite convention is often used).

The potential $V$ at the point $P$ is given by

$$V = \left(\frac{q}{R} - \frac{q}{r}\right)/4\pi\epsilon_0\,, \tag{A18.21}$$

or

$$V = (q/4\pi\epsilon_0) \times \text{ change in } (1/r) \text{ from } Q \text{ to } S\,, \tag{A18.22}$$

which is
$$V = \frac{qa}{4\pi\epsilon_0}\frac{\partial}{\partial s}\left(\frac{1}{r}\right); \qquad \text{(A18.23)}$$

$\partial/\partial s$ denotes differentiation at $Q$ in the direction of $QS$ in three-dimensional space. From previous arguments, and since $qa$ is the dipole moment $p$, we can write

$$V = \frac{1}{4\pi\epsilon_0}\,\mathbf{p}.\mathrm{grad}_Q\left(\frac{1}{r}\right), \qquad \text{(A18.24)}$$

or
$$V = -\mathbf{p}.\mathbf{r}/4\pi\epsilon_0 r^3 . \qquad \text{(A18.25)}$$

Generally, we shall need to define the gradient at $P$, the point where $V$ is to be calculated. Since $P$ and $Q$ lie on the same vector, $\mathrm{grad}_Q\left(\frac{1}{r}\right) = -\mathrm{grad}_P\left(\frac{1}{r}\right)$, and

$$V_P = \mathbf{p}.\mathbf{r}/4\pi\epsilon_0 r^3 . \qquad \text{(A18.26)}$$

If the dipole is assumed to lie along the $z$ Cartesian axis with its tail at $Q$,

$$V_P = pz/4\pi\epsilon_0 r^3 = p\cos\theta/4\pi\epsilon_0 r^2 . \qquad \text{(A18.27)}$$

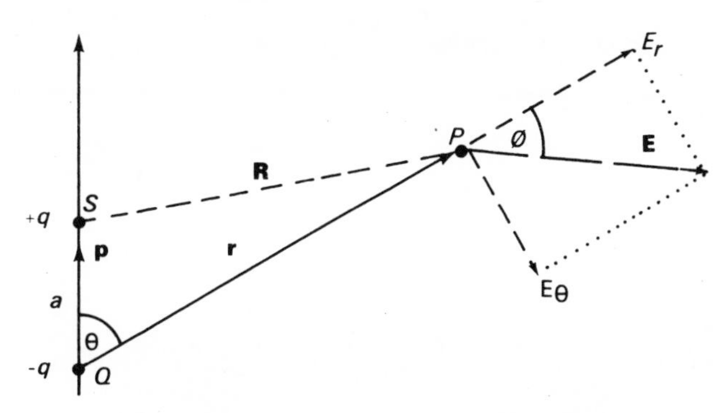


Figure A18.2 – Potential due to an ideal dipole.

In molecules, dipoles have a finite length and so are not ideal. We shall be content to use the same formulae but we must be cognisant of the errors which occur. From Figure A18.3 we have, from (A18.21),

$$V_X = \frac{q}{4\pi\epsilon_0}(1/a - 1/2a) = 0.5\,q/4\pi\epsilon_0 a . \qquad \text{(A18.28)}$$

In using (A18.26), we choose a single $r$ by measuring from $X$ to the centre of the dipole. Thus,

$$V_X = \frac{qa}{4\pi\epsilon_0(3a/2)^2} = 0.44\, q/4\pi\epsilon_0 a \ , \qquad \text{(A18.29)}$$

which gives an error of about 11%. For the point $Y$, the error is reduced to 4%.

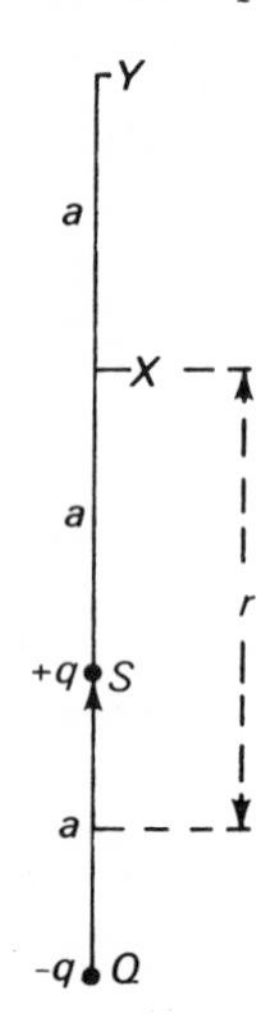


Figure A18.3 – Potential due to a molecular dipole.

**A18.4 Field due to an ideal dipole**

Referring again to Figure A18.2, the electrostatic field at $P$ is given, through (A18.14) and (A18.26), by

$$\mathbf{E} = -\frac{1}{4\pi\epsilon_0}\,\text{grad}(\mathbf{p.r}/r^3) \ , \qquad \text{(A18.30)}$$

whence

$$\mathbf{E} = -\frac{1}{4\pi\epsilon_0}\left\{\mathbf{p.r}\,\text{grad}\left(\frac{1}{r^3}\right) + \left(\frac{1}{r^3}\right)\text{grad}\,(\mathbf{p.r})\right\} . \qquad \text{(A18.31)}$$

It is readily shown, by resolution into Cartesian components, that

$$\mathbf{p.r} = p_x\, x + p_y\, y + p_z\, z \ , \qquad \text{(A18.32)}$$

whence

$$\text{grad}(\mathbf{p.r}) = \mathbf{p} \ , \qquad \text{(A18.33)}$$

and
$$\text{grad}\left(\frac{1}{r^3}\right) = -\left(\frac{3}{r^4}\right)\text{grad}(r) = -3\mathbf{r}/r^5 \;; \tag{A18.34}$$

hence,
$$\mathbf{E} = \frac{1}{4\pi\epsilon_0 r^5}\{3(\mathbf{p.r})\mathbf{r} - r^2\mathbf{p}\} \tag{A18.35}$$

It is sometimes convenient to resolve **E** into components along $r$ ($E_r$) and normal to this direction† ($E_\theta$), as shown in Figure A18.2. Using (A18.14) and (A18.27),

$$E_r = -\left(\frac{\partial V_P}{\partial r}\right)_\theta = 2p\cos\theta/4\pi\epsilon_0 r^3 \;, \tag{A18.36}$$

$$E_\theta = -\frac{1}{r}\left(\frac{\partial V_P}{\partial \theta}\right)_r = p\sin\theta/4\pi\epsilon_0 r^3 \;. \tag{A18.37}$$

Thus, we can see that the field varies with $1/r^3$, and is wholly along $r$ only when $\theta = 0$ or $\pi$.

## APPENDIX 19 – VOLUME OF MOLECULES IN A GAS

The constant $b$ in the van der Waals' equation, (4.27), is related to the volume occupied by the gas molecules. More precisely, it is an *excluded* volume, as the following analysis shows.

Let the molecules of a gas be spherical and of diameter $\sigma$. Two molecules cannot approach more closely than the sum of their van der Waals' radii, $2 \times \sigma/2$. Figure A19.1 shows the situation of two molecules in closest contact. The spherical volume of space in which the centres of the molecules cannot move is shaded, and the radius of *this* sphere is $\sigma$.

Thus the volume *excluded* per pair of molecules is $4\pi\sigma^3/3$, and that per single molecule $4\pi\sigma^3/6$. The actual volume of a single molecule is $4\pi(\sigma/2)^3/3$, or $\pi\sigma^3/6$. Hence, the excluded volume for a single molecule is four times its own volume.

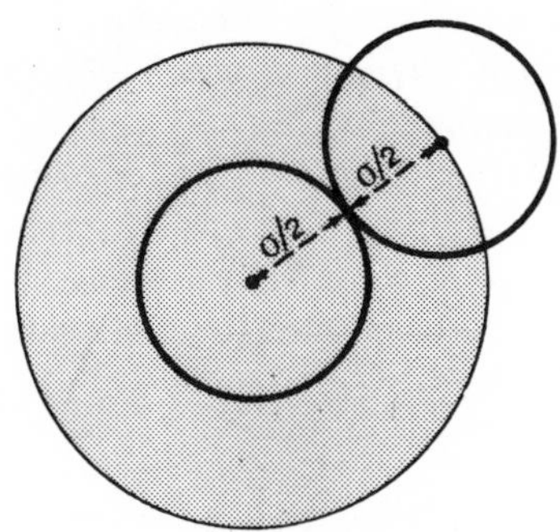


Figure A19.1 – Excluded volume in a gas.

†A small increment in the direction of $E_\theta$ is of magnitude $r\delta\theta$.

## APPENDIX 20 MEASUREMENT OF RELATIVE PERMITTIVITY

The determination of the relative permittivity of a substance may be carried out by comparing the capacitance of a condenser when filled with the substance under investigation with its capacitance in a vacuum. The measurement is carried out by a heterodyne-beat technique. Two similar oscillators, of frequency conveniently about $10^6$ Hz, are coupled to an amplifier and output device. One oscillator has its frequency fixed, and the other is coupled to a standard variable condenser connected in parallel with the experimental cell. By adjusting the variable condenser, the frequency of the variable oscillator can be brought into coincidence with that of the fixed oscillator. The beat frequency, equal to the difference in frequency of the two oscillators, is detectable to at least 1 Hz.

The readings of the standard condenser are taken with the experimental cell both filled with the dielectric and in a vacuum: the ratio of the capacitances is the relative permittivity of the substance. The beat-frequency technique results in a precision of about 1 in $10^6$. In practice corrections must be applied for the capacity of the leads and end-effects of the capacitor. A calibration technique may be employed using carbon dioxide and benzene as standards.

## PROBLEMS TO CHAPTER 4

1. The refractive index of gaseous chlorine at 293 K is 1.000768, and its density is 3.21 kg $m^{-3}$. Calculate the electron polarizability and the relative permittivity of this gas. What is the probable effective diameter of the chlorine molecule?
2. The crtical density of gaseous methane is 162 kg $m^{-3}$. Calculate the dielectric constant for this gas at 298 K and 1 atm pressure.
3. The dipole moment for the sulphur dioxide molecule is $5.40 \times 10^{-30}$ C m at 298 K. Calculate the orientation contribution to the static dielectric constant of sulphur dioxide.

*4. The dipole moment of the hydrogen fluoride molecule is $6.07 \times 10^{-30}$ C m. What are the electrostatic potential and the magnitude and direction of the electrostatic field strength at a point $P$, 500 Å from the fluorine atom, with an H–F–$P$ angle of 30°?

5. Lange (1925) obtained the following data for solutions of nitrobenzene in benzene. Determine the dipole moment, $P_{m,e}$ and $P_{m,a}$ for nitrobenzene at 297 K. The values for $P_m(C_6H_5NO_2)$ have been corrected for the polarization of the solvent and extrapolated to infinite dilution.

| $T$/K | $10^4 P_m(C_6H_5NO_2)/m^3\ mol^{-1}$ |
|---|---|
| 297 | 3.32 |
| 318 | 3.12 |
| 338 | 2.96 |
| 373 | 2.71 |

$M_r(C_6H_5NO_2)$ = 0.12311 kg mol$^{-1}$
$n(C_6H_5NO_2)$ = 1.556 at 297 K
$\rho(C_6H_5NO_2)$ = 1210 kg m$^{-3}$ at 297 K

*6. The cohesive energy of an ionic crystal, including a first-order correction for polarization, may be written as

$$U(r) = Ae^2/4\pi\epsilon_0 r + B\exp(-r/\rho) - C/r^6.$$

Following sections 2.4.2 and 4.5.3, determine equations for $\rho/r_e$ and $U(r_e)$.

*7. Taking $\alpha(Na^+) = 0.54 \times 10^{-40}$ and $\alpha(Cl^-) = 3.22 \times 10^{-40}$ F m$^2$, and using the data of sections 2.4.2 and 4.5.3, calculate $U(r_e)$ for NaCl. Compare the results with those obtained earlier.

8. If the bond moment of C-Cl from $C_6H_5Cl$ is $5.1 \times 10^{-30}$ C m, calculate the dipole moments of the tetra- and penta-substituted chlorobenzenes.

9. Compare the structures of hexane (Figure P4.1) and dicyanoethyne (Figure P4.2): the shortest intermolecular contact distances are 3.6 Å (C . . . . C) in hexane and 3.3 Å (C . . . . N) is dicyanoethyne. Suggest a reason why the molecules of dicyanoethyne do not pack in the simple manner shown by hexane.

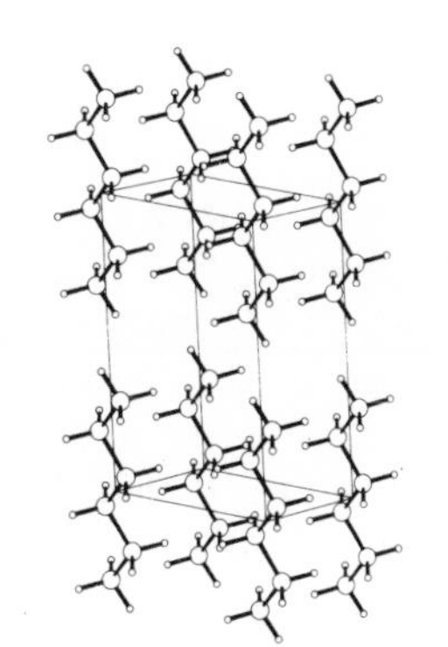
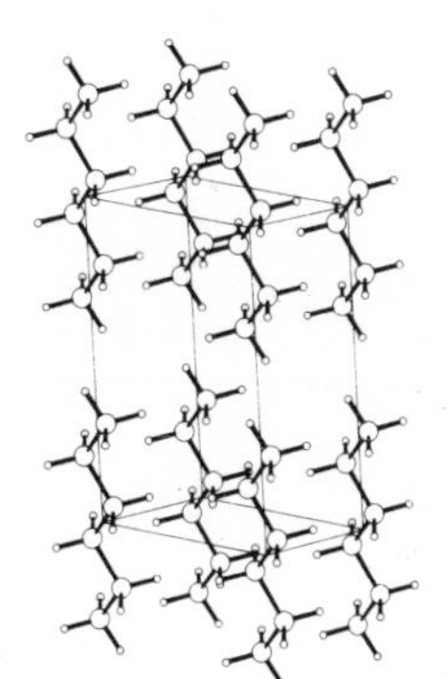

Figure P4.1

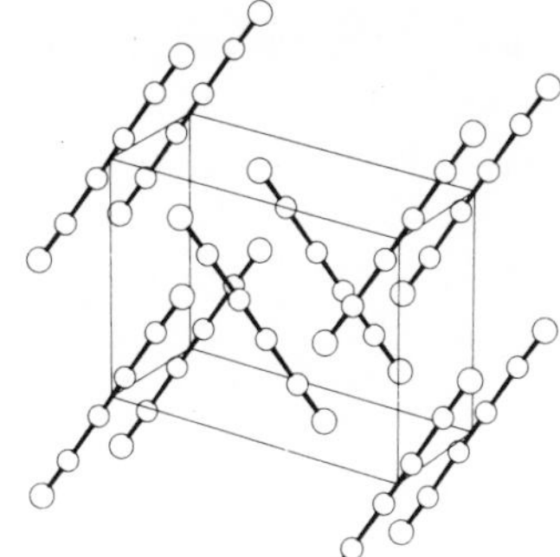
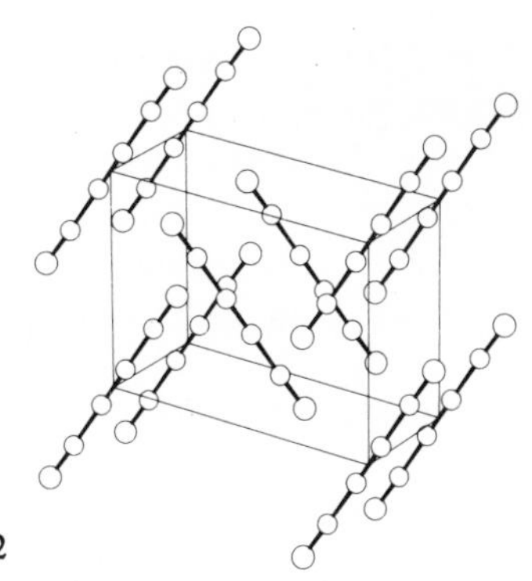

Figure P4.2

Chapter 5

# Metallic Compounds

## 5.1 INTRODUCTION

Metals are distinguished from other solids by several physical properties, among which their high thermal and electrical conductivities and their opacity to visible light are well known. About three-quarters of the known elements are metals, yet their structures are few in number and geometrically quite simple. We shall endeavour to show how metallic properties can be understood, indeed, what we mean by the term metallic character, and how the metallic bond correlates with the molecular orbital picture of the covalent bond.

## 5.2 CLASSICAL FREE-ELECTRON THEORY

Classical free-electron theory was summarised and reviewed in 1900 by Drude, who was commissioned to write a survey of the optical properties of matter. In 1905 Lorentz developed the 'Drude theory' to include ideas from Maxwell's kinetic theory and the resulting work is often called the Drude-Lorentz theory.

Electrons in a metal are effectively of two kinds. Those in the closed inner shells, or *core*, belong to a lattice array of positive ions, whereas the *valence* electrons permeate the metal and belong to it as a whole. It is the valence electrons that are potentially able to be influenced by external (and internal) electric, magnetic or thermal fields; the term conduction electrons is often used to describe the mobile valence electrons. The cohesive energy of a metal was considered to arise from the attraction between the ions and the 'sea' of valence electrons (see Figure 1.25), without considering any interaction between the conduction electrons and those in the ion cores. The theory pre-dates wave mechanics: it was successful in explaining conduction and optical properties, but failed completely to explain the heat capacities of metals.

### 5.2.1 Electrical conductivity

Experimental observations show that the resistance $R$ of a conductor of uniform cross-sectional area $A$ and length $l$, is directly proportional to $A$ and inversely proportional to $l$. We write

$$\rho = RA/l \ , \tag{5.1}$$

where $\rho$ is the electrical resistivity of the metal. The electrical conductivity, $\sigma$, is the reciprocal of the resistivity, so that

$$\sigma = l/RA \ . \tag{5.2}$$

Remembering that $R = V/I$, where $V$ is the potential difference across a conductor of resistance $R$ through which a current $I$ is flowing, we have

$$\sigma = \mathbf{j} \cdot \mathbf{E}^{-1} \ , \tag{5.3}$$

where **j** is the current density and **E** is the electric field intensity. We can look upon the potential difference, $V$, as representing the energy involved in moving an electron from point to point; it is an energy per unit charge. For an electron in a field **E**, the mobility, $\mu$, of the electron is defined as the drift velocity, $\mathbf{v}_d$, under unit field strength, or

$$\mu = \mathbf{v}_d \cdot \mathbf{E}^{-1} \ ; \tag{5.4}$$

If the electron concentration is $(N/V)$†, then the current density can be written

$$\mathbf{j} = (N/V)e\mathbf{v}_d = (N/V)e\mu\mathbf{E} \ , \tag{5.5}$$

where $e$ is the elementary charge. From (5.3) and (5.5),

$$\sigma = (N/V)e\mu \ . \tag{5.6}$$

The mean velocity of electron drift $\bar{\mathbf{v}}_d$, in the direction of the applied field, during a time $t$ before collision takes place with an ion, is given by

$$\bar{\mathbf{v}}_d = \tfrac{1}{2} \times \text{mean acceleration} \times \text{time} = e\mathbf{E}t/2m \ , \tag{5.7}$$

where $m$ is the mass of the electron; hence, the mean electron mobility is given by

$$\bar{\mu} = et/2m \ , \tag{5.8}$$

and

$$\sigma = (N/V)e^2t/2m \ . \tag{5.9}$$

We shall introduce a relaxation time, $\tau$, as the time in which an electron attains its average drift velocity, that is, $\tau = t/2$, and so

† See page 276.

$$\sigma = (N/V)e^2\tau/m \ . \tag{5.10}$$

The mean free path $\Lambda$ of an electron is its average distance of travel between collisions, and it may be formulated as

$$\Lambda = \overline{v_d}\,\tau \ . \tag{5.11}$$

If the electrons in a metal behave like a gas, then the average kinetic energy is given by (see Appendix 21)

$$\tfrac{1}{2}m\overline{v_d^2} = \tfrac{3}{2}kT \ . \tag{5.12}$$

However application of (5.12) gives a value for $\overline{v_d}$ which is much too small. The average kinetic energy of the electrons in a metal is better equated to three-fifths of a quantity called the Fermi energy (see Section 5.3), which for copper is $1.13 \times 10^{-18}$ J. Hence, $\overline{v_d}$ is $1.2 \times 10^6$ m s$^{-1}$, and we can calculate $\tau$ if we have a value for $\Lambda$. From measurements of electrical resistivity of thin films of copper, $\Lambda$ has been estimated at about 400 Å. Hence, $\tau = 3.3 \times 10^{-14}$s, and, from (5.10), $\sigma = 7.9 \times 10^7$ ohm$^{-1}$ m$^{-1}$. The experimental value of $\sigma$ at 293 K is $6.00 \times 10^7$ ohm$^{-1}$ m$^{-1}$, and the agreement between the two results for $\sigma$ is satisfactory.

Although (5.12) does not predict $\overline{v_d}$ correctly, it does show that $\overline{v_d}$ increases as the temperature increases. From (5.11) and (5.10), we see that $\sigma$ decreases or $\rho$ increases with increasing temperature, as found by experiment.

### 5.2.2 Heat capacity of solids

#### 5.2.2.1 *Classical particles, including electrons*

We show in Appendix 21 that the average kinetic energy of a system of particles, in thermal equilibrium at a temperature $T$, is given by

$$\overline{U}_K = \tfrac{3}{2}kT \ , \tag{5.13}$$

and the corresponding heat capacity, $C_V$, by

$$C_V = \left(\frac{\partial \overline{U}_K}{\partial T}\right)_V = \tfrac{3}{2}k \ . \tag{5.14}$$

The electron gas in a metal is assumed, in the classical theory, to follow the Maxwell-Boltzmann distribution of energies: the fraction, $\delta N/N$, of electrons in a system having energies lying between $E$ and $E + \delta E$ is given by

$$\delta N/N = \frac{2\pi}{(\pi kT)^{\frac{3}{2}}} E^{\frac{1}{2}} \exp(-E/kT)\delta E \ . \tag{5.15}$$

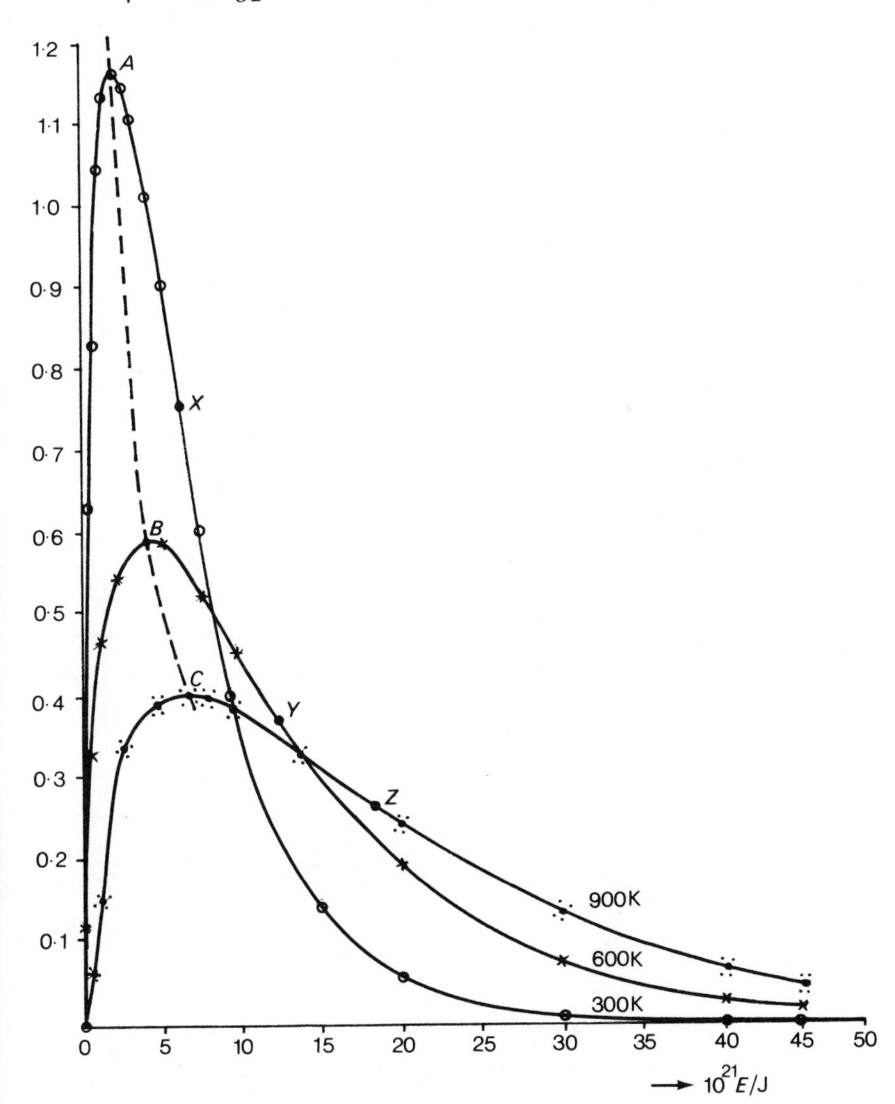


Figure 5.1 – Maxwell-Boltzmann distribution of energies: the maxima occur at ½$kT$, and follow the curve $ABC$ with increasing temperature; the average energies, $\frac{3}{2}kT$, are marked $X$, $Y$, $Z$.

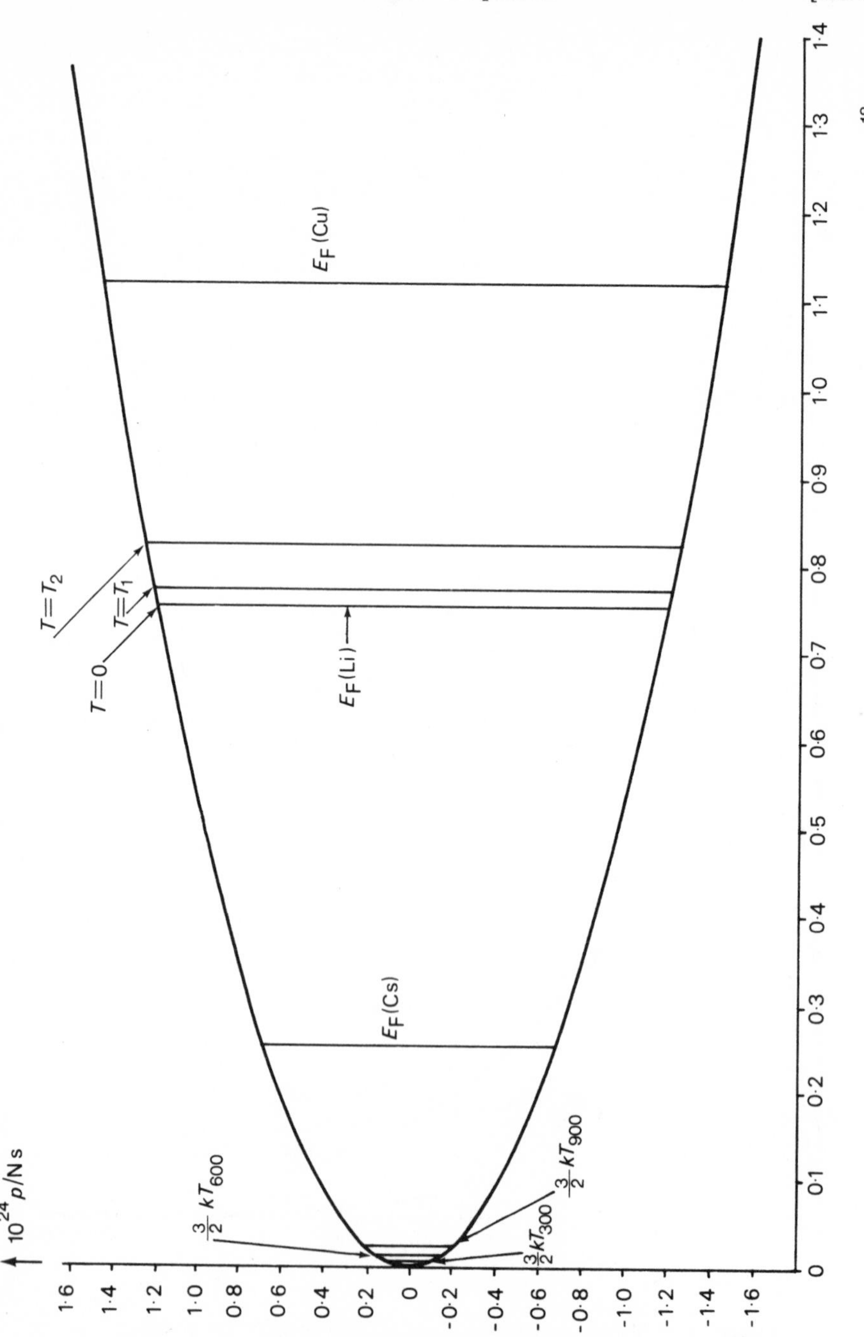


Figure 5.2 – Free-electron theory: variation of momentum, $p$, with energy $E$; some values of $E_{\mathrm{F}}$ and of $\frac{3}{2}kT$ are shown.

Figure 5.1 shows the variation of $\frac{1}{N}\left(\frac{\delta N}{\delta E}\right)$ with $E$ at three temperatures. Since the kinetic energy of a free electron of speed $v$ is $\frac{1}{2} mv^2$, we can write

$$E = p^2/2m \ , \tag{5.16}$$

where $p$ is the momentum of the electron. Figure 5.2 shows the variation of $p$ with $E$. An increase in $E$ causes the electron to move up the parabola, the final position depending on the balance between the thermal energy supplied and that lost in collision with the positive ions.

5.2.2.2 *Classical Solids*

The heat capacity $C_V$ of a typical monatomic solid is approximately 25 J mol$^{-1}$, and the variation of $C_V$ with temperature has been shown in Figure 2.20 (the value of any ordinate must be halved for a monatomic solid). Since the geometrical basis for any crystal is a lattice, we can consider that the atoms in a metal crystal are vibrating harmonically, each about its mean position, with a common frequency, $\nu$.

Each atom can then be treated as a simple harmonic oscillator, from which the total energy, kinetic and potential, is given by

$$U = \tfrac{1}{2}mv^2 + 2\pi^2\nu^2mx^2 \ , \tag{5.17}$$

where $m$ is the mass of the vibrating particle and $v$ and $x$ are, respectively, its speed and displacement from its equilibrium position, at any instant. The average thermal energy of a classical, one-dimensional, harmonic oscillator is, from Appendix 21, given by

$$\bar{U} = kT \ . \tag{5.18}$$

Generalizing to $N$ oscillators in three dimensions, we have for the mean total energy now,

$$\bar{U} = 3NkT \ . \tag{5.19}$$

If $N$ is put equal to $L$, the Avogadro constant,

$$\bar{U} = 3RT \ , \tag{5.20}$$

and

$$C_V = 3R \ . \tag{5.21}$$

From this equation we see that the molar heat capacity is 24.9 J mol$^{-1}$. This result is of course enshrined in Dulong and Petit's law: it is obeyed reasonably well at high temperatures but, as Figure 2.20 shows, it fails dramatically at low temperatures.

5.2.2.3 *Einstein and Debye solids*

Einstein treated the vibrations of $N$ atoms in a lattice array as $3N$ independent oscillators, of frequency $\nu$, in one dimension, but quantized the energy in accordance with Planck's theory (see Section 3.2.1). The distribution of energies is no longer continuous, and we write for the average energy

$$\bar{U} = 3N \sum_{n=0}^{\infty} nh\nu \exp(-nh\nu/kT) / \sum_{n=0}^{\infty} \exp(-nh\nu/kT) \ . \tag{5.22}$$

Putting $x = -h\nu/kT$, we see that (5.22) can be written as

$$\bar{U} = 3Nh\nu \frac{\mathrm{d}}{\mathrm{d}x} \{\ln [1 + \exp(x) + \exp(2x) + \ldots]\} = \frac{h\nu}{\{\exp(-x) - 1\}} \ , \tag{5.23}$$

or

$$\bar{U} = \frac{3N\,h\nu}{\{\exp(h\nu/kT) - 1\}} \ . \tag{5.24}$$

At high temperatures, $\{\exp(h\nu/kT) - 1\} \approx kT$, and $\bar{U} \to 3NkT$†, the classical value, but at low temperatures, where $\exp(h\nu/kT) \gg 1$,

$$\bar{U} \approx 3Nh\nu \exp(-h\nu/kT) \ , \tag{5.25}$$

and

$$C_V \approx 3Nk \left(\frac{h\nu}{kT}\right)^2 \exp(-h\nu/kT) \ . \tag{5.26}$$

From (5.26) we see that $C_V \to 0$ as $T \to 0$, but it does so exponentially at low temperatures. Experiment shows however that $C_V \to 0$ as a $T^3$ variation.

A satisfactory correction was made by Debye. He proposed that not all oscillators had the same frequency. At long wavelengths (low frequencies) relative to the interatomic spacings, large volume regions in the crystal may couple in their vibrational motion. In a finite crystal at low temperatures, there will be some vibrations for which $h\nu \ll kT$, and for them, the long wavelength vibrations are particularly significant. Thus these vibrations will make a classical contribution to the energy, thereby modifying the exponential dependence on $T$ in Einstein's equation (5.26): only for the low temperature region is a modification needed.

† $3RT$ if $N$ is set equal to the Avogadro constant.

Debye's treatment is complex and we state, without proof here, his result for $U$:

$$U = 9NkT(T/\Theta)^3 \int_{\Theta/T}^{0} \frac{(h\nu/kT)^4}{T[\exp(h\nu/kT) - 1]} \mathrm{d}T \ . \tag{5.27}$$

The Debye temperature $\Theta$ is chosen to give the correct number of vibrational modes. It varies from 110 K (lead) to 1160 K (beryllium); for copper, $\Theta$ is 343 K.

At very low temperatures, the limit $\Theta/T$ in the integral (5.27) may be set to infinity. Then, in molar terms,

$$U = \frac{3\pi^4 RT^4}{5\Theta^3} , \tag{5.28}$$

and

$$C_V = 1944(T/\Theta)^3 \ \mathrm{kJ\,mol^{-1}} \ . \tag{5.29}$$

5.2.2.4 *Heat capacity paradox*

If a mole of conduction electrons in a metal behaved as entirely free particles, their contribution to the heat capacity would be, from (5.14), 1.5R, and the total heat capacity, including (5.21), would then be 4.5R, or 37.5 J mol$^{-1}$. Experiment shows that metals, like other solids, show no tendency to deviate far from the Dulong and Petit limit. Since there can be little doubt about the nature of atomic vibrations or their contribution to the heat capacity, we are left with the situation that the electronic motion makes a negligibly small contribution to the heat capacity, a result clearly at variance with classical theory.

## 5.3 WAVE-MECHANICAL FREE-ELECTRON THEORY

In the classical theory, the free electrons interact with the lattice array of positive ions only insofar as to use them to arrest their motion. In the wave-mechanical treatment, electrons are still regarded as free from any particular atom, but are bound collectively to the positive ions, and interact with them uniformly over the whole crystal.

According to the Pauli principle, the energy states of the conduction electrons must be specified by sets of different quantum numbers, just like all other electrons. We shall consider that a conduction electron in a metal behaves like a particle in a box. We have sudied this problem already in Section 3.4.1: there is now an important difference to consider. We assumed in the three-dimensional electron in a box problem that the potential was infinite outside the box, zero inside it and that the box was a cube. In its application to a metal crystal, we require the potential to be periodic. The two situations are summarised in the following conditions:

*Electron in a box*

$$\psi(x,y,z) = 0 \text{ for } \begin{cases} x = 0, A \\ y = 0, A \\ z = 0, A \, ; \end{cases} \tag{5.30}$$

*Electron in a metal*

$$\psi(x,y,z) = \begin{cases} \psi(x + A,y,z) \\ \psi(x,y + A,z) \\ \psi(x,y,z + A) \; . \end{cases} \tag{5.31}$$

The solution follows the same lines as previously discussed. We can rewrite (3.52) in the form

$$E_k = h^2k^2/8\pi^2m \; . \tag{5.32}$$

From (3.52), we can see that the square of the wave vector **k** has replaced the sum $(n_x{}^2 + n_y{}^2 + n_z{}^2)$, and we can allot Cartesian components to **k**, such that

$$k^2 = k_x{}^2 + k_y{}^2 + k_z{}^2 \; , \tag{5.33}$$

where

$$k_x = 2\pi n_x/A \; , \tag{5.34}$$

and so on: $A$ is the edge length of the box, and $E_k$ depends on $A$. We can see that the conditions of (5.31) are satisfied by (5.34) as follows. By analogy with the electron in a box, let the eigenfunction for this problem be of the form

$$\psi_{\mathbf{k}}(\mathbf{r}) = \exp(\mathrm{i}\mathbf{k} \cdot \mathbf{r}) \; , \tag{5.35}$$

where

$$\mathbf{k} \cdot \mathbf{r} = k_x x + k_y y + k_z z \; . \tag{5.36}$$

Then,

$$\exp\{\mathrm{i}k_x(x + A)\} = \exp(\mathrm{i}k_x x) \exp(\mathrm{i}k_x A) = \exp(\mathrm{i}k_x x) \exp(\mathrm{i}2\pi n) = \exp(\mathrm{i}k_x x) \; , \tag{5.37}$$

and so on.

The electrons are still free, and so the curve of $E$ against $p$ remains parabolic: by introducing the de Broglie equation, (3.16), we can show easily that (5.32) becomes

$$E_k = p^2/2m \; . \tag{5.38}$$

The significance of (5.32) is that only discrete values for the energies are permissible. These energy levels are very closely spaced in a solid, and may be considered to approximate to a band, or continuum. At 0 K, electrons occupy energy levels up to a finite value; this value is the Fermi energy, $E_F$. If we take $3E_F/5$ as an average thermal energy†, as we did in Section 5.2.1, then the equivalent temperature, the Fermi temperature, $T_F$, or $E_F/k$, is in the region of 50000 K. This result must not be interpreted as the temperature of the electron gas: it means that heat supplied to a metal crystal will, normally, be very much less than $T_F$, and so will have negligible effect on the energy distribution of electrons.

The wave vectors **k** may be considered to exist in **k** space. This space is rather like the crystallographer's reciprocal space; **k** has the dimensions of reciprocal length. In crystallography, a reciprocal distance $d^*$ is given as $1/d$. In this subject, it is more convenient to use $2\pi/d$ and we shall follow this convention. From (5.33) and (5.34), we have

$$\mathbf{k} = \mathbf{i}\left(\frac{2\pi}{a}\right)(n_x + n_y + n_z) \; . \tag{5.39}$$

Let us consider next introducing $N$ electrons into our metal crystal, in accordance with the Pauli principle. For each energy state dictated by the combination of the quantum numbers, one electron with a spin component of +½ or −½ can be accommodated.

As **k** increases, so does the energy and momentum, represented by the lattice point $n_x$, $n_y$, $n_z$. We may note that although quantum numbers are positive integers, the quadratic form of $E$ can be satisfied for $n_x$, $n_y$ and $n_z$ taking values of ±(integer) or zero.

A system of $N$ free electrons in the ground state may be represented by the points of lattice lying inside a sphere of radius $k_F$ (Figure 5.3). The energy at the surface of this sphere is called the Fermi energy, $E_F$, given by

$$E_F = \frac{h^2 k_F^2}{8\pi^2 m} \; ; \tag{5.40}$$

values of $E_F$ for some metals are shown in Figure 5.2.

The volume element $(2\pi/a)^3$, where $a$ is the dimension of the real cubic unit cell, defines a primitive unit cell in **k** space and is identified with a single energy state: one lattice point occupies the volume of a primitive unit cell. Hence, in a sphere of radius $k_F$, the total number, $N$, of electron energy states is given by

$$N = 2 \times \tfrac{4}{3}\pi k_F^3/(2\pi/a)^3 \; ; \tag{5.41}$$

†See problem 5.5.

the factor of 2 arises because for each lattice point two energy states, corresponding to spins of ±½, are possible. The electron concentration $(N/V)$, where $V = a^3$, is given by

$$(N/V) = k_F^3/3\pi^2 \ . \tag{5.42}$$

Alternatively, $(N/V)$ may be obtained from $\eta L/a^3$, where $\eta$ is the number of valence electrons in a cubic unit cell of side $a$, and $L$ is the Avogadro constant.

From (3.52)

$$E_F = \frac{h^2}{8\pi^2 m}\{3\pi^2(N/V)\}^{2/3} \ : \tag{5.43}$$

for metallic copper, $a = 3.61 \times 10^{-10}$ m for the face-centred unit cell containing four atoms, each atom containing one valence electron. Hence, $(N/V) = 8.5 \times 10^{28}\ m^{-3}$, and $E_F = 1.13 \times 10^{-18}$ J, or 7.05 electronvolt (eV).

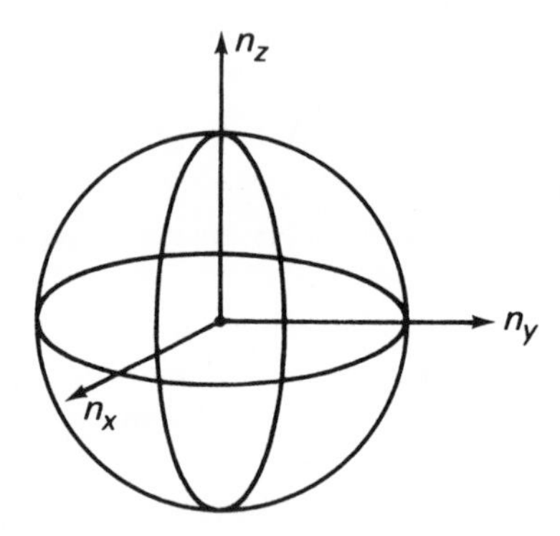


Figure 5.3 – Sphere of radius $k_F$ enclosing a portion of the lattice in **k** space.

5.3.1 **Density of states**

The density of states function, $g(E)$, is defined as the rate of change of the number of states with energy: $g(E)$ d$E$ represents the number of states in the energy range $E$ to $E$ + d$E$. Thus using (5.43) as an expression for $E$,

$$g(E) = \mathrm{d}N/\mathrm{d}E = 8\sqrt{2}\pi V(m/h^2)^{3/2}E^{1/2} \ . \tag{5.44}$$

Strictly, $m$ in (5.44) is not the rest mass of the electron for all values of $E$. An electron in a periodic potential crystal field is accelerated relative to the crystal, as though it had an effective mass, $m^*$, defined by

$$m^* = (h^2/4\pi^2)\mathrm{d}^2E/\mathrm{d}k^2 \ . \tag{5.45}$$

From Figure 5.8b (see later), we note from the shape of the curve that $m^*$ can vary from $-\infty$ to $+\infty$.

Figure 5.4 is a plot of (5.44); it is parabolic, like Figure 5.2. At 0 K, the energy levels are filled to the sharp cut-off at the Fermi energy, $E_F(0)$; higher energy states are empty. As the temperature is increased, thermal agitation moves electrons from some states below $E_F(0)$ to states above it, as shown by the rounded curve for $T > 0$ K, but for which $kT \ll E_F$. Only those electrons within an energy range of approximately $kT$ from the Fermi energy are excited in this way. We may note, in passing, that the Pauli principle ensures that nearly all electrons in a metal are unavailable for thermal excitation at low temperatures.

The heat capacities of metals at very low temperatures may be written as the sum of lattice ($\alpha$) and electronic ($\gamma$) contributions as

$$C_V = \alpha T^3 + \gamma T \ , \tag{5.46}$$

and experiments confirm that $C_V/T$ varies linearly with $T^2$. The term $\alpha$ is obtained from (5.29), and $\gamma$, a function of the electrons, has the form

$$\gamma = \pi^2 R/2T_F \ . \tag{5.47}$$

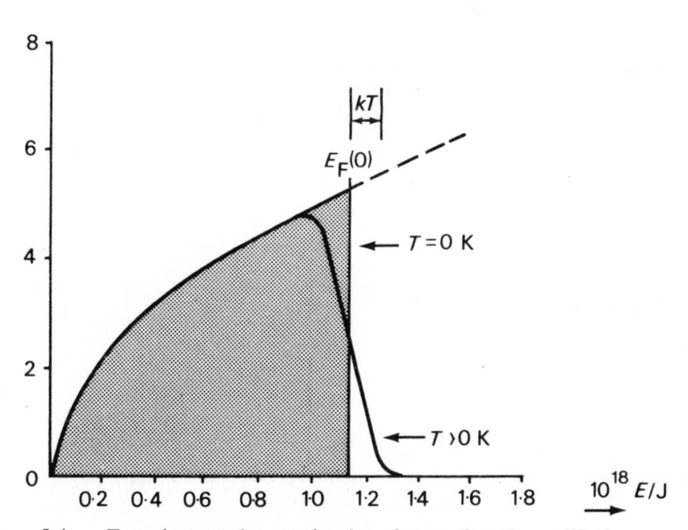


Figure 5.4 – Free-electron theory: density of states function, $g(E)$, for copper.

For copper, $T_F$ is above 81000 K: Figure 5.5 shows the lattice and electron contributions to the heat capacity of copper: only at $T < 5$ K is the electronic

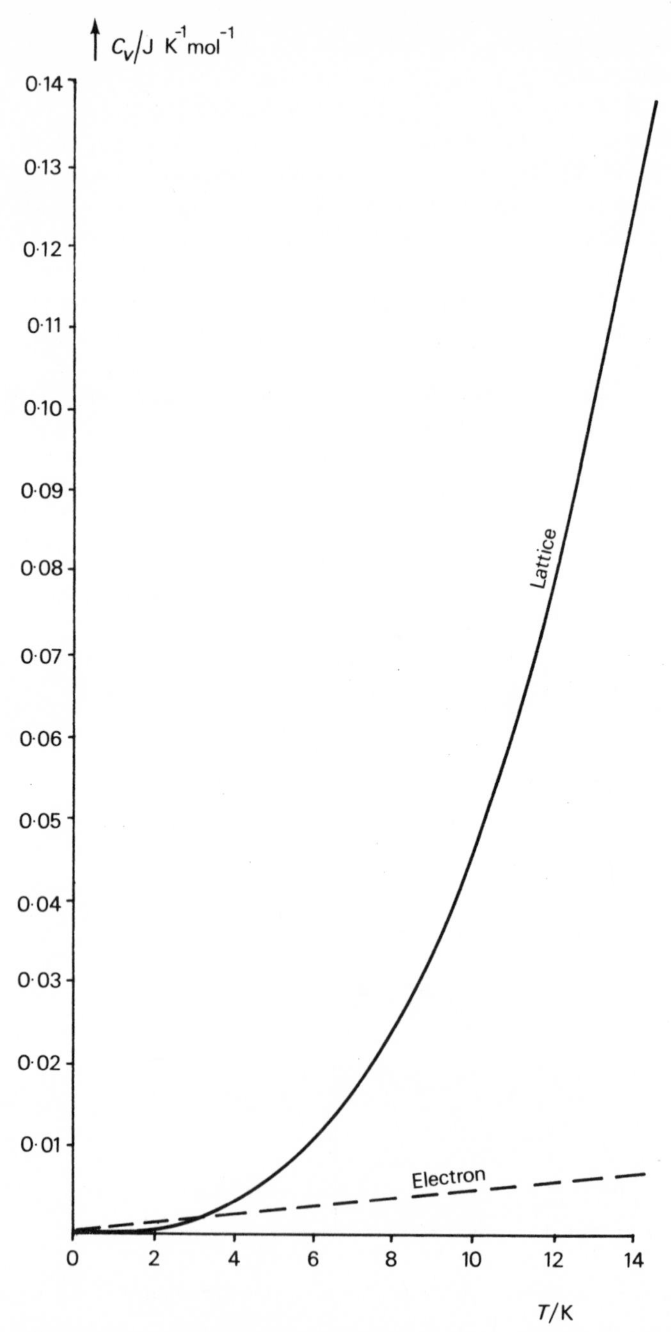


Figure 5.5 – Lattice and electronic contributions to the heat capacity of copper between 0 and 14 K.

contribution significant. At 298 K this contribution is about 0.015 J $K^{-1}$ $mol^{-1}$, less than one thousandth of the Dulong and Petit limiting value. We can be satisfied that the heat capacity 'paradox' is thus resolved.

### 5.3.2 **Fermi-Dirac distribution**

The concentration of valence electrons in a metal is about $10^4$ times greater than that of molecules in unit volume of a gas at stp, and the average spacing between these electrons is commensurate with their de Broglie wavelength. In these circumstances, the Maxwell-Boltzmann distribution equation is inadequate: any number of particles can have identical energy and momentum. Fermi-Dirac quantum statistics treats all electrons as indistinguishable, and requires that each energy state is either empty or fully occupied by a single electron and specified by the values of the quantum numbers $n_x$, $n_y$, $n_z$ and $m_s$. A derivation of the Fermi-Dirac distribution function is given in Appendix 22. For convenience we write (A22.24) as

$$f(E) = \frac{1}{\exp\{(E - E_F)/kT\} + 1} \ . \tag{5.48}$$

The function is plotted in Figure 5.6: $f(E)$ should be interpreted as the probability that a state of energy $E$ will be occupied in an ideal electron gas in thermal equilibrium with its surroundings.

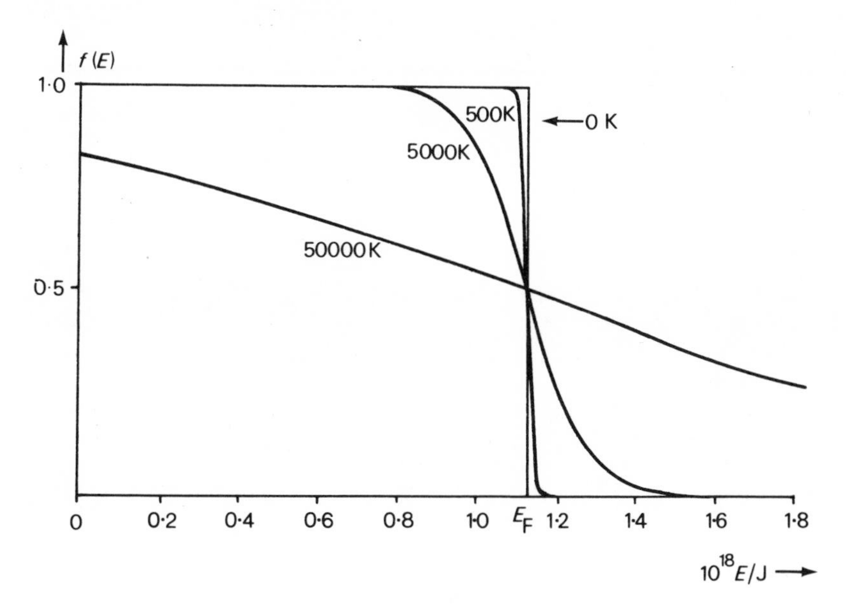


Figure 5.6 – Fermi-Dirac distribution function, $f(E)$, showing $E_F$ for copper.

At 0 K, $f(E) = 1$ for all $E < E_F$, and $f(E) = 0$ for all $E > E_F$. As the temperature is increased, $f(E)$ becomes less than unity for $E < E_F$, and greater than zero for $E > E_F$. The high energy 'Maxwellian' tail of the distribution for $T > 0$ K corresponds to $(E - E_F) \gg kT$ or $\exp{(E - E_F)/kT} \gg 1$. Then,

$$f(E) \approx \exp\{(E_F - E)/kT\} \propto \exp(-E/kT) \; , \tag{5.49}$$

which is the classical distribution. The curve for $T > 0$ K in Figure 5.4 is just the density of states function at 0 K multiplied by the Fermi-Dirac distribution at a given temperature.

We can identify the Fermi energy with the chemical potential of the electrons, as follows. Using (A22.7) and (A22.20), we have

$$S = k\{\sum_i g_i \ln g_i - g_i \ln(g_i - N_i) + N_i \ln(g_i - N_i) - N_i \ln N_i\} \; . \tag{5.50}$$

Remembering that $g_i/N_i = f(E)$, and letting $(E - E_F)/kT$ be $x$, rearrangement of (5.50) gives

$$S = k\left\{\sum_i -g_i \ln[1 - f(E)] + N \ln\left(\frac{1}{f(E)}\right) - 1\right\} , \tag{5.51}$$

or

$$S = k\left\{\sum_i g_i \ln\left[\frac{1 + \exp(x)}{\exp(x)}\right] + N \ln \exp(x)\right\} , \tag{5.52}$$

which simplifies to

$$S = k\{\sum_i g_i \ln[\exp(-x) + 1] + Nx\} \; . \tag{5.53}$$

Using (A22.6), we obtain

$$S = k\left\{\sum_i g_i \ln[1 + \exp(E_F - E_i)/kT] + \frac{E}{T} + \frac{NE_F}{T}\right\} . \tag{5.45}$$

From thermodynamics, we know that the chemical potential, $\mu$, is defined by

$$\mu = -T\left(\frac{\partial S}{\partial N}\right)_{E,V} , \tag{5.55}$$

whence†

$$\mu = E_F \; . \tag{5.56}$$

†Strictly at 0 K, since there is a small dependence of $\mu$ on $T$.

The theory which we have developed so far is capable of explaining many properties of metals. We cannot see yet how the extreme variations in electrical resistivity among elements (see page 205) arise.

## 5.4 BAND THEORY

In order to explain better the electrical properties of solids, we have to determine the circumstances in which a certain fraction of the electrons behave as though they were free. We shall find that electron energy levels in solids are arranged in bands, which are very closely spaced energy levels approximating to a continuum. Between the bands there are forbidden regions in which no energy states are allowed. Both the occupancy of the bands and the width of the gaps serve to determine the electrical properties of solids. Figure 5.7 shows schematically the situation for three main classes of electrical conductors. If a band is partially filled the solid behaves as a metal. If all bands are filled except for one or two bands which are either nearly filled or nearly empty the solid is an impurity-semiconductor, and if all occupied bands are filled the solid is an insulator. If all occupied bands are filled but the energy difference between the uppermost filled band (valence band) and the band of next highest energy (conduction band) is small, then the solid is an intrinsic-semiconductor. Clearly there will be gradations throughout these classes, as measurements of electrical resistivity show. We shall consider next how the energy gaps between the bands arise.

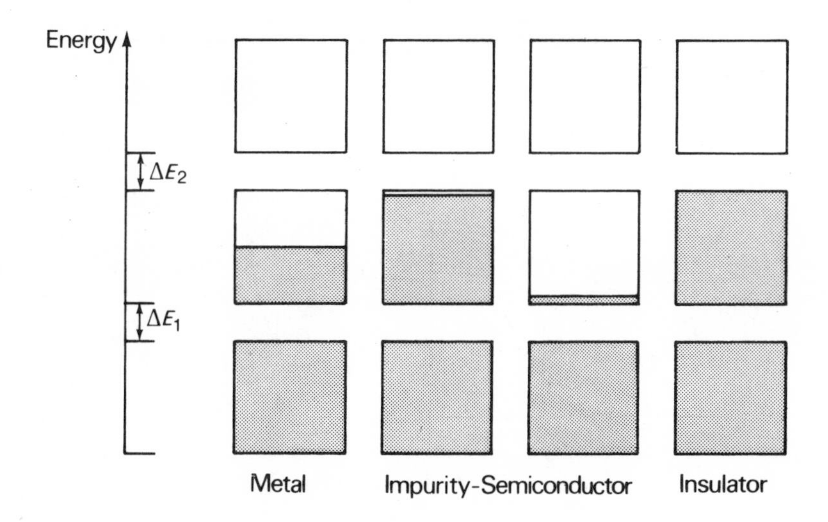


Figure 5.7 – Schematic arrangement of energy bands in solids: shading indicates electron occupancy; $\Delta E_1$ and $\Delta E_2$ are forbidden energy ranges.

### 5.4.1 Energy bands and Brillouin zones

On the free-electron model, the permitted energy states are distributed quasi-continuously according to (5.32): the free-electron eigenfunctions are, in one dimension, of the form $\exp(ikx)$, and represent travelling waves of momentum $kh/2\pi$. The modern theory of the electronic structure of solids starts from the Schrödinger equation, and incorporates the periodic potential field that arises from a lattice array of atoms.

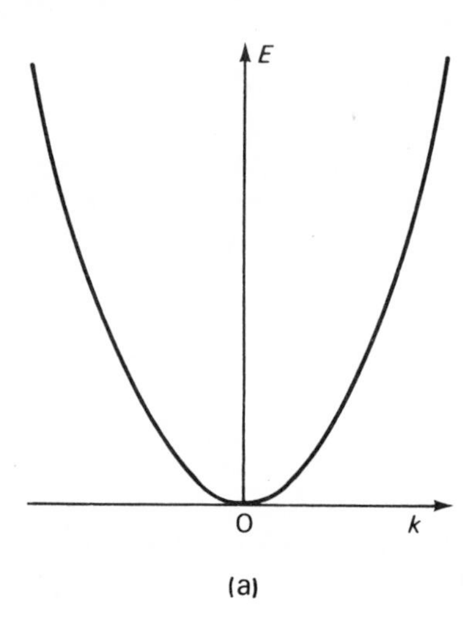


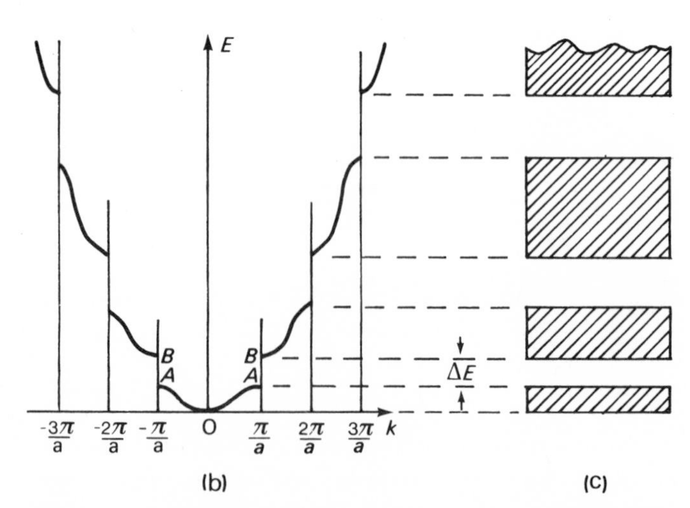


Figure 5.8 – Electron energies in solids: (a) free-electron theory; (b) electrons in a periodic potential field; (c) energy bands, showing gaps such as $\Delta E$.

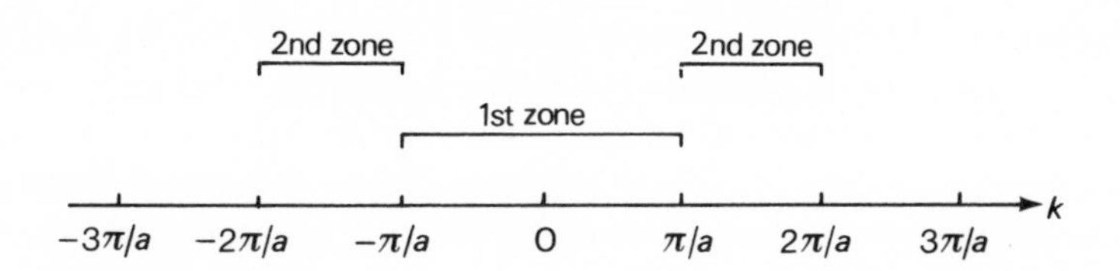


Figure 5.9 – The first two Brillouin zones for a one-dimensional lattice of periodicity $a$.

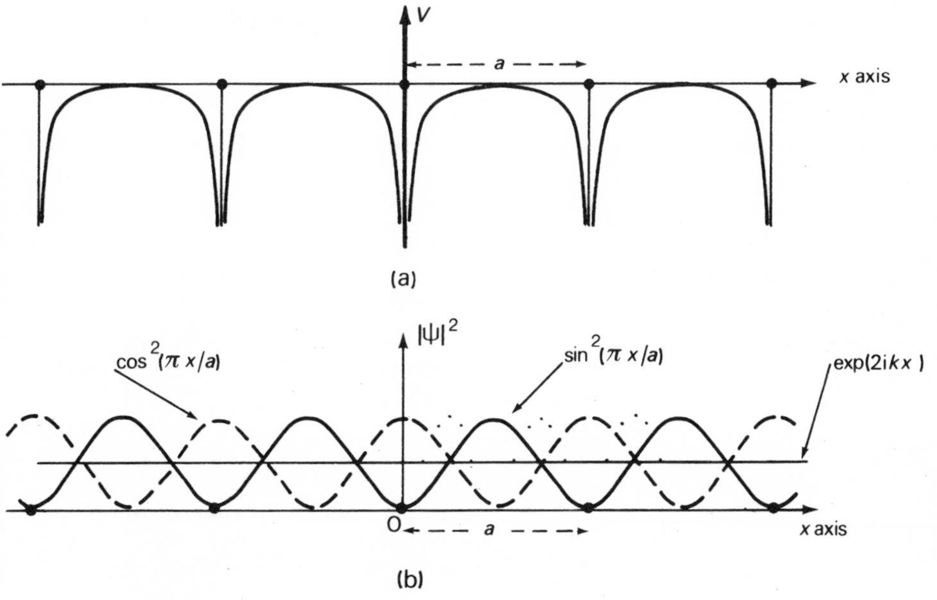


Figure 5.10 – One-dimensional lattice of periodicity $a$: (a) periodic potential field $V(x)$; (b) probability densities $\cos^2(\pi x/a)$, $\sin^2(\pi x/a)$ and $\exp(\mathrm{i}2kx)$.

The solution of the one-dimensional problem has been given by Bloch as

$$\psi_k = \exp(\mathrm{i}kx)u_k(x) \tag{5.57}$$

where $u_k(x)$ is a function with the periodicity, $a$, of the lattice. The energy dependence on $k$ is still quadratic, as in (5.40), but for $k = \pm n\pi/a$ discontinuities appear in the energy, so giving rise to a band structure (Figure 5.8). For most values of $k$ and, hence, the electron wavelength, the electrons behave very much like free electrons. At $k = \pm n\pi/a$, the conditions for Bragg reflexion of electron waves in one-dimension are realized: $k = \pm n\pi/a$ is equivalent to the Bragg equation $2a \sin\theta = n\lambda$, where $k = 2\pi/\lambda$ and $\sin\theta = 1$. The first order reflexion at $k = \pm\pi/a$ arises because the waves reflected from adjacent atoms interfere constructively, the phase difference being just $2\pi$. The region in **k** space between $\pm\pi/a$ is referred to as the first Brillouin zone (Figure 5.9). The energy is quasi-continuous within a zone, according to (5.32), and discontinuous at the zone boundaries. As $k$ increases towards $n\pi/a$, the eigenfunctions contain increasing amounts of the Bragg-reflected wave. At $k = \pi/a$, for example, the wave $\exp(i\pi x/a)$ reflects as $\exp(-i\pi x/a)$, and the resultant combinations are standing waves $\psi_1$ and $\psi_2$ of the forms $\cos(\pi x/a)$ and $\sin(\pi x/a)$, respectively. The probability densities of the two standing waves are $|\psi_1|^2$ and $|\psi_2|^2$, whereas that for the travelling wave is $\exp(2ikx)$. Figure 5.10 illustrates a one-dimensional periodic potential field and the wave probability functions.

The travelling wave distributes charge uniformly along the $x$ axis, $\psi_1$ has its peaks at $ma$, where $m$ is an integer, and $\psi_2$ at $(m + \frac{1}{2})a$. It is clear therefore that the potential energies of the three distributions are in the order $|\psi_1|^2 < \exp(2ikx) < |\psi_2|^2$. Hence an energy gap $\Delta E$ arises; the waves $\psi_1$ and $\psi_2$ correspond to the points $A$ and $B$ in Figure 5.8. The combination of the waves $\exp(\pm ikx)$ at $k = \pm n\pi/a$, that is, at the boundaries of the Brillouin zones, leads to an energy gap of $2V_k$, where $V_k$ is the potential energy function at the position corresponding to $k$ in reciprocal space. This result may be compared with the bonding/antibonding situation in MO theory, which also depends upon a core potential energy. Band theory modifies the density of the states function shown in Figure 5.4; a more correct plot is given by Figure 5.11.

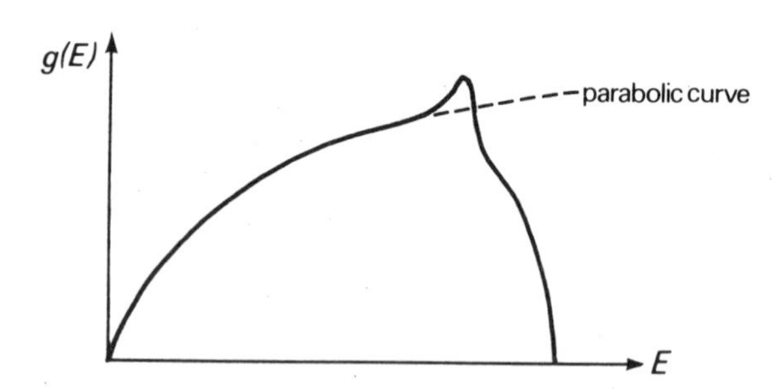


Figure 5.11 – Band theory: density of states function, $g(E)$.

Brillouin zones can be extended to two and three dimensions. The zone boundaries are determined by the regions in reciprocal space where the Bragg equation is satisfied. Thus Brillouin zones are determined by the crystal structure of a solid, rather than by its chemical composition.

The Bragg equation can be written as

$$2 \times \frac{2\pi}{\lambda}\cos\phi = d^* \tag{5.58}$$

where $\phi = (90 - \theta)$. Multiplying both sides by $d^*$, we obtain

$$2 \times \frac{2\pi}{\lambda} d^* \cos\phi = d^{*2} \tag{5.59}$$

which may be written as

$$\mathbf{k} \cdot \mathbf{d} = d^{*2}/2 \tag{5.60}$$

Thus, the Bragg equation is satisfied if **k** terminates on the plane (line in two dimensions) normal to **d*** at the mid-point of $d^*$, and these terminations determine the boundaries of the Brillouin zones.

### 5.4.2 Energy bands and molecular orbital theory

The Bloch theory is essentially a molecular orbital description of the metallic bond. The important feature of the MO theory is that each electron moves in the potential field created by all other atoms in the molecule, the core field. It uses a one-electron Hamiltonian, and the solutions of the corresponding wave functions are obtained from an LCAO approximation. We saw that in a compound such as benzene, strong overlap of the $\pi$ orbitals led to electron delocalization over the whole molecule. In the case of a metal, the number of atoms is infinite or, at least, very large, and we can envisage an extension of the delocalization over the many atoms present. The molecular orbitals are now the conduction orbitals and the metallic properties depend upon the degree of their overlap. In lithium, for example, the overlap integral $\int \psi_1(2s)\psi_2(2s)d\tau$ for adjacent atoms is about 0.5. We can see that the extension of this overlap to all atoms in a metal crystal must lead to considerable delocalization, and so establish metallic character.

In order to highlight an important difference between the MO's in a metal and those which we have described earlier in this book, we will consider the building up of a crystal of lithium. This element has the electronic configuration $(1s)^2(2s)$. When we discussed the one-electron species $H_2^+$, we saw that when the two nuclei were brought together, two diatomic orbitals, $\chi_\pm$ (Figure 3.21), were obtained: from the *single* energy level of each atom *two* energy levels were obtained in the molecule. We have described $Li_2$ in Section 3.9.5: if we add a third atom, there will be a total of three MO's; with four atoms, four MO's are obtained, and so on. Figure 5.12 shows the situation for the 2s and 2p energy bands in a metal. A similar bond is formed for the 1s electrons but inner, or core, bands are not normally involved in such characteristic metallic properties as electrical and thermal conductivity.

A cubical crystal of lithium of side 1 mm contains about $5 \times 10^{19}$ atoms. The energy levels in each band are therefore so closely spaced that the band is almost continuous in energy; a continuum is the limiting situation for infinite degeneracy. The number of 2s energy states actually present in this crystal of lithium is $9.33 \times 10^{19}$.

If the energies of two AO's are close, the bands formed will not remain discrete. They will merge and so contain more than one character, that is, a sort of hybrid band will be formed. If the AO's forming a band are degenerate, like p orbitals, then the resulting band could, in principle, have three parts, but these too could merge in forming a band structure.

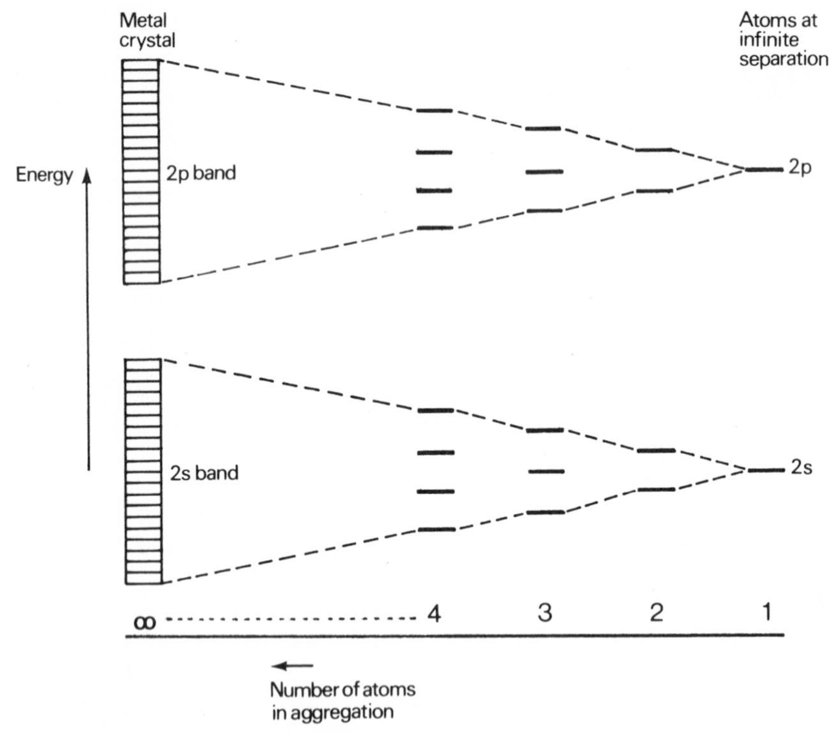


Figure 5.12 – Diagrammatic representation of the 2s and 2p energy bands in a metal.

We can now see that there are great similarities in the construction of both a metal and a covalently-bonded molecule. As the atoms are brought together from infinite separation, or zero energy, the AO levels split and form bands which themselves overlap. The electronic energy is determined after feeding electrons into the bands, following the *aufbau* scheme and taking account of the Pauli exclusion principle. The total number of energy levels in each band is given by the density of states function, and the distribution of energies within each band follows Fermi-Dirac statistics.

Our description of energy bands can be applied to all solids as their potential energy is a periodic function. Not every property of a solid need be discussed in terms of bands. However the electrical conductivity of solids is nicely explained by band theory, as the following simplified discussion shows.

In lithium there is one valence electron per atom, so the lowest band is only half-full. The energy states in each band occur in pairs ($n$, $l$, $m_l$, $\pm m_s$), and could represent waves travelling in opposite directions ($\pm m_s$) in the crystal. In the absence of an electric field, the net momentum of the electrons is zero. However if an external potential difference is applied to the metal there will be a net resultant electron flow and some electrons, those within about $kT$ of the Fermi energy, may be raised from the half-filled band to an empty, higher level band in the process. The current-limiting process involves collision between the electrons and the positive ions, to which reference has already been made. In an insulating material, like diamond, which has four valence electrons per atom, each energy level is filled by two electrons with opposite spins. Hence the band is filled, and there can be no net electron flow under an applied potential difference. A fuller treatment involves lattice vibrations and phonon scattering, which lie outside the present scope.† In the case of either impurity-semiconductors or intrinsic-semiconductors the concentration of mobile electrons is very small: the situation corresponds to the Maxwellian tail of Figure 5.4, where $E$ is greater than $E_F$ and $\exp\{(E - E_F)/kT\}$ decreases with an increase in temperature.

## 5.5 STRUCTURE OF METALS

The metallic bond does not have the directional character of the covalent bond. In consequence, metals take up structures which are determined to a large extent by space-filling criteria. Most metals adopt one or more of three relatively simple structures. They are the close-packed cubic (Figure 5.13), the close-packed hexagonal (Figure 5.14) and the body-centred cubic (Figure 5.15).

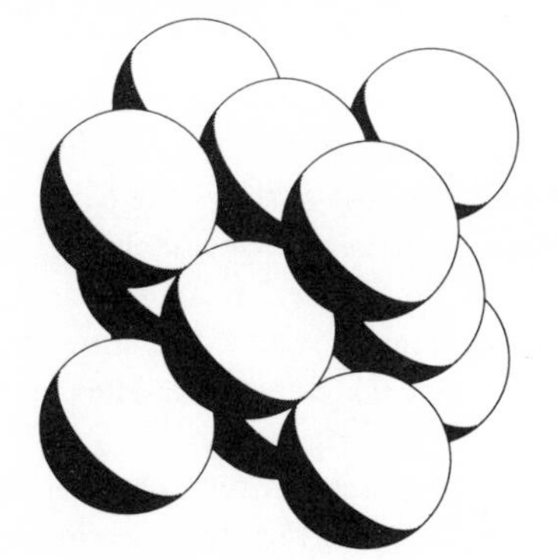

Figure 5.13 – Unit cell of the close-packed cubic (CPC) structure of identical spheres.

†See Suggested Reading: *Solid-State Physics* (Kittel).

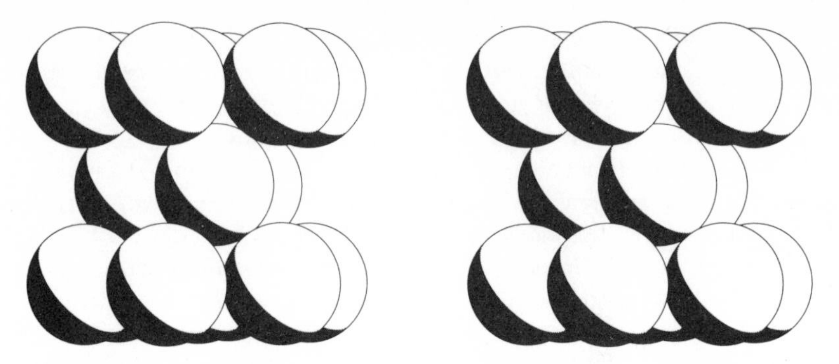

Figure 5.14 – Close-packed hexagonal (CPH) structure of identical spheres; three unit cells are included in the diagram.

Figure 5.15 – Unit cell of the body-centred cubic (BCC) structure of identical spheres.

The close-packed structures represent equally efficient methods of filling space with identical spherical atoms. A first layer of the structure is obtained by placing spheres in contact such that their centres form the apicies of equilateral triangles (Figure 5.16a). A second similar layer is placed on the first so that the spheres of the second layer rest in the depressions of the first (Figure 5.16b). A third layer can be added in one of two ways. If it is arranged such that the spheres in the third layer lie above spaces in *both* the first and second layers, the close-packed cubic (CPC) structure is obtained (Figure 5.16c). If a third layer lies, sphere for sphere, exactly above the first layer, the close-packed hexagonal (CPH) structure is obtained (Figure 5.16d).

The CPC structure is conveniently referred to a face-centred cubic unit cell (Figures 1.23 and 5.13). Let the sphere have radius $r$ and the unit cell side be $a$. Since the spheres are close-packed, we have

$$4r = a\sqrt{2} \qquad (5.61)$$

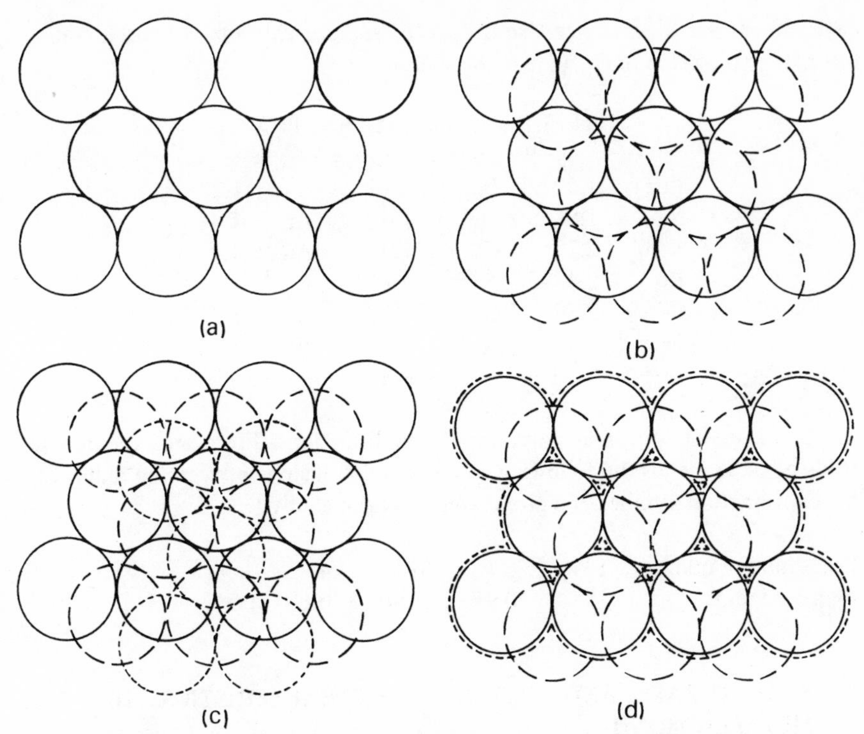


Figure 5.16 – Close-packing of identical spheres: (a) a first layer ($A$), full lines; (b) a second layer ($B$), dashed lines; (c) CPC, with layer sequence ... *ABCABCA* ...; (d) CPH, with sequence ... *ABABA* ... In (c) and (d), third layers are dotted lines. In (d) the third layer circles are, for clarity, shown just larger than the first layer circles; in reality, they are exactly superimposed in this view.

The volume of the cubic unit cell is $a^3$, or $16\sqrt{2}r^3$. Since it contains four atoms (spheres), the volume occupied per atom is $4\sqrt{3}r^3$. The volume of an atom is $4\pi r^3/3$, and hence the packing efficiency is 0.74. The same value obtains for the CPH structure; the coordination number is 12 in both of these structures.

The BCC structure is less closely packed; the coordination number is 8. By analogy with (5.61) we now have

$$4r = a\sqrt{3} \tag{5.62}$$

from which the packing efficiency is 0.68.

Close-packing leads to high density. The close-packed planes are $\{111\}$ in the CPC structure and $\{0001\}$ in the CPH structure. From (5.61) and (5.62) we can develop the idea of an atomic radius for metals (Table 5.1), since $a$ is an experimentally measurable parameter. It is worthwhile to recall the effect of environment on atomic size. Thus $r_{Na^+}$ in sodium chloride is approximately one half of the value of $r_{Na}$ in metallic sodium.

**Table 5.1** – Some metallic radii/Å

| | | | | | |
|---|---|---|---|---|---|
| Li | 1.52 | Be | 1.12 | Cu | 1.28 |
| Na | 1.86 | Mg | 1.60 | Ag | 1.44 |
| K | 2.27 | Ca | 1.97 | Au | 1.44 |
| Rb | 2.48 | Sr | 2.15 | Fe | 1.24 |
| Cs | 2.65 | Ba | 2.22 | Co | 1.25 |
| Fr | 2.93 | Ra | 2.29 | Ni | 1.25 |

Some metals occur in polymorphic modifications. From an experimental study of these metals the following empirical relationship between the radius and the coordination number of a given atom has been evolved.

| Coordination number | 12 | 8 | 6 | 4 |
|---|---|---|---|---|
| Relative radius | 1.00 | 0.97 | 0.96 | 0.88 |

## 5.6 STRUCTURAL AND PHYSICAL CHARACTERISTICS OF THE METALLIC BOND

The metallic bond is spatially undirected, and metal structures have high coordination numbers and high densities. Metals are opaque and have high reflective power. Electrons near the Fermi level can absorb energy, according to (3.4), and are raised to higher states. If there is only slight interaction between these electrons and the lattice ions, the energy gained is radiated away without change of phase and the crystal is transparent. Metals interact in this manner with radiation of wavelength less than that of the ultraviolet region of the spectrum.

In recent years, improvements have been made in the determination of the shapes and energies of bands in metals by the technique of electron spectroscopy for chemical analysis (ESCA). A metal specimen is exposed to ultraviolet or X-ray radiation of known energy $E_\nu$, sufficient to induce electron emission from the metal. The kinetic energy $E_K$ of the emitted electrons (about 1 keV) is measured accurately in a $\beta$-ray spectrometer. The energy $E$ needed to remove an electron from an atom is given by analogy with (3.11) as

$$E = E_\nu - E_K \; . \qquad (5.63)$$

Figure 5.17 shows the overlapping 5d/6s bands determined in ESCA experiments on a single crystal of gold, using Al $K\alpha$ radiation. The results apply strictly to the outer layers, since electron emission can take place through only approximately 20 Å; the corresponding energies deep within the metal may well be different in value.

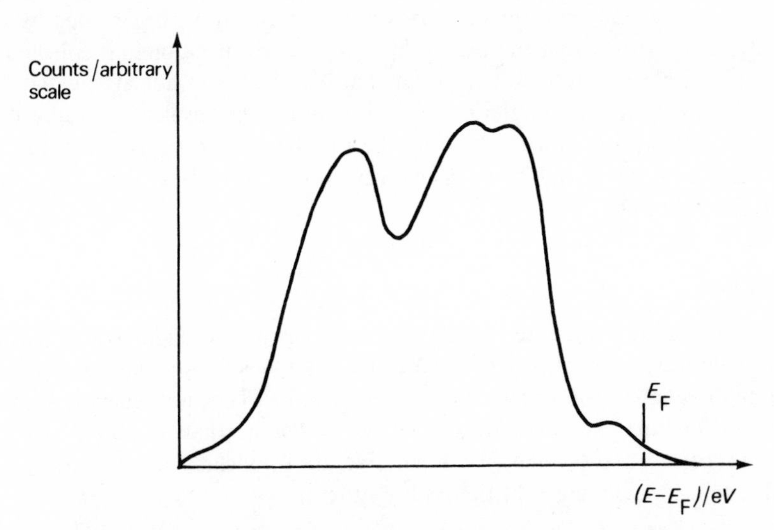


Figure 5.17 – Overlapping 5d/6s energy bonds in gold (after Shirley, and reproduced by courtesy of the American Institute of Physics).

Metals have variable strength. Deformation by gliding is common in metals and it takes place most easily in directions parallel to close-packed planes. In the CPC metals there are four such planes, {111}, whereas there is only one, {0001}, in the CPH metals and none in the BCC metals. Consequently we find that metals such as copper and silver are more malleable and ductile than metals such as beryllium or tungsten. Metals have sharp melting points, but they vary widely (Hg 234 K, W 3683 K) and the liquid interval is often very long (Ga 2370 K, Hf 3400 K).

Perhaps the most distinctive properties of metals are their electrical and thermal conductivities, to which we have referred. The electrical resistivity of a metal increases almost linearly with increasing temperature (see Section 5.2.1). In conducting solids, an increase in temperature excites electrons to higher energy states. At high temperatures, collisions between electrons and the vibrating lattice ions are very important in metals because of the high electron concentration. It is this effect which causes electrical conductivity to decrease with increasing temperature. In semi-conducting solids, the promotion of electrons to excited states is most significant and their conductivity increases with increasing temperature. The presence of defects in solids often has a profound effect on their electrical properties, and the characteristics of many semi-conductors depend upon the nature and concentration of impurities.

The electrical resistivity of many metals falls dramatically to zero when the specimen is cooled to a very low (critical) temperature $T_c$. This phenomenon is called superconductivity; mercury becomes superconducting at approximately 4 K. A forbidden energy gap of width about $3kT_c$ exists just above the Fermi level at 0 K. In a superconductor, the whole of the Fermi gas is moved by an applied electric field, from the region $|\mathbf{k}| = 0$ to a new position in the Brillouin zone; this displacement produces a supercurrent which is not attenuated by normal thermal electron scattering. The situation in an insulator is quite different. An energy gap still exists, but whereas the energy gap is linked to the Fermi gas in a superconductor, in an insulator it is tied to the lattice and therefore immutable.

## 5.7 ALLOYS

Metals readily form alloys, and their study has played an important part in the development of the theory of the metallic state. Alloys are numerous in type and exhibit variable, non-stoichiometric compositions. They are generally formed by simply melting the constituents together, a feature that is quite consistent with our picture of the metallic bond. We shall conclude our discussion of metallic solids by looking at just one alloy system.

### 5.7.1 Copper-gold alloys

Copper and gold are chemically similar and have similar radii (Table 5.1). If copper is added progressively to gold and the molten alloy quenched rapidly to room temperature, a face-centred cubic structure is obtained (Figure 5.18) which shows random replacement of gold by copper. The decrease in the unit-cell dimension is directly proportional to the concentration of copper added, and a complete range of solid solutions is obtainable.

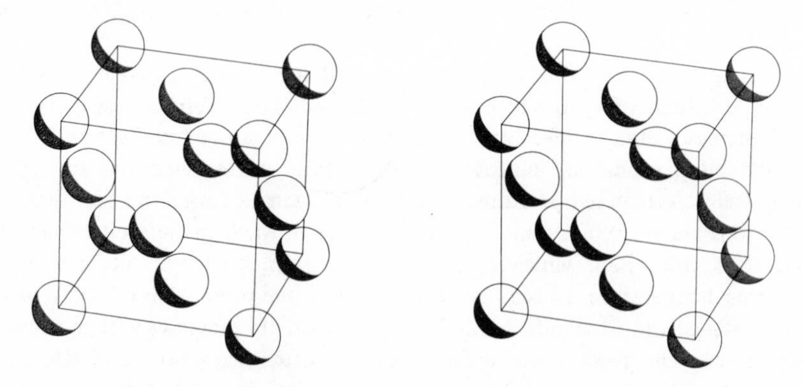

Figure 5.18 – Unit cell of the structure of a random solid solution of copper in gold: each sphere represents, statistically, a certain fraction, $c$ of copper and $1-c$ of gold.

A different situation arises if the specimen is annealed. Random replacement of gold by copper occurs until the composition reaches CuAu. Then the atoms segregate into layers in a tetragonal structure (Figure 5.19). Random replacement of gold, in the tetragonal structure, continues to the composition $Cu_3Au$ which crystallizes in a pseudo face-centred cubic structure (Figure 5.20). The CuAu and $Cu_3Au$ ordered structures are termed superstructures. The electrical resistivity of quenched copper-gold alloys shows a smooth variation from copper to gold. The annealed samples however show pronounced minima for the two superstructures.

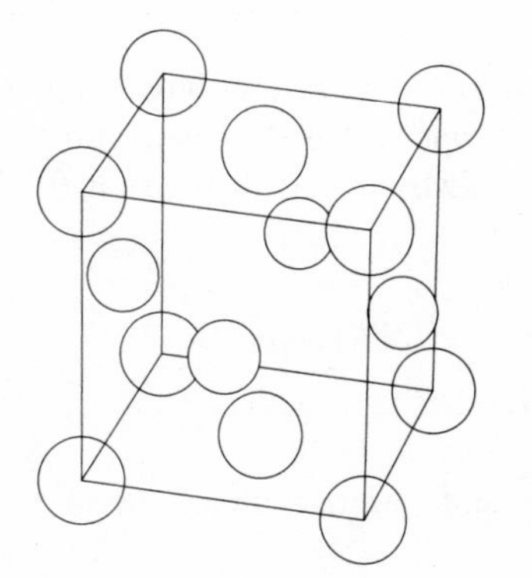
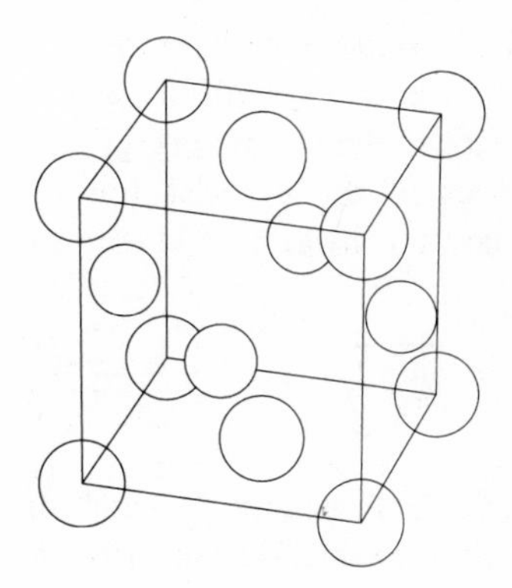

Figure 5.19 – Unit cell of the ordered tetragonal structure of CuAu; circles in decreasing order of size represent Au and Cu.

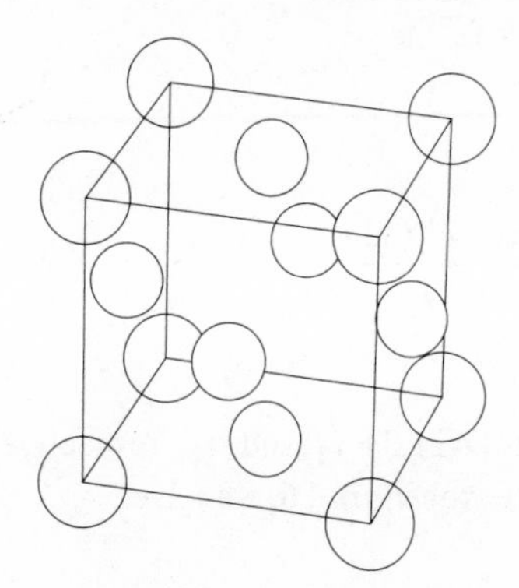
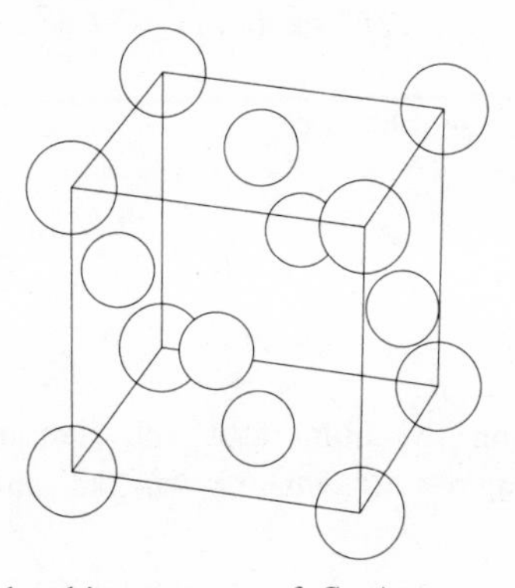

Figure 5.20 – Unit cell of the ordered cubic structure of $Cu_3Au$; circles in decreasing order of size represent Au and Cu.

If the metallic radii are very similar, the strain introduced by replacement will be small. The superstructure will then not be more stable than the random solid solution. Thus silver and gold of equal metallic radii will form a continuous series of solid solutions, but no superstructures are formed. If there is no strain for a given possible composition, such as AgAu, then the free energy charge for the hypothetical reaction

$$\text{AgAu (superstructure)} \rightarrow \text{AgAu (solid solution)} \tag{5.64}$$

will be governed by the entropy change, and a random solid solution would be expected.

## APPENDIX 21 AVERAGE THERMAL ENERGIES

### A21.1 Average kinetic energy

Consider a system of particles, each of mass $m$ but with different values of velocity $\mathbf{v}$; the kinetic energy of any particle is $mv^2/2$. We shall assume that the energies of the particles follow a Boltzmann distribution; then, from (4.39), we have for the average kinetic energy $\bar{U}_K$,

$$\bar{U}_K = \overline{mv^2/2} = \frac{\int_{-\infty}^{\infty}\int_{-\infty}^{\infty}\int_{-\infty}^{\infty} (mv^2/2)\exp(-mv^2/2kT)\mathrm{d}v_x\mathrm{d}v_y\mathrm{d}v_z}{\int_{-\infty}^{\infty}\int_{-\infty}^{\infty}\int_{-\infty}^{\infty} \exp(-mv^2/2kT)\mathrm{d}v_x\mathrm{d}v_y\mathrm{d}v_z} \tag{A21.1}$$

Following Appendix 15, and since positive and negative directions of $\mathbf{v}$ are equally probable, we can write (A21.1) as

$$\bar{U}_K = \frac{2\int_0^{\infty} (mv^2/2)\exp(-mv^2/2kT)v^2\mathrm{d}v \int_0^{\pi}\sin\theta\,\mathrm{d}\theta \int_0^{2\pi}\mathrm{d}\phi}{2\int_0^{\infty}\exp(-mv^2/2kT)v^2\mathrm{d}v \int_0^{\pi}\sin\theta\,\mathrm{d}\theta \int_0^{2\pi}\mathrm{d}\phi}\,, \tag{A21.2}$$

which simplifies to

$$\bar{U}_K = \frac{m/2\int_0^{\infty} v^4\exp(-mv^2/2kT)\mathrm{d}v}{\int_0^{\infty} v^2\exp(-mv^2/2kT)\mathrm{d}v}\,. \tag{A21.3}$$

Calling the numerator and denominator of (A21.3) $I_1$ and $I_2$, respectively, letting $\alpha = pv^2$, where $p = m/2kT$, and following Appendix 16, we have

$$I_1 = (m/4)p^{-5/2}\int_0^{\infty}\alpha^{5/2-1}\exp(-\alpha)\mathrm{d}\alpha = (m/4)p^{-5/2}\,\Gamma(\tfrac{5}{2})$$

$$= (m/4)p^{-5/2}(\tfrac{5}{2})\,\Gamma(\tfrac{3}{2})\,,$$

and $$I_2 = 2^{-1}p^{-3/2}\int_0^{\infty}\alpha^{3/2-1}\exp(-\alpha)\mathrm{d}\alpha = 2^{-1}p^{-3/2}\,\Gamma(\tfrac{3}{2})\,. \tag{A21.5}$$

Hence $$\bar{U}_K = I_1/I_2 = (\tfrac{3}{2})\,kT\,. \tag{A21.6}$$

This value can be derived by other methods, but this procedure shows the usefulness of both (4.39) and the results in Appendices 15 and 16.

### A21.2 Average vibrational energy

In Section 5.2.2.2, we treated the vibrational energy of a monoatomic solid in terms of that of a one-dimensional simple harmonic oscillator. Using (5.17), and following the treatment in A21.1, we can write the average vibrational energy $\bar{U}_V$ of the monoatomic solid as

$$\bar{U}_V = \frac{m/2 \int_{-\infty}^{\infty}\int_{-\infty}^{\infty}(v^2 + 4\pi^2\nu^2x^2)\exp\{-m(v^2 + 4\pi^2\nu^2x^2)/2kT\}dvdx}{\int_{-\infty}^{\infty}\int_{-\infty}^{\infty}\exp\{-m(v^2 + 4\pi^2\nu^2x^2)/2kT\}dvdx} \tag{A21.7}$$

It is easily confirmed that (A21.7) is equivalent to

$$\bar{U}_V = \frac{m\int_0^{\infty} v^2\exp(-mv^2/2kT)dv}{2\int_0^{\infty}\exp(-mv^2/2kT)dv} + \frac{4\pi^2\nu^2m\int_0^{\infty} x^2\exp(-4\pi^2\nu^2mx^2/2kT)dx}{2\int_0^{\infty}\exp(-4\pi^2\nu^2mx^2/2kT)dx} \tag{A21.8}$$

Each term on the right-hand side of (A21.8) solves to $kT/2$; hence

$$\bar{U}_V = kT \tag{A21.9}$$

The reader may compare this derivation with the less rigorous account in Section 2.9.1. It may be noted that both terms on the right-hand side of (5.19) are squared terms: their average contributions to $\bar{U}_V$ would be expected to be equal.

## APPENDIX 22 SOME QUANTUM STATISTICS

### A22.1 Fermi-Dirac distribution

Each energy state, $g_i$, of energy $E_i$, in a system of electrons is determined by the four quantum numbers $n$, $l$, $m_l$ and $m_s$, or the equivalent $n_x$, $n_y$ and $n_z$ of (5.39) together with $m_s$. Electrons are of course, indistinguishable, but any electron of given $n_x$, $n_y$, $n_z$ and $m_s$ occupies the corresponding energy state: thus, each state can be either empty or completely filled by one electron. Two states of the same values of $n_x$, $n_y$ and $n_z$, occupied by electrons with $m_s = \pm ½$ constitutes an electron pair, or fully occupied AO.

In classical (Boltzmann) statistics, the state of each particle in a system is determined solely by the energy that it possesses, that is to say its energy states are non-degenerate. In quantum statistics, energy states are usually degenerate as the following calculation shows:

Consider an electron at 300 K. From Appendix 21 the mean kinetic energy is $\frac{3}{2}kT$, or $6.21 \times 10^{-21}$ J. From (5.39) and (5.40) we have

$$E = \frac{h^2}{2ma^2}(n_x{}^2 + n_y{}^2 + n_z{}^2) \; . \qquad \text{(A22.1)}$$

The mass ($m$) of an electron is $9.11 \times 10^{-31}$ kg, and if we confine the electron to a cubical box of side ($a$) 1 mm then, by equating $E$ to $6.21 \times 10^{21}$ J, we find

$$n_x{}^2 + n_y{}^2 + n_z{}^2 \approx 2.6 \times 10^{10} \; . \qquad \text{(A22.2)}$$

Thus an electron possesses the mean classical kinetic energy, $\frac{3}{2}kT$, if it selects any quantum numbers satisfying (A22.2): in other words the energy levels are highly degenerate.

Consider a set of energy states or cells, $g_i$ ($i = 1,2, \ldots ,s$), of similar energy $E_s$, given by (5.40), and let the number of electrons in the set be $N_s$, so that $N_s$ cells are occupied, and $g_s - N_s$ are empty. Following Section 2.10 and in particular (2.98), we can write for the number of different ways ($W_s$) in which the $N_s$ electrons can occupy $g_s$ cells, allowing for the indistinguishability of electrons, as

$$W_s = \frac{g_s}{(g_s - N_s)!\,N_s!} \; . \qquad \text{(A22.3)}$$

To a given range of energies, there corresponds for each range an equation like (A22.3). Hence the total number of distinguishable arrangements ($W$) for an entire system of $p$ sets is given by

$$W = \prod_{i=1}^{p} W_i = \prod_{i=1}^{p} \frac{g_i!}{(g_i - N_i)!\,N_i!} \; . \qquad \text{(A22.4)}$$

The most probable distribution is that which maximises $W$. However the maximization is subject to the constraints

$$\prod_{i=1}^{p} N_i = N \; , \qquad \text{(A22.5)}$$

where $N$ is the number of electrons in the entire system, and

$$\sum_{i=1}^{p} E_i N_i = E \; , \qquad \text{(A22.6)}$$

where $E$ is the total energy of the entire system. Using Stirling's approximation, (2.103), with (A22.4) gives

$$\ln W = \sum_{i=1}^{p} \{g_i \ln g_i - (g_i - N_i)\ln(g_i - N_i) - N_i \ln N_i\} \tag{A22.7}$$

For a maximum†, $\mathrm{d}\ln W = 0$; thus,

$$\mathrm{d}\ln W = \frac{\partial \ln W}{\partial N_1}\mathrm{d}N_1 + \frac{\partial \ln W}{\partial N_2}\mathrm{d}N_2 + \ldots + \frac{\partial \ln W}{\partial N_i}\mathrm{d}N_i + \ldots + \frac{\partial \ln W}{\partial N_p}\mathrm{d}N_p = 0. \tag{A22.8}$$

From (A22.5) and (A22.6), we have

$$\sum_{i=1}^{p} \mathrm{d}N_i = 0 \ , \tag{A22.9}$$

and

$$\sum_{i=1}^{p} E_i \,\mathrm{d}N_i = 0 \ . \tag{A22.10}$$

To solve (A22.8) subject to the constants (A22.9) and (A22.10), we use Lagrange's method of undetermined multipliers (see section A22.2). Using (A22.7) in (A22.8), we obtain

$$\mathrm{d}\ln W = \sum_{i=1}^{p} \{\ln(g_i - N_i) - \ln N_i\}\mathrm{d}N_i = 0 \ . \tag{A22.11}$$

We assign multipliers, $\alpha$ and $\beta$, such that

$$\sum_{i=1}^{p} \{\ln(g_i - N_i) - \ln N_i + \alpha + \beta E_i\}\mathrm{d}N_i = 0 \ ; \tag{A22.12}$$

only two multipliers are needed since there are only two constraining equations. Equation (A22.12) must hold for any small change in $N_i$, and we shall consider the simplest general variation. More than one term of $N_i$ must be involved, so as to conform to (A22.9). If we vary only two different terms $N_i$, and $N_j$, then since $\mathrm{d}N_i = -\mathrm{d}N_j$ to satisfy (A22.9), (A22.10) cannot be satisfied simultaneously, since $E_i \neq E_j$. We conclude that the simplest general variation involves three $N_i$ terms.

Let $\mathrm{d}N_i$ ($i = 1,2,3$) be non-zero. Then, the coefficients of these quantities must be zero, from (A22.12). Hence

$$\ln(g_1 - N_1) - \ln N_1 + \alpha + \beta E_1 = 0 \ , \tag{A22.13}$$

†A consideration of entropy, through (A22.20), will convince us that we are discussing a maximum rather than a minimum.

and $$\ln(g_2 - N_2) - \ln N_2 + \alpha + \beta E_2 = 0 \ . \tag{A22.14}$$

Since we have chosen $dN_3 \neq 0$,

$$\ln(g_3 - N_3) - \ln N_3 + \alpha + \beta E_3 = 0 \ . \tag{A22.15}$$

We can repeat this procedure for $dN_i \neq 0$ $(i = 2,3,4)$, giving

$$\ln(g_4 - N_4) - \ln N_4 + \alpha + \beta E_4 = 0 \ , \tag{A22.16}$$

and so on. In general, for the $i$th state, we have

$$\ln(g_i - N_i) - \ln N_i + \alpha + \beta E_i = 0 \ . \tag{A22.17}$$

This equation may be written in the form

$$\frac{g_i}{N_i} - 1 = \exp(\alpha)\exp(\beta E_i) \ ; \tag{A22.18}$$

(A22.18) is one form of the Fermi-Dirac distribution function. We need next to identify $\alpha$ and $\beta$: $\alpha$ can be determined through (A22.5), but for the moment we shall write $\exp(\alpha) = A$. At high temperatures, the number of states which are energetically accessible is very large; then $g_i/N_i \gg 1$, and

$$N_i \approx g_i A^{-1} \exp(-\beta E_i) \ . \tag{A22.19}$$

In the high temperature limit, Boltzmann statistics apply, and from (1.9) we can equate $\beta$ to $1/kT$. A more rigorous proof can be obtained through (1.7), in the form

$$S = k\ln W_{max} \ . \tag{A22.20}$$

We have now $$N_i/g_i = \frac{1}{A\exp(E_i/kT) + 1} \ . \tag{A22.21}$$

It is convenient to define an energy, $E_F$, such that

$$A = \exp(-E_F/kT) \ . \tag{A22.22}$$

Hence, $$\frac{N_i}{g_i} = \frac{1}{\exp(E - E_F)/kT + 1} \ . \tag{A22.23}$$

We write $f(E) = N_i/g_i$, where $f(E)$ gives the probability that a state of energy $E$ is occupied. Thus,

$$f(E) = \frac{1}{\exp(E - E_F)/kT + 1} \tag{A22.24}$$

is a convenient form of the Fermi-Dirac distribution function.

## A22.2 Lagrange undetermined multipliers

In the previous section we needed to determine the maximum value of a function of several variables, subject to constraining equations relating the variables themselves. The problem was solved by Lagrange's method of undetermined multipliers and we consider it useful to describe the method in a little detail.

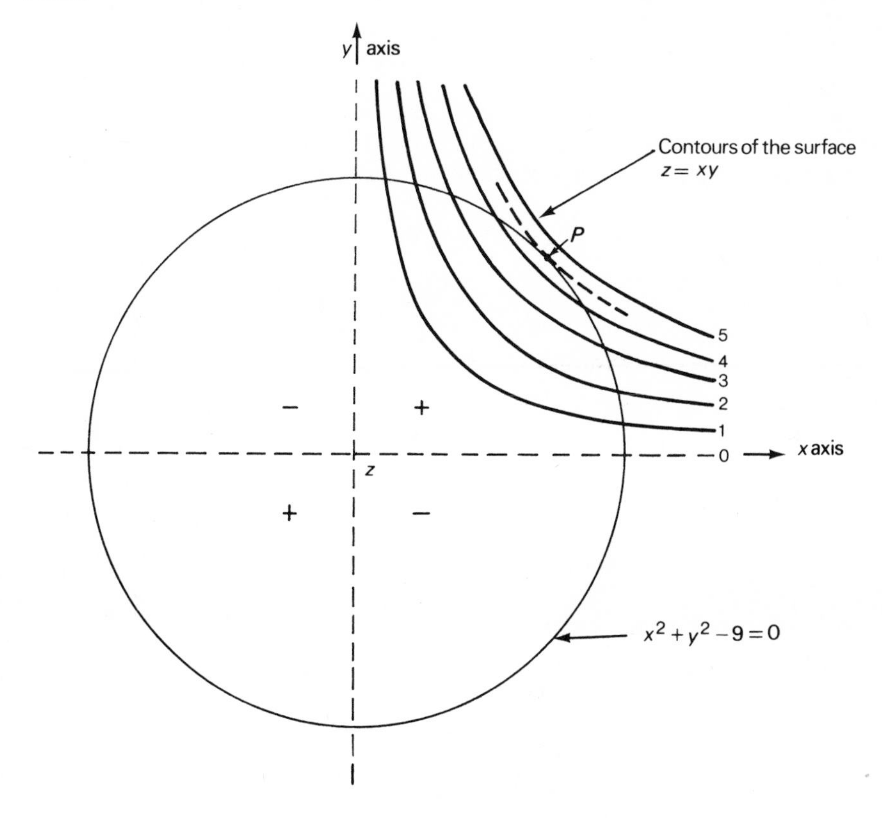


Figure A22.1 – A portion of the surface $z = xy$, shown by contours of interval 1, and the constraint $g(x,y)$, the circle $x^2 + y^2 - 9 = 0$. The dashed curve is a portion of one region of the $xy$ surface passing through a required maximum.

Consider Figure A22.1: it shows a portion of the surface $z = xy$ in $xyz$ space. We consider also a constraining equation $g(x,y) = 0$, which is the circle $x^2 + y^2 - 9 = 0$ in the example. Clearly, the extrema for the function $xy$ are $\pm\infty$, but we need to constrain the result to lie on $g(x,y)$. We shall let the position $P$ of the maximum in the positive $xy$ quadrant be $x = \mu$, $y = \nu$, that is on the curve $z = c$, where $c$ is a constant.

Since the two curves $z = c$ and $g = 0$ touch, they have a common tangent. Thus, at the point $(\mu,\nu)$, $\mathrm{d}z = 0$, that is

$$\mathrm{d}z = \frac{\partial z}{\partial x}\mathrm{d}x + \frac{\partial z}{\partial y}\mathrm{d}y = 0 \,, \tag{A22.25}$$

or

$$\mathrm{d}y/\mathrm{d}x = -\frac{\partial z/\partial x}{\partial z/\partial y} \,. \tag{A22.26}$$

Hence,

$$\frac{\partial z/\partial x}{\partial z/\partial y} = \frac{\partial g/\partial x}{\partial g/\partial y} \,, \tag{A22.27}$$

or, introducing a proportionality constant, $\lambda$, the two equations

$$\partial z/\partial x + \lambda \partial g/\partial x = 0 \,, \tag{A22.28}$$

and

$$\partial z/\partial y + \lambda \partial g/\partial y = 0 \tag{A22.29}$$

are satisfied. Since we have also

$$g(x,y) = 0 \,, \tag{A22.30}$$

we can determine $\mu$, $\nu$ and $\lambda$. Let us apply these results to our problem. We can form the function

$$h(x,y;\lambda) = z(x,y) + \lambda g(x,y) \,, \tag{A22.31}$$

and treat $x$ and $y$ as independent variables. Thus,

$$h = xy + \lambda(x^2 + y^2 - 9) \,, \tag{A22.32}$$

whence

$$\frac{\partial h}{\partial x} = y + 2\lambda x = 0 \,, \tag{A22.33}$$

and

$$\frac{\partial h}{\partial y} = x + 2\lambda y = 0 \,, \tag{A22.34}$$

which, together with

$$x^2 + y^2 - 9 = 0 \;, \tag{A22.35}$$

enables us to solve for four points:

$$\begin{aligned} \mu &= 3/\sqrt{2} & \nu &= 3/\sqrt{2} \\ \mu &= -3/\sqrt{2} & \nu &= -3/\sqrt{2} \\ \mu &= 3/\sqrt{2} & \nu &= -3/\sqrt{2} \\ \mu &= -3/\sqrt{2} & \nu &= 3/\sqrt{2} \end{aligned} \tag{A22.36}$$

The first two results give maxima in positive regions of Figure A22.1, with $xy = 4.5$: the second two results give minima, in the negative regions of the figure, with $z = -4.5$.

The same result can be reached by the following alternative route. We can use (A22.35) to solve for $y = y(x)$. Then $z = x,y(x)$, and $x$ may be treated as a single variable. Thus using (A22.26) we have

$$\frac{\mathrm{d}z}{\mathrm{d}x} = \frac{\partial z}{\partial x} + \frac{\partial z}{\partial y}\frac{\partial y}{\partial x} = \frac{\partial z}{\partial x} - \frac{\partial z}{\partial y}\frac{\partial g/\partial x}{\partial g/\partial y} \;. \tag{A22.37}$$

If we now set $\mathrm{d}z/\mathrm{d}x = 0$, the results of (A22.36) follow.

We shall now extend the discussion to its application in the Fermi-Dirac distribution. Let the function to be maximised be

$$z = f(x_1,x_2, \ldots, x_n) \;, \tag{A22.38}$$

subject to two conditions,

$$g_1(x_1,x_2, \ldots, x_n) = 0 \;, \tag{A22.39}$$

and

$$g_2(x_1,x_2, \ldots, x_n) = 0 \;. \tag{A22.40}$$

At a maximum, $\mathrm{d}z = 0$, that is

$$\frac{\partial z}{\partial x_1}\mathrm{d}x_1 + \frac{\partial z}{\partial x_2}\mathrm{d}x_2 + \ldots + \frac{\partial z}{\partial x_n}\mathrm{d}x_n = 0 \;. \tag{A22.41}$$

From the constraining equations, we have

$$\frac{\partial g_1}{\partial x_1}\mathrm{d}x_1 + \frac{\partial g_1}{\partial x_2}\mathrm{d}x_2 + \ldots + \frac{\partial g_1}{\partial x_n} = 0 \ , \qquad \text{(A22.42)}$$

and

$$\frac{\partial g_2}{\partial x_1}\mathrm{d}x_1 + \frac{\partial g_2}{\partial x_2}\mathrm{d}x_2 + \ldots + \frac{\partial g_2}{\partial x_n} = 0 \ . \qquad \text{(A22.43)}$$

Multiplying (A22.42) and (A22.43) by $\lambda_1$ and $\lambda_2$ respectively and adding to (A22.41) gives

$$\sum_{i=1}^{n} \frac{\partial z}{\partial x_i} + \lambda_1\frac{\partial g_i}{\partial x_i} + \lambda_2\frac{\partial g_2}{\partial x_i}\mathrm{d}x_i = 0 \ . \qquad \text{(A22.44)}$$

$\lambda_1$ and $\lambda_2$ are chosen such that the coefficients of $\mathrm{d}x_1$ and $\mathrm{d}x_2$ in (A22.44) are zero, that is

$$\frac{\partial z}{\partial x_1} + \lambda_1\frac{\partial g_1}{\partial x_1} + \lambda_2\frac{\partial g_2}{\partial x_1} = 0 \ , \qquad \text{(A22.45)}$$

and

$$\frac{\partial z}{\partial x_2} + \lambda_1\frac{\partial g_1}{\partial x_2} + \lambda_2\frac{\partial g_2}{\partial x_2} = 0 \ . \qquad \text{(A22.46)}$$

Since the $n$ variables are subject only to two conditions, $n - 2$ variables are independent. Therefore, the coefficients of $\mathrm{d}x_i$ ($i = 3,4,5, \ldots n$) in (A22.44) must vanish. Hence,

$$\frac{\partial z}{\partial x_i} + \lambda_1\frac{\partial g_1}{\partial x_i} + \lambda_2\frac{\partial g_2}{\partial x_i} = 0 \ (i = 1,2 \ldots,n) \qquad \text{(A22.47)}$$

These $n$ equations and the two constraining equations determine the $n + 2$ unknowns.

## PROBLEMS TO CHAPTER 5

1. The electrical resistivity for lithium at 273 K is $8.55 \times 10^{-8}$ ohm m, the atomic mass is 0.00694 kg mol$^{-1}$, the density is 537.5 kg m$^{-3}$, and the mean electron velocity is $1.30 \times 10^{6}$ m s$^{-1}$. Calculate the mean free path and mean mobility of the valence electrons in lithium.

*2. The anharmonic vibration of atoms about their mean position in a solid may be represented approximately by the potential energy function $V(x) = ax^2 - bx^3$, where $a$ and $b$ are constants. By determining an expression for the mean displacement, $\bar{x}$, show that $\bar{x}$ is directly proportional to temperature, that is, it is consistent with the solid expanding on heating. *Note:* since $x$ is small. $\exp(bx^3) \approx (1 + bx^3)$.

3. Calculate, from free-electron theory, the Fermi energy and the Fermi temperature for lithium. The unit cell constant for the BCC structure of Li is $3.50 \times 10^{-10}$ m.
4. Determine $g(E_F)$ for lithium.
5. Use the density of states function to calculate the average thermal energy of electrons between 0 and $E_F$.

*6. (a) Draw a two-dimensional, square lattice in **k** space and mark on it the boundaries of the first three Brillouin zones.
(b) What is the first Brillouin zone for a lattice referred to a primitive cubic unit cell of side $a$?

7. Calculate the packing efficiency for the CPH structure.
8. The CPC structure contains tetrahedral holes (surrounded by 4 atoms) and octahedral holes (surrounded by 6 atoms). How many holes of each kind are unique to one face-centred unit cell, and what are the fractional coordinates of their centres?

Chapter 6

# Postamble

The study of the forces acting between atoms and molecules in different states of aggregation is one of the most important and fascinating fields in the physical sciences. We have made but an introductory excursion into it, and the reader is, I hope, now sufficiently interested in this subject to delve into the bibliography of more advanced writings in order to extend his knowledge of structure and bonding.

The division of the main subject matter of this book into four sections based on extreme bond types is not unusual. We appreciate that no compound shows just one of these bond types, but in the cohesion of its chemical units makes use of several of the types of forces which we have discussed. The actual bonding is possibly some combination of two or more of the extreme types which are recognised. It must be remembered that the divisions which we have used are arbitrary: they represent a way in which we can think about bonding within the limits of the theories available to us.

In the development of scientific ideas we usually find that an earlier theory which, in order to be raised to that level of significance, did explain certain properties adequately, emerges as a special case of a more general theory. We saw this feature take shape with the Planck theory of black-body radiation. The earlier results of Wien and of Rayleigh-Jeans, adequate in their own ranges of applicability, were nothing more than limiting cases of Planck's theory in two different circumstances. So in a more generalised theory of bonding, yet to be developed, the bond types which we have separated out for convenience will be recognizable as special cases of the more general theory. These ideas are somewhat speculative, but inasmuch as the present theories are imperfect so will improvements be forthcoming, for such is the nature of the progress of scientific discovery.

In order to draw together some of the work which we have studied, let us take sodium chloride again and look further into the nature of its bonding. We know that in the solid state it is composed of ions. The electrons remain fairly well localized in the atomic orbitals of the ions and it is very unlikely that the bond contains any significant contribution of the metallic type, for the feature of this bond is complete delocalization of the valence electrons. We may consider the other possibilities: ionic forces, covalent forces and contributions from van der Waals' forces.

The ionic contribution to NaCl is readily understood, and if we set up a model for NaCl which includes polarization terms arising from dipole and quadrupole interactions, and the equation of state correction (see Appendix 8), we obtain for the cohesive energy –769 kJ mol$^{-1}$. The best experimental value obtained through a Born-Haber cycle is –774 kJ mol$^{-1}$. The discrepancy is small and not generally significant within the probable errors of the parameters involved in the calculations. But can we do any better? The p orbitals of $Na^+$ and $Cl^-$ are directed towards one another in the solid, which is the most favourable direction for electron overlap, and we can study the overlap integrals of the type $\int \psi(Na^+)\psi(Cl^-)d\tau$, where $\psi(Na^+)$ and $\psi(Cl^-)$ are one-electron AO's for the ions. This integral has a value of about 0.06, showing clearly the high degree of localization of the electrons which we have already suggested. However the contribution is not zero, and if we include its effect the calculated cohesive energy becomes –774 kJ mol$^{-1}$. So we can be well satisfied with this calculation insofar as our present theory is concerned (Table 6.1).

**Table 6.1 – Cohesive energy of sodium chloride**

| | kJ mol$^{-1}$ |
|---|---|
| Electrostatic energy | – 861 |
| Repulsion energy | + 101 |
| Polarization energy (dipole-dipole and higher terms) | – 9 |
| Covalent energy | – 5 |
| Total calculated energy | – 774 |
| Experimental energy | – 774 |

The classical calculation of the polarization terms has certain unsatisfactory features. A wave mechanical approach is sounder and has been carried out recently for LiF. The result obtained for $U(r_e)$, including only dipole-dipole terms for the dispersion energy, is –1027 kJ mol$^{-1}$, and the experimental Born-Haber cycle value is –1027 kJ mol$^{-1}$. The correspondence is encouraging and no doubt we shall see advances in these calculations in future years.

## APPENDIX 23 PROPAGATION OF ERRORS

The number of significant figures in a result is not necessarily similar to the number of significant figures in the data. Consider $y = p^n$, where $p = 2.0 \pm 0.1$.

For $n = 0.1$, $y$ lies between 1.066 and 1.077, whereas for $n = 4$, $y$ lies between 13.0 and 19.4.

Consider any function $y = f(p)$ (Figure A23.1). In the small interval $\delta p$, the change $\delta y$ in $y$ is given with good accuracy by

$$\delta y = \left(\frac{\mathrm{d}y}{\mathrm{d}p}\right) \delta p \; . \tag{A23.1}$$

Consider next any function $y = f(p_1,p_2)$ where $p_1$ and $p_2$ are independent variables. For two small independent changes $\delta p_1$ and $\delta p_2$, the changes in $y$ are given by analogy with (A23.1) by

$$(\delta y)_{p_1} = \left(\frac{\partial y}{\partial p_1}\right) \delta p_1 \tag{A23.2}$$

and

$$(\delta y)_{p_1} = \left(\frac{\partial y}{\partial p_2}\right) \delta p_2 \; . \tag{A23.3}$$

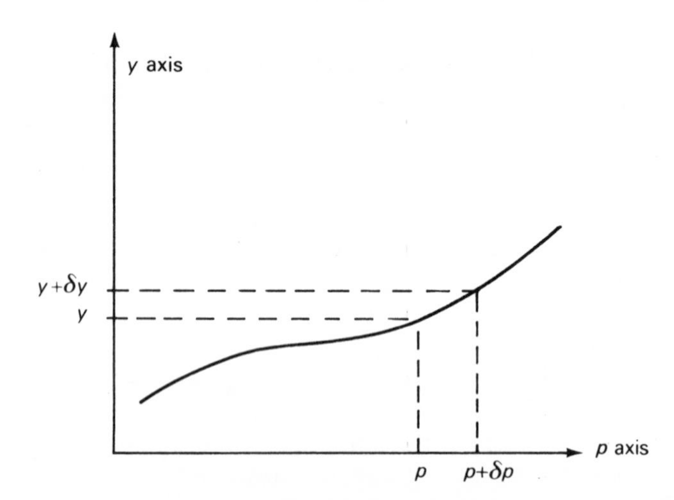


Figure A23.1 – A function $y = f(p)$.

Since we have assumed that these two variations in $y$ are uncorrelated, they can be represented along two rectangular axes (Figure A23.2). Hence

$$(\delta y)^2 = (\delta y)^2_{p1} + (\delta y)^2_{p2} = \left(\frac{\delta y}{\partial p_1}\right)^2 (\delta_{p_1})^2 + \left(\frac{\partial y}{\partial p_2}\right)^2 (\delta_{p_2})^2 \tag{A23.4}$$

Generalizing for a function $y = f(p_j)$ $(j = 1,2,3, \ldots n)$:

$$(\delta y)^2 = \sum_{i=1}^{n} \left(\frac{\partial y}{\partial p_j}\right)^2 (\delta p_j)^2 \ . \qquad \text{(A23.5)}$$

The quantity $\delta y$ can be equated to the standard deviation in $y$, $\sigma(y)$.

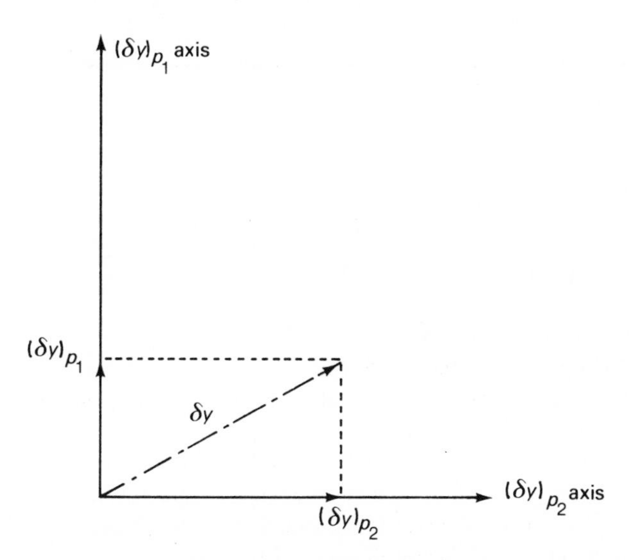


Figure A23.2 – Representation of the uncorrelated errors $\delta y_{p_1}$ and $\delta y_{p_2}$.

## PROBLEMS TO CHAPTER 6

1. The main source of error in the value for the cohesive energy of an ionic solid arises through the thermochemical terms in the case of the Born-Haber cycle and through the compressibility in the case of the electrostatic model. In NaCl we have the following data:

| | Standard deviation $\sigma$ | |
|---|---|---|
| $S$ | 2.5 kJ mol$^{-1}$ | (Table 2.1) |
| $\Delta H_f$ | 0.8 kJ mol$^{-1}$ | (Table 2.6) |
| $\kappa$ | $0.1 \times 10^{-11}$ N$^{-1}$ m$^2$ | |

   Estimate the standard deviation in $U$ from the Born-Haber (2.16) and electrostatic (2.56)–(2.57) models for NaCl. The propagation of errors is considered in Appendix 23.

*2. Compose a table showing the characteristic physical and structural properties associated with the four main types of interatomic forces.

# Suggested Reading List

**Covalent Compounds**

Coulson, C. A., *Valence,* O.U.P. (1961).

Evans, R. C., *Crystal Chemistry,* C.U.P. (1964).

**Crystallography and Crystal Structures**

Ladd, M. F. C. and Palmer, R. A., *Structure Determination by X-ray Crystallography,* Plenum (1977).

Sutton, L. E. (Editor), *Tables of Interatomic Distances and Configurations in Molecules and Ions,* Chemical Society, London (1958, 1965).

Wyckoff, R. W. G., *Crystal Structures, Vols. 1–6,* Interscience (1963–1971).

**Ionic Compounds and Cohesive Energies of Ionic Solids**

Adams, D. M., *Inorganic Solids*, Wiley (1974).

Cotton, F. A. and Wilkinson, G., *Advanced Inorganic Chemistry,* Wiley-Interscience (1973).

Evans, R. C., *Crystal Chemistry,* C.U.P. (1964).

Ladd, M. F. C. and Lee, W. H., in *Progress in Solid State Chemistry, Vols. 1, 2, 3,* Pergamon (1964, 1965, 1966).

Tosi, M. P., in *Solid State Physics, Vol. 16,* Academic Press (1964).

Waddington, T. C., in *Advances in Inorganic Chemistry and Radiochemistry*, Vol. 1, Academic Press (1959).

**Mathematics, Statistics and Statistical Mechanics**

Courant, R., *Differential and Integral Calculus, Vol. 2,* Blackie (1949).

Guggenheim, E. A., *Boltzmann's Distribution Law,* North Holland (1955).

Margenau, H. and Murphy, G. M., *The Mathematics of Physics and Chemistry,* Van Nostrand (1943).

Topping, J., *Errors of Observation and Their Treatment,* Chapman and Hall (1971).

Woodward, L. A., *Molecular Statistics for Students of Chemistry,* Clarendon (1975).

**Metallic Compounds**

Evans, R. C., *Crystal Chemistry,* C.U.P. (1964).

Kittel, C., *Introduction to Solid State Physics,* Wiley (1971).

**Molecular Compounds**

Evans, R. C., *Crystal Chemistry,* C.U.P. (1964).

Hirschfelder, J. O. (Editor), *Advances in Chemical Physics, Vol. 12: Intermolecular Forces,* Interscience (1967).

Prock, A. and McConkey, G., *Topics in Chemical Physics,* Elsevier (1962).

**Molecular Structures**

Brand J. C. D. and Speakman, J. C., *Molecular Structure,* Arnold (1975).

Karplus, M. and Porter, R. N., *Atoms and Molecules,* Benjamin (1970).

Sutton, L. E. (Editor), *Tables of Interatomic Distances and Configurations in Molecules and Ions,* Chemical Society, London (1958, 1965).

**Physical Chemistry**

Atkins, P. W., *Physical Chemistry,* O.U.P. (1978).

**Quantum Mechanics and Quantum Chemistry**

Anderson, J. M., *Introduction to Quantum Chemistry,* Benjamin (1960).

Atkins, P. W., *Molecular Quantum Mechanics,* Clarendon (1973).

Coulson, C. A., *Valence,* O.U.P. (1961).

Eyring, H., Walter, J. and Kimball, G. E., *Quantum Chemistry,* Wiley (1944).

Murrell, J. N., Kettle, S. F. and Tedder, J. M., *Valence Theory,* Wiley (1970).

Murrell, J. N., Kettle, S. F. and Tedder, J. M., *The Chemical Bond,* Wiley (1978).

Pauling, L and Wilson, E. B., *Introduction to Quantum Mechanics,* McGraw-Hill (1935).

Roberts, J. D., *Molecular Orbital Calculations,* Benjamin (1962).

Slater, J. C., *Quantum Theory of Matter,* McGraw-Hill (1968).

**Solid-State Chemistry and Solid-State Physics**

Ladd, M. F. C. and Lee, W. H., in *Progress in Solid State Chemistry, Vols. 1, 2, 3,* Pergamon (1964, 1965, 1966).

Kittel, C., *Introduction to Solid State Physics,* Wiley (1971).

Prock, A. and McConkey, G., *Topics in Chemical Physics,* Elsevier (1962).

Smith, R. A., *Wave Mechanics of Crystalline Solids,* Chapman and Hall (1961).

Tosi, M. P., in *Solid-State Physics,* Vol. 16, Academic Press (1964).

**Spectroscopy**

Atkins, P. W., *Molecular Quantum Mechanics,* Clarendon (1973).

Barrow, G. M., *Introduction to Molecular Spectroscopy,* McGraw-Hill (1962).

Kuhn, H., *Atomic Spectra,* Longmans (1962).

**Thermodynamics and Thermodynamic Data**

Denbigh, K., *The Principles of Chemical Equilibrium,* C.U.P. (1966).

Rossini, F. D. *et al, Circular No. 500,* National Bureau of Standards (1952).

Smith, E. B., *Basic Chemical Thermodynamics,* O.U.P. (1973).

**Valence Theory**

Coulson, C. A., *Valence,* O.U.P. (1961).

Murrell, J. N. and Harget, A. J., *Semi-empirical Self-consistent-field Molecular Orbital Theory of Molecules,* Wiley (1972).

Murrell, J. N., Kettle, S. F. A. and Tedder, J. M., *Valence Theory,* Wiley (1970).

Murrell, J. N., Kettle, S. F. A. and Tedder, J. M., *The Chemical Bond,* Wiley (1978).

Orgel, L. E., *An Introduction to Transition Metal Chemistry,* Methuen (1960).

# Solutions to Problems

## SOLUTIONS TO CHAPTER 1

1. The $(NH_4)^+$ and $(NO_3)^-$ ions are both in free rotation, so that their envelopes of motion averaged over time are effectively spherical.
2. 4.40 Å.
3. NaCl, KCl — $CaF_2$, $\beta$-$PbF_2$
   $BaSO_4$, $PbSeO_4$ — NaBr, MgO
4. KCl is mainly an ionic solid and its melt consists of free ions. $BeCl_2$ is mainly a molecular solid containing tetrahedrally coordinated Be atoms doubly-bridged by Cl:

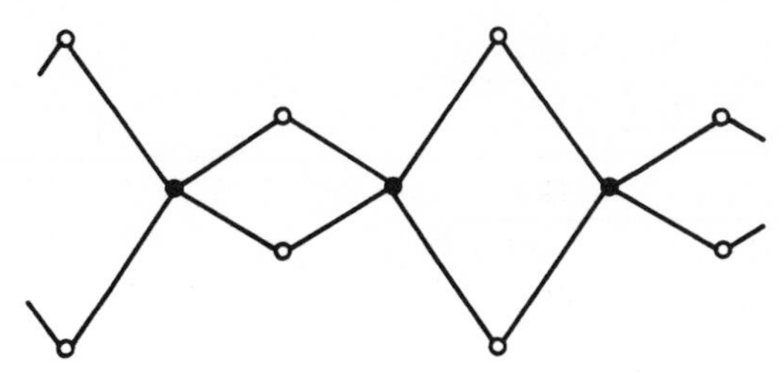

In the liquid state of $BeCl_2$, some of this structure persists and consequently it is more ordered than that of liquid KCl.

5.

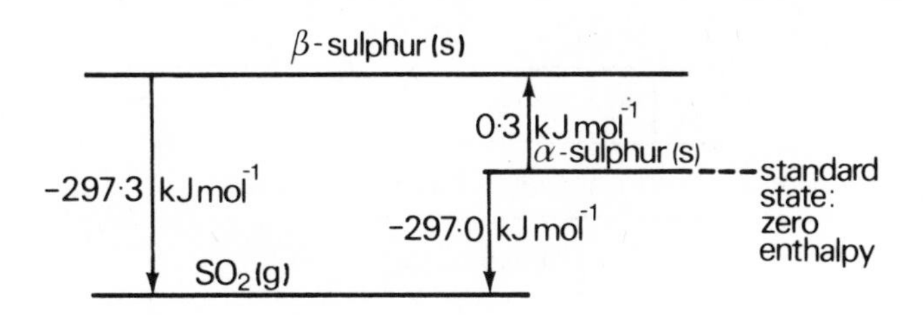


$S$ ($\beta$-sulphur at 386K) 32.66 J $K^{-1}$ $mol^{-1}$

6.

| *Ionic* | *Covalent* | *Van der Waals'* | *Metallic* |
|---|---|---|---|
| RbF | AlN | $P_4$ | Pb |
| $P_2Cl_{10}$ | SiC | $CO_2$ | $Cu_3Au$ |
| $KClO_3$ | | $C_6H_6$ | |
| $Na_2SO_4$ | | Ne | |

7. $(21\bar{3})$ $(203)$ $(\bar{3}00)$ $(6\bar{9}8)$ $(06\bar{4})$ $(0\bar{2}1)$.
8. In the correct reflexion position $2d_{010} \sin \theta_{010} = \lambda$. In this position X-rays incident at the same angle on the interleaving planes are reflected with the same intensity but with a path difference of $\lambda/2$. Their combination interferes destructively and the 010 reflexion does not occur with this type of structure.

## SOLUTIONS TO CHAPTER 2

1. 589.17 kJ $mol^{-1}$.
2. $\ln P = (22723 \pm 8)(1/T) + (22.9 \pm 0.7)$.
   $\Delta H = (189 \pm 7)$ kJ $mol^{-1}$.
   *Least-squares sums*
   $[1/T] = 5.1850 \times 10^{-3}$. $[1/T^2] = 5.3941 \times 10^{-6}$.
   $[\ln P] = -3.1898$. $[(\ln P)/T] = -3.7004 \times 10^{-3}$.
3. *Cycle*

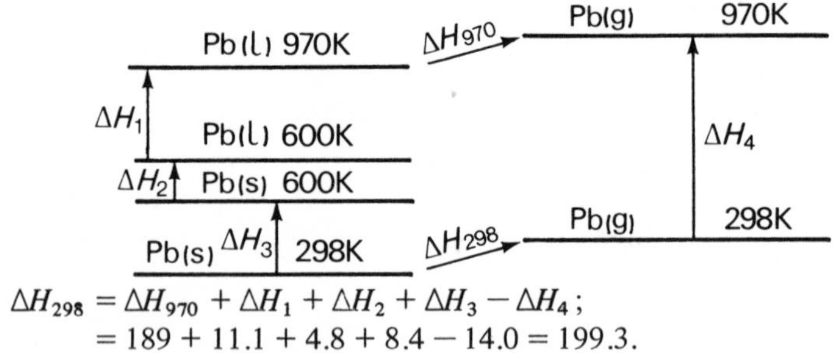


$$\Delta H_{298} = \Delta H_{970} + \Delta H_1 + \Delta H_2 + \Delta H_3 - \Delta H_4;$$
$$= 189 + 11.1 + 4.8 + 8.4 - 14.0 = 199.3.$$

$S_M(\text{Pb}) = 199$ kJ $mol^{-1}$.

4. $U_C(\text{MgCl}) = -726$ kJ $mol^{-1}$ and $U_C(\text{MgCl}_2) = -2057$ kJ $mol^{-1}$. Although more energy is needed to produce $Mg^{2+}$ than $Mg^+$, it is more than balanced by the lower crystal energy of $MgCl_2$ compared to that of MgCl.
5. *Cycle*

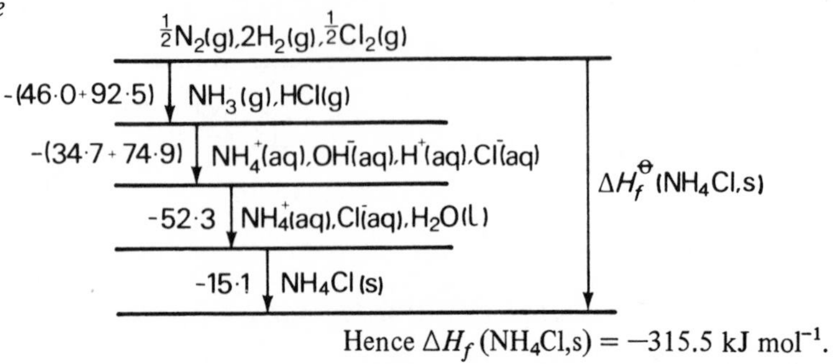


Hence $\Delta H_f(\text{NH}_4\text{Cl,s}) = -315.5$ kJ $mol^{-1}$.

6. From (2.39) the energy lost per ion $= 0.806e^2/(4\pi\epsilon_0 V^{\frac{1}{3}})$. The volume $V$ occupied per ion, $Na^+$ or $Cl^-$, in the NaCl structure type is $r^3$. Hence taking the conducting shell centred on any ion and of radius $r$ equal to $r_e$, the energy lost per ion is $0.806e^2/(4\pi\epsilon_0 r)$. Hence $A = 1.612$.
7. $\rho/r_e = 0.126$ and $U(r_e) = -3488$ kJ mol$^{-1}$; hence $E(O^{2-}) = 713$ kJ mol$^{-1}$.
8. $-1437$ kJ mol$^{-1}$.
9. $u_S = \{e^2/(4\pi\epsilon_0)\}\{6(0.7\times0.7)/2.12 - 4(0.7\times0.8)/1.30\}\times10^{10}$.
   $= -7.76 \times 10^{-19}$ J (per ion).
10. $a = 4.56 \times 10^{-10}$ m whence $\rho = 4450$ kg m$^{-3}$.
11.

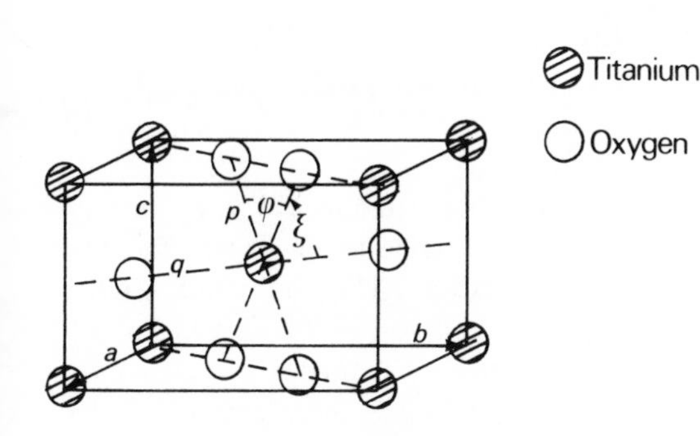


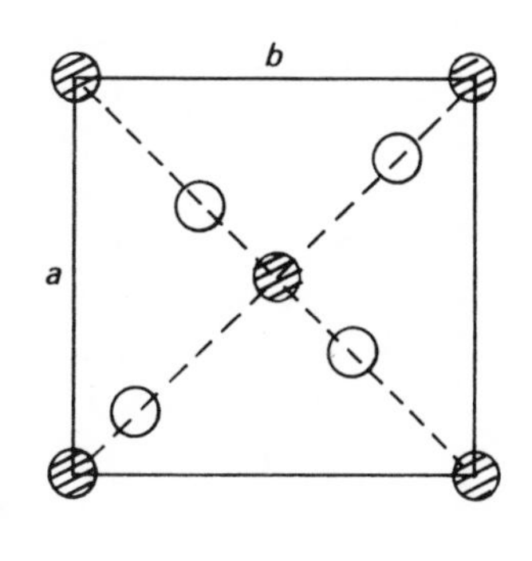


The unique bond lengths and angles in the $\{TiO_6\}$ unit are labelled $p$, $q$, $\phi$ and $\xi$: $p = 1.945$ Å, $q = 1.985$ Å, $\phi = 80.96°$, $\xi = 90.01°$.

12.

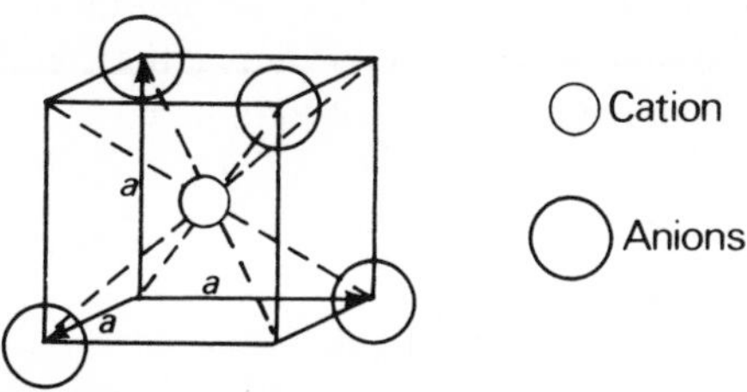


From the diagram $a\sqrt{2} = 2r_-$ and $a\sqrt{3} = 2r_+ + 2r_-$ when the atoms are in maximum contact. Hence $R(=r_+/r_-) = \sqrt{(3/2)} - 1 = 0.225$.

| | $\Delta H$/kJ mol$^{-1}$ |
|---|---|
| 13. $AgI(s) + K^+(aq) + I^-(aq) \rightarrow K^+(aq) + [AgI_2]^-(aq)$ | $-9.54$ |
| $AgI(s) + Na^+(aq) + I^-(aq) \rightarrow Na^+(aq) + [AgI_2]^-(aq)$ | $-7.46$ |
| $Ag(s) + I_2(s) + K^+(aq) + I^-(aq) \rightarrow K^+(aq) + [AgI_2]^-(aq)$ | $-72.0$ |
| $Ag(s) + I_2(s) + Na^+(aq) + I^-(aq) \rightarrow Na^+(aq) + [AgI_2]^-(aq)$ | $-70.3$ |

Hence the average $\Delta H_f(AgI,s) = 62.7$ kJ mol$^{-1}$.

14. 39.6 kJ mol. The negative $\Delta H_d$ promotes solubility; this effect is negated by the large negative $\Delta S_d$. The small highly-charged ions $Mg^{2+}$ and $F^-$ are structure-making in water; this effect is enhanced by F . . . H . . . F hydro-

gen bonds. The entropy of the ions+water systems is less than that of the sum of the separate components.

15. Following (2.98) $W_1$ (for the number of defects) is $N!/\{(N\text{-}n)!\}$ and $W_2$(for the number of interstices) is $N'!/\{(N'\text{-}n)!n!\}$. Hence the total probability $W$ is $W_1W_2$. Continuing the analogy with (2.99) to (2.104) $[\partial(\Delta G)/\partial n]_T$ becomes $\Delta u - \ln[(N\text{-}n)/n] - \ln[(N'\text{-}n)/n]$. Assuming that both $n \ll N$ and $n \ll N'$ we obtain $n = \sqrt{NN'}\exp(-\Delta u/2kT)$. For AgCl at 500 K, $n/\sqrt{NN'} = 9.1 \times 10^{-6}$.
16. $W$ in (2.98) is cubed and $n/N = \exp(-\Delta u/3kT)$.
17. From (P2.11) $\ln\beta_t = x^2/(Dt)$. By least squares, $A = 623$ arbitrary units and $D = 9.35 \times 10^{-7}\,\mathrm{mm^2\,s^{-1}}$.
18. From (P2.12), $\ln D = \ln D_0 - U/RT$. By least squares, $U = 191$ kJ mol$^{-1}$ and $D_0 = 3.43 \times 10^{-7}\,\mathrm{m^2\,s^{-1}}$. The significance of $D_0$ might be said to be the rate of superficial diffusion in the absence of any activation energy barrier.
19. The curve of $C_P/T$ against $T$ is drawn and extrapolated to 298 K. By cutting and weighing against a standard area of the same graph paper, the area between 15.05 and 298 K was 3726 squares. The calibration gave 1 square $= 8.04 \times 10^{-3}$ J mol$^{-1}$ K$^{-1}$; hence $S$ (15.05 to 298 K) is 29.96 J mol$^{-1}$ K$^{-1}$. Below 15.05 K the Debye approximation may be used: $C_P = aT^3$ where $a$ is a constant. Hence $\int_0^{15.05} aT^2 dT = aT^3/3$ or $C_P/3$, where $C_P$ may be taken as the minimum value recorded. Hence $S$ (15.05 to 0 K) $= 0.06$ J mol$^{-1}$ K$^{-1}$. Thus the entropy of nickel at 298 K is 30.0 J mol$^{-1}$ K$^{-1}$.
20. $\theta_{300} = 30.47°$. The intensity of the 300 reflexion would be expected to be weak: the species $Cs^+$ and $I^-$ are isoelectronic, and the contribution from $I^-$ at the centre of the unit cell effectively cancels the contribution from $Cs^+$ for 300. Compare the case of the 111 reflexion for KCl.

## SOLUTIONS TO CHAPTER 3

1.

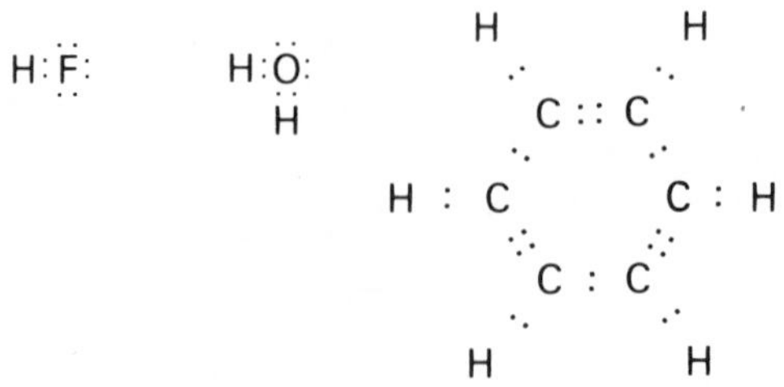

```
                                    H           H
                                     .·         ·.
  H:F̤̈:          H:Ö:                    C :: C
                  H                    ·.        .·
                              H : C              C : H
                                       .·.       .·.
                                         C : C
                                       ·.        .·
                                    H           H
```

2. (i)

| $\lambda$/Å | 5461 | 3650 | 3126 |
|---|---|---|---|
| $-V_0$/volt | 2.05 | 0.92 | 0.32 |

(ii) $V_0 e = (6.74 \times 10^{-34})\,\nu - 6.99 \times 10^{-19}$
$k$ (Planck's constant) is $6.74 \times 10^{-34}$ J s.

3. $\Delta p_x \approx 6.6 \times 10^{-19}$, and average kinetic energy $\approx (3\Delta p_x^2/2m_e) \approx 7 \times 10^{-7}$ J
Potential energy $\approx -\dfrac{(1.6022 \times 10^{-19})^2}{4\pi\epsilon_0 \times 10^{-15}} \approx -2 \times 10^{-13}$ J. The potential energy does not balance the kinetic energy associated with the uncertainty principle, and the system is therefore unstable.

4. $K = 4/R_{\mathrm{H}}$; $3.026 \times 10^{-19}$ J.

5. $\frac{2}{a}\int_0^a \sin(m\pi x/a)\sin(n\pi x/a)\mathrm{d}x = \delta_{m,n}$; use the identity $\sin A \sin B = \frac{1}{2}\{\cos(A-B)-\cos(A+B)\}$. When $n \neq m$ $\delta_{m,n} = 0$: for $n = m$ re-write the integral as $\frac{2}{a}\int_0^a \sin^2(n\pi x/a)\mathrm{d}x$ in order to avoid the indeterminate quotient 0/0. Now $\delta_{m,n} = 1$.

6. $E_{1,1,1} = 1.8 \times 10^{-7}$ J.

7. $\psi_{1,0,0}^2 = \dfrac{1}{\pi a_0^3}\exp(-2r/a_0)$.

$$\overline{\left(\frac{1}{r}\right)} = \int_0^\infty \left(\frac{1}{r}\right)\psi_{1,0,0}^2\, r^2\mathrm{d}r \int_0^\pi \sin\theta\, \mathrm{d}\theta \int_0^{2\pi} \mathrm{d}\phi = \frac{4}{a_0^3}\int_0^\infty r^2 \exp(-2r/a_0)\mathrm{d}r\ 1/a_0.$$

Average potential energy $(\bar{V}) = -e^2/(4\pi\epsilon_0 a_0)$. From (3.84) and (3.85), for $n = 1$, $E = -e^2/(8\pi\epsilon_0 a_0)$; hence the average kinetic energy $(\bar{T})$ is $E - \bar{V}$, or $e^2/(8\pi\epsilon_0 a_0)$, and $\bar{T} = -E$.

8. $$\int\psi^2(1\mathrm{s})\mathrm{d}\tau = \frac{1}{\pi a_0^3}\int_{1.10a_0}^{1.11a_0} r^2 \exp(-2r/a_0)\mathrm{d}r \int_{0.20\pi}^{0.21\pi}\sin\theta\,\mathrm{d}\theta \int_{0.60\pi}^{0.61\pi}\mathrm{d}\phi$$
$$= 0.00127\int_{1.10a_0}^{1.11a_0} r^2 \exp(-2r/a_0)\mathrm{d}r.$$
Use $\int x^n \exp(ax)\mathrm{d}x = \left(\dfrac{x^n}{a}\right)\exp(ax) - \left(\dfrac{n}{a}\right)\int x^{n-1}\exp(ax)\mathrm{d}x$.
Probability $= 2.53 \times 10^{-7}$.

9. $\mathrm{N}(1\mathrm{s})^2(2\mathrm{s})^2(2\mathrm{p}_x)(2\mathrm{p}_y)(2\mathrm{p}_z)$
$\mathrm{Cl}(1\mathrm{s})^2(2\mathrm{s})^2(2\mathrm{p}_x)^2(2\mathrm{p}_y)^2(2\mathrm{p}_z)^2(3\mathrm{s})^2(3\mathrm{p}_x)^2(3\mathrm{p}_y)^2(3\mathrm{p}_z)$
$\mathrm{Cl}^-(1\mathrm{s}^2)(2\mathrm{s})^2(2\mathrm{p}_x)^2(2\mathrm{p}_y)^2(2\mathrm{p}_z)^2$ or (Ar).
$\mathrm{K}^+$(Ar).

10. $Be_2$ VB: No electrons for pairing so no canonical forms: $Be_2$ is not a stable entity.
MO: $(1\sigma_g)^2(1\sigma_u)^2(2\sigma_g)^2(2\sigma_u)^2$, an antibonding configuration.
LiH VB: Possible canonical forms are Li-H, $Li^+H^-$, $Li^-H^+$. $Li^-H^+$ is of least significance. The structure will be a resonance hybrid of mainly these three forms.
MO: $(1\sigma)^2(2\sigma^2)$; a $\sigma$-type bond is formed.

11. $\dfrac{q^- \quad q^+}{1.18\ \text{Å}}$; total fractional ionic character $q = 0.32$. From (3.145) $q = 0.29$.

12. s:p = 1:2. In (3.151) $\lambda = 1/\sqrt{2}$ and so $\theta = 120°$.

13. $\psi(\text{sp}^x) = \text{s} + \lambda\text{p}$; s:p = 1:$x$ and so $\lambda = x^{1/2}$.
$N^2\int(\text{s}+x^{1/2}\text{p})(\text{s}+x^{1/2}\text{p})\text{d}\tau = 1 = N^2\int\text{s}^2\text{d}\tau + x\int\text{p}^2\text{d}\tau + 2x^{1/2}\int\text{spd}\tau = N^2(1+x)$, since $\int\text{spd}\tau = 0$. Hence $N = 1/(1+x)^{1/2}$ and $\psi(\text{sp}^x) = (1+x)^{-1/2}\{\psi(s) + x^{1/2}\psi(\text{p})\}$.

14. $\text{p}_x = \text{p}\cos(109.47/2) = \dfrac{1}{\sqrt{3}}\,\text{p}$, and similarly for $\text{p}_y$ and $\text{p}_z$. Then it follows that

$$\begin{aligned}
\psi(\text{sp}^3)_{1,1,1} &= \tfrac{1}{2}(\text{s} + \text{p}_x + \text{p}_y + \text{p}_z),\\
\psi(\text{sp}^3)_{-1,-1,1} &= \tfrac{1}{2}(\text{s} - \text{p}_x - \text{p}_y + \text{p}_z),\\
\psi(\text{sp}^3)_{1,-1,-1} &= \tfrac{1}{2}(\text{s} + \text{p}_x - \text{p}_y - \text{p}_z)\\
\psi(\text{sp}^3)_{-1,1,-1} &= \tfrac{1}{2}(\text{s} - \text{p}_x + \text{p}_y - \text{p}_z).
\end{aligned}$$

$\int\psi(\text{sp}^3)_{1,1,1}\psi(\text{sp}^3)_{1,1,1}\text{d}\tau = \tfrac{1}{4}\{\int\text{s}^2\text{d}\tau + \int\text{p}_x^2\text{d}\tau + \int\text{p}_y^2\text{d}\tau + \int\text{p}_z^2\text{d}\tau\} = 1$, and so on. $\int\psi(\text{sp}^3)_{1,1,1}\psi(\text{sp}^3)_{-1,-1,1}\text{d}\tau = \tfrac{1}{4}\{\int\text{s}^2\text{d}\tau - \int\text{p}_x^2\text{d}\tau - \int\text{p}_y^2\text{d}\tau + \int\text{p}_z^2\text{d}\tau\}$ + other integrals like $\int\text{spd}\tau$ and $\int\text{p}_x\text{p}_y\text{d}\tau$ which are both zero. Hence, $\int\psi(\text{sp}^3)_{1,1,1}\psi(\text{sp}^3)_{-1,-1,1}\text{d}\tau = 0$, and so on. Hence the $\text{sp}^3$ hybrid orbitals are mutually orthogonal.

15. From section 3.9 it follows that $E_\pi$ for ethene is $2\alpha + 2\beta$. If buta-1,3-diene were localized, $E_\pi$ would be equivalent to that for two molecules of ethene, that is $4\alpha + 4\beta$. For delocalized buta-1,3-diene, we can write, following section 3.7.2 and Appendix 17, the determinantal equation

$$\begin{vmatrix}
H_{11}\text{-}ES_{11} & H_{12}\text{-}ES_{12} & H_{13}\text{-}ES_{13} & H_{14}\text{-}ES_{14}\\
H_{12}\text{-}ES_{12} & H_{22}\text{-}ES_{22} & H_{23}\text{-}ES_{23} & H_{24}\text{-}ES_{24}\\
H_{13}\text{-}ES_{13} & H_{23}\text{-}ES_{23} & H_{33}\text{-}ES_{33} & H_{34}\text{-}ES_{34}\\
H_{14}\text{-}ES_{14} & H_{24}\text{-}ES_{24} & H_{34}\text{-}ES_{34} & H_{44}\text{-}ES_{44}
\end{vmatrix} = 0\ .$$

Taking $H_{ij} = \alpha$, $H_{ij}$(adjacent) $= \beta$, $H_{ij}$(non-adjacent) $= 0$, $S_{ii} = 1$ and $S_{ij} = 0$, we obtain

$$\begin{vmatrix}
\alpha\text{-}E & \beta & 0 & 0\\
\beta & \alpha\text{-}E & \beta & 0\\
0 & \beta & \alpha\text{-}E & \beta\\
0 & 0 & \beta & \alpha\text{-}E
\end{vmatrix} = 0$$

Dividing throughout by $\beta$ and putting $\alpha - E/\beta = x$, we have

$$\begin{vmatrix} x & 1 & 0 & 0 \\ 1 & x & 1 & 0 \\ 0 & 1 & x & 1 \\ 0 & 0 & 1 & x \end{vmatrix}$$

Solution by factorization gives

$$x^4 - 3x^2 + 1 = 0.$$

Solving for $x^2$:

$$x^2 = \frac{3 \pm\sqrt{9-4}}{2} = 2.618 \text{ or } 0.382. \text{ Hence,}$$

$$x = \pm 1.618, \pm 0.618.$$

We can represent the orbitals diagrammatically:

| | | $E$ per electron |
|---|---|---|
| □ | Antibonding | $\alpha - 1.618\beta$ |
| □ | | $\alpha - 0.618\beta$ |
| ↑↓ | Bonding | $\alpha + 0.618\beta$ |
| ↑↓ | | $\alpha + 1.618\beta$ |

Total energy $= 2(\alpha + 0.618\beta) + 2(\alpha + 1.618\beta) = 4\alpha + 4.472\beta$. Hence, $E_{\text{deloc}} = 0.472\beta$, or 30 kJ mol$^{-1}$.

16. 6; permute 2,1,0 among $n_x$, $n_y$, $n_z$.
17. Use $hc/\lambda = h^2\{(N+2)^2 - (N+1)^2\}/8mL^2$; $L = 1.4 \times 10^{-10} \times 8$ m and $\lambda =$ $4.595 \times 10^{-7}$ m, or 4595 Å; yellow.
18. $\int\psi(2p_x)\,\psi(2p_y)d\tau$ includes $\int_0^{2\pi} \cos\phi\sin\phi d\phi$; this definite integral is zero.

## SOLUTIONS TO CHAPTER 4

1. $4.9 \times 10^{-40}$ F m$^2$; $3.3 \times 10^{-10}$ m.
2. 1.0010
3. 1.0066
4. 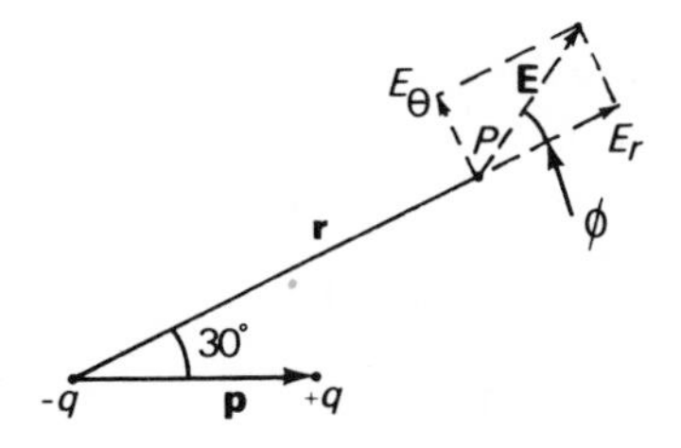

$1.89 \times 10^{-5}$ V;
$E = 787$ V m$^{-1}$,
$\phi = 16.1°$(from (A18.35), $\mathbf{E}_0$ must be drawn as shown).
5. $12.7 \times 10^{-30}$ C m; $P_{m,e} = 3.27 \times 10^{-5}$ m$^3$ mol$^{-1}$ $P_{m,a} = 0.05 \times 10^{-5}$ m$^3$ mol$^{-1}$.

6. $\rho/r_e = \{Ae^2/4\pi\epsilon_0 r_e + 6C/r_e^6\}/\{9V/\kappa + 2Ae^2/4\pi\epsilon_0 r_e + 42C/r_e^6\}$
$U(r_e) = L\{(Ae^2/4\pi\epsilon_0 r_e)(\rho/r_e - 1) + (C/r_e^6)(6\rho/r_e - 1)\}$

7. This result: $c = 471 \times 10^{-79}$ J m$^6$
$\rho/r_e = 0.120$
$u(r_e) = -773$ kJ mol$^{-1}$
Simpler model: $\rho/r_e = 0.113$
$u(r_e) = -764$ kJ mol$^{-1}$
Born-Haber cycle: $u(r_e) = -774$ kJ mol$^{-1}$

8.

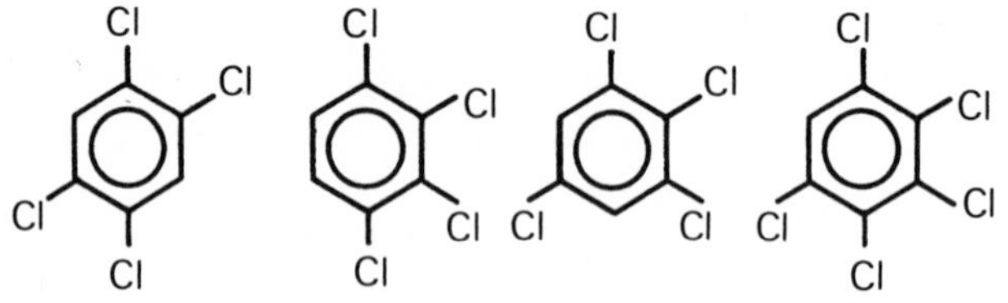


$10^{30}p$/C m 0 8·8 5·1 5·1

9. In dicyanoethyne, partial overlap of the $\pi$-orbitals of the carbon and nitrogen atoms of adjacent molecules is facilitated by the packing adopted; the C . . . N distance is less than the sum of van der Waals' radii by about 0.2 Å.

## SOLUTIONS TO CHAPTER 5

1. $1.16 \times 10^{-8}$ m; $1.57 \times 10^{-3}$ m V$^{-1}$ s$^{-1}$.

2. $$\bar{x} = \int_{-\infty}^{\infty} x \exp \beta(-ax^2 + bx^3)\mathrm{d}x \Big/ \int_{-\infty}^{\infty} \exp \beta(-ax^2 + bx^3)\mathrm{d}x,$$
where $\beta = 1/kT$.

$$\bar{x} = \int_{-\infty}^{\infty} \{x \exp \beta(-ax^2) + \beta b x^4 \exp\beta(-ax^2)\}\mathrm{d}x \Big/$$
$$\int_{-\infty}^{\infty} \{\exp \beta(-ax^2) + \beta b x^3 \exp \beta(-ax^2)\mathrm{d}x$$

The integrals involving odd powers of $x$ are zero and those involving even powers are twice the integral from 0 to $\infty$. Hence,

$$\bar{x} = (b/a^{\frac{5}{2}})\,\beta^{-\frac{3}{2}}(3\pi^{\frac{1}{2}}/4)/(\pi/a\beta)^{\frac{1}{2}} = \frac{3b}{4a^2\beta} = \frac{3bkT}{4a^2}.$$

3. $7.57 \times 10^{-19}$ J; $5.48 \times 10^4$ K.
4. $2.0 \times 10^{18}$;
5. $$\bar{E} = \int_0^{\bar{E}_\mathrm{F}} E\, g(E)\mathrm{d}E \Big/ \int_0^{\bar{E}_\mathrm{F}} g(E)\mathrm{d}E = 3E_\mathrm{F}/5.$$

6. (a)

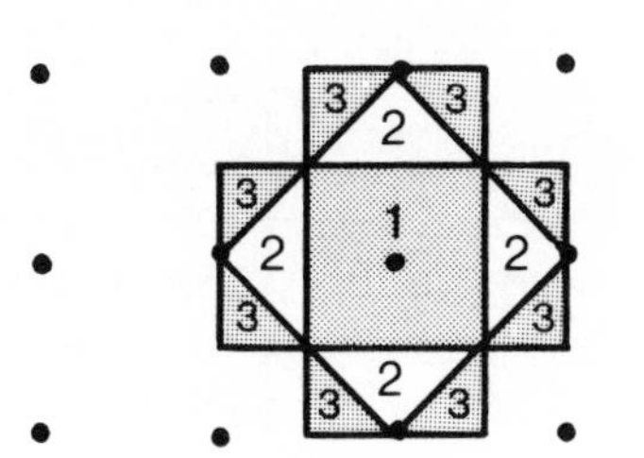


(b) A cube of side $2\pi/a$.

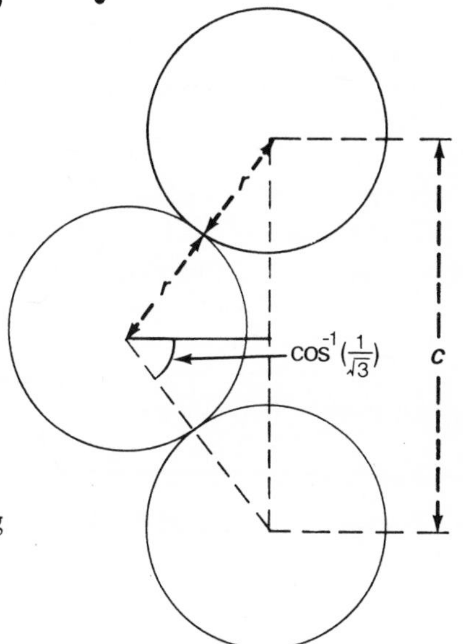


7. From the geometry of the structure we have the construction shown opposite. Hence $c = 2r\sqrt{8/3}$. Since $a = 2r$ the volume occupied per sphere is $2\sqrt{8}\,r^3$ and the packing efficiency is 0.74.

8. 8 tetrahedral holes: ¼,¼,¼; ¼,¾,¼; ¾,¼,¼; ¾,¾,¼
   ¼,¼,¾; ¼,¾,¾; ¾,¼,¾; ¾,¾,¾
   4 octahedral holes: ½,0,0; 0,½,0; 0,0,½; ½,½,½
   (It might help to refer to the $CaF_2$ structure type, Figure 1.11).

## SOLUTIONS TO CHAPTER 6

| | $\sigma(U)$/kJ mol$^{-1}$ |
|---|---|
| 1. Born-Haber cycle | 2.6 |
| Electrostatic model | 1.8 |

2. The following table is not unique. It has been composed from the text: try to enlarge on it as you continue your study of this subject with more advanced texts.

**Some physical and structural characteristics of interatomic bonding forces**
Mp, melting point; $\beta$, coefficient of cubical thermal expansion;
$\epsilon_r$, relative permittivity; $n$ refractive index; *CN*, coordination number.

| Property | Ionic | Covalent |
|---|---|---|
| Mechanical | Strong, hard crystals. | Strong, hard crystals. |
| Thermal | High Mp.<br>Low $\beta$.<br>Ions in melt (may decompose). | High Mp.<br>Low $\beta$.<br>Molecules in melt (may decompose). |
| Electrical | Insulators.<br>Conduction in melt by ion transport. Often soluble in solvent of high $\epsilon_r$. | Insulators in the solid and in the melt. Generally insoluble. |
| Optical and Magnetic | Absorption and similar properties are those of individual ions (similar in solution). | High $n$; absorption different in solid and solution. |
| Structural | Non-directional.<br>High *CN* ($\geqslant$6) | Strongly directional.<br>Low *CN* ($\leqslant$4). |

| Property | van der Waals' | Metallic |
|---|---|---|
| Mechanical | Weak, soft crystals. | Variable strength and gliding. |
| Thermal | Low Mp.<br>High $\beta$. | Variable Mp with long liquid interval.<br>High $\beta$. |
| Electrical | Insulators. | Conduction by electron transport. |
| Optical and Magnetic | Properties are those of molecules and are similar in solution. | Opaque.<br>Properties similar in liquid. |
| Structural | Non-directional. | Non-directional.<br>Very high *CN* ($\geqslant$8). |

# Index

## D



## E

F



G



H



I

# Periodic Table of the Elements

Each box contains the chemical symbol of the element, its atomic number, atomic weight (see below) and outermost electronic configuration. The elements are arranged by classical group number and period (principal quantum number of outermost electron/s). Some of the electronic configurations, particularly for elements of high atomic number, are the subject of current investigation and debate.

The atomic weights are those recommended by IUPAC 1975 §: they are relative values, being scaled to $^{12}C = 12$. The atomic weights of certain elements depend on the history of the material The values given apply to terrestrial elements and to certain artificial elements. The precision is ±1 in the last digit quoted, or ±3 where the element is marked with an *. Values in parentheses refer to the mass number of the isotope of longest half-life for certain radioactive elements whose atomic weights cannot be quoted precisely without reference to their origin.

§ *Pure & Applied Chemistry* Vol. 47, pp.75-95, (1975).

*NOTES ON ATOMIC WEIGHTS*

- *a* Variations in isotopic compostion of terrestrial material limits the precision to that quoted.
- *b* Geological specimens reveal such variations in isotopic composition that the difference in implied atomic weight exceeds the precision given in the table.
- *c* Variations in atomic weight may occur in commerical material because of unknown or unsuspected changes in isotopic composition.
- *d* Atomic weight is for radioisotope of longest half-life.

| Period \ Group | IA | IIA | IIIA | IVA | VA | VIA | VIIA | VIII | VIII | VIII | IB | IIB | IIIB | IVB | VB | VIB | VIIB | O |
|---|---|---|---|---|---|---|---|---|---|---|---|---|---|---|---|---|---|---|
| 1 | 1 1.0079<br>*a*<br>**H**<br>$(1s)^1$ | | | | | | | | | | | | | | | | | 2 4.00260<br>*b*<br>**He**<br>$(1s)^2$ |
| 2 | 3 6.941<br>*abc*<br>**Li** *<br>$(2s)^1$ | 4 9.01218<br>**Be**<br>$(2s)^2$ | | | | | | | | | | | 5 10.81<br>*ac*<br>**B**<br>$(2s)^2(2p)^1$ | 6 12.011<br>*a*<br>**C**<br>$(2s)^2(2p)^2$ | 7 14.0067<br>**N**<br>$(2s)^2(2p)^3$ | 8 15.9994<br>*a*<br>**O** *<br>$(2s)^2(2p)^4$ | 9 18.998403<br>**F**<br>$(2s)^2(2p)^5$ | 10 20.179<br>*c*<br>**Ne** *<br>$(2s)^2(2p)^6$ |
| 3 | 11 22.98977<br>**Na**<br>$(3s)^1$ | 12 24.305<br>*b*<br>**Mg**<br>$(3s)^2$ | | | | | | | | | | | 13 26.98154<br>**Al**<br>$(3s)^2(3p)^1$ | 14 28.0855<br>**Si** *<br>$(3s)^2(3p)^2$ | 15 30.97376<br>**P**<br>$(3s)^2(3p)^3$ | 16 32.06<br>*a*<br>**S**<br>$(3s)^2(3p)^4$ | 17 35.453<br>**Cl**<br>$(3s)^2(3p)^5$ | 18 39.948<br>*ab*<br>**Ar** *<br>$(3s)^2(3p)^6$ |
| 4 | 19 39.0983<br>**K** *<br>$(4s)^1$ | 20 40.08<br>*b*<br>**Ca**<br>$(4s)^2$ | 21 44.9559<br>**Sc**<br>$(3d)^1(4s)^2$ | 22 47.90<br>**Ti** *<br>$(3d)^2(4s)^2$ | 23 50.9414<br>**V** *<br>$(3d)^3(4s)^2$ | 24 51.996<br>**Cr**<br>$(3d)^5(4s)^1$ | 25 54.9380<br>**Mn**<br>$(3d)^5(4s)^2$ | 26 55.847<br>**Fe** *<br>$(3d)^6(4s)^2$ | 27 58.9332<br>**Co**<br>$(3d)^7(4s)^2$ | 28 58.70<br>**Ni**<br>$(3d)^8(4s)^2$ | 29 63.546<br>*a*<br>**Cu** *<br>$(3d)^{10}(4s)^1$ | 30 65.38<br>**Zn**<br>$(3d)^{10}(4s)^2$ | 31 69.72<br>**Ga**<br>$(4s)^2(4p)^1$ | 32 72.59<br>**Ge** *<br>$(4s)^2(4p)^2$ | 33 74.9216<br>**As**<br>$(4s)^2(4p)^3$ | 34 78.96<br>**Se** *<br>$(4s)^2(4p)^4$ | 35 79.904<br>**Br**<br>$(4s)^2(4p)^5$ | 36 83.80<br>*bc*<br>**Kr**<br>$(4s)^2(4p)^6$ |
| 5 | 37 85.4678<br>*b*<br>**Rb** *<br>$(5s)^1$ | 38 87.62<br>*b*<br>**Sr**<br>$(5s)^2$ | 39 88.9059<br>**Y**<br>$(4d)^1(5s)^2$ | 40 91.22<br>*b*<br>**Zr**<br>$(4d)^2(5s)^2$ | 41 92.9064<br>**Nb**<br>$(4d)^4(5s)^1$ | 42 95.94<br>**Mo**<br>$(4d)^5(5s)^1$ | 43 (97)<br>**Tc**<br>$(4d)^5(5s)^2$ | 44 101.07<br>*b*<br>**Ru** *<br>$(4d)^7(5s)^1$ | 45 102.9055<br>**Rh**<br>$(4d)^8(5s)^1$ | 46 106.4<br>*b*<br>**Pd**<br>$(4d)^{10}(5s)^0$ | 47 107.868<br>*b*<br>**Ag**<br>$(4d)^{10}(5s)^1$ | 48 112.41<br>*b*<br>**Cd**<br>$(4d)^{10}(5s)^2$ | 49 114.82<br>*b*<br>**In**<br>$(5s)^2(5p)^1$ | 50 118.69<br>**Sn** *<br>$(5s)^2(5p)^2$ | 51 121.75<br>**Sb** *<br>$(5s)^2(5p)^3$ | 52 127.60<br>*b*<br>**Te** *<br>$(5s)^2(5p)^4$ | 53 126.9045<br>**I**<br>$(5s)^2(5p)^5$ | 54 131.30<br>*bc*<br>**Xe**<br>$(5s)^2(5p)^6$ |
| 6 | 55 132.9054<br>**Cs**<br>$(6s)^1$ | 56 137.33<br>*b*<br>**Ba**<br>$(6s)^2$ | 57 138.9055<br>*b*<br>**La** †*<br>$(4f)^1(5d)^0(6s)^2$ | 72 178.49<br>**Hf** *<br>$(5d)^2(6s)^2$ | 73 180.9479<br>**Ta** *<br>$(5d)^3(6s)^2$ | 74 183.85<br>**W** *<br>$(5d)^4(6s)^2$ | 75 186.207<br>**Re**<br>$(5d)^5(6s)^2$ | 76 190.2<br>*b*<br>**Os**<br>$(5d)^6(6s)^2$ | 77 192.22<br>**Ir** *<br>$(5d)^7(6s)^2$ | 78 195.09<br>**Pt** *<br>$(5d)^9(6s)^1$ | 79 196.9665<br>**Au**<br>$(5d)^{10}(6s)^1$ | 80 200.59<br>**Hg** *<br>$(5d)^{10}(6s)^2$ | 81 204.37<br>**Tl** *<br>$(6s)^2(6p)^1$ | 82 207.2<br>*ab*<br>**Pb**<br>$(6s)^2(6p)^2$ | 83 208.9804<br>**Bi**<br>$(6s)^2(6p)^3$ | 84 (209)<br>**Po**<br>$(6s)^2(6p)^4$ | 85 (210)<br>**At**<br>$(6s)^2(6p)^5$ | 86 (222)<br>**Rn**<br>$(6s)^2(6p)^6$ |
| 7 | 87 (223)<br>**Fr**<br>$(7s)^1$ | 88 226.0254<br>*bd*<br>**Ra**<br>$(7s)^2$ | 89 227.0278<br>*d*<br>**Ac** ‡<br>$(5f)^0(6d)^1(7s)^2$ | | | | | | | | | | | | | | | |
| 6 | † Lanthanides | | 57 138.9055<br>*b*<br>**La** *<br>$(4f)^1(5d)^0(6s)^2$ | 58 140.12<br>*b*<br>**Ce**<br>$(4f)^1(5d)^1(6s)^2$ | 59 140.9077<br>**Pr**<br>$(4f)^3(5d)^0(6s)^2$ | 60 144.24<br>*b*<br>**Nd** *<br>$(4f)^4(5d)^0(6s)^2$ | 61 (145)<br>**Pm**<br>$(4f)^5(5d)^0(6s)^2$ | 62 150.4<br>*b*<br>**Sm**<br>$(4f)^6(5d)^0(6s)^2$ | 63 151.96<br>*b*<br>**Eu**<br>$(4f)^7(5d)^0(6s)^2$ | 64 157.25<br>*b*<br>**Gd** *<br>$(4f)^7(5d)^1(6s)^2$ | 65 158.9254<br>**Tb**<br>$(4f)^8(5d)^1(6s)^2$ | 66 162.50<br>**Dy** *<br>$(4f)^{10}(5d)^0(6s)^2$ | 67 164.9304<br>**Ho**<br>$(4f)^{11}(5d)^0(6s)^2$ | 68 167.26<br>**Er** *<br>$(4f)^{12}(5d)^0(6s)^2$ | 69 168.9342<br>**Tm**<br>$(4f)^{13}(5d)^0(6s)^2$ | 70 173.04<br>**Yb** *<br>$(4f)^{14}(5d)^0(6s)^2$ | 71 174.97<br>**Lu**<br>$(4f)^{14}(5d)^1(6s)^2$ | |
| 7 | ‡ Actinides | | 89 227.0278<br>*d*<br>**Ac**<br>$(5f)^0(6d)^1(7s)^2$ | 90 232.0381<br>*bd*<br>**Th**<br>$(5f)^0(6d)^2(7s)^2$ | 91 231.0359<br>*d*<br>**Pa**<br>$(5f)^2(6d)^1(7s)^2$ | 92 238.029<br>*bc*<br>**U**<br>$(5f)^3(6d)^1(7s)^2$ | 93 237.0482<br>*d*<br>**Np**<br>$(5f)^4(6d)^1(7s)^2$ | 94 (244)<br>**Pu**<br>$(5f)^6(6d)^0(7s)^2$ | 95 (243)<br>**Am**<br>$(5f)^7(6d)^0(7s)^2$ | 96 (247)<br>**Cm**<br>$(5f)^7(6d)^1(7s)^2$ | 97 (247)<br>**Bk**<br>$(5f)^8(6d)^1(7s)^2$ | 98 (251)<br>**Cf**<br>$(5f)^{10}(6d)^0(7s)^2$ | 99 (254)<br>**Es**<br>$(5f)^{11}(6d)^0(7s)^2$ | 100 (257)<br>**Fm**<br>$(5f)^{12}(6d)^0(7s)^2$ | 101 (258)<br>**Md**<br>$(5f)^{13}(6d)^0(7s)^2$ | 102 (259)<br>**No**<br>$(5f)^{14}(6d)^0(7s)^2$ | 103 (260)<br>**Lr**<br>$(5f)^{14}(6d)^1(7s)^2$ | |